# Springer Series in Operations Research and Financial Engineering

*Series Editors:*
Thomas V. Mikosch
Sidney I. Resnick
Stephen M. Robinson

# Springer Series in Operations Research and Financial Engineering

John A. Muckstadt

# Analysis and Algorithms for Service Parts Supply Chains

 Springer

John A. Muckstadt
Cornell University
School of Operations Research and Industrial Engineering
Ithaca, NY 14853
U.S.A.
jack@orie.cornell.edu

*Series Editors:*
Thomas V. Mikosch
University of Copenhagen
Laboratory of Actuarial Mathematics
DK-1017 Copenhagen
Denmark
mikosch@act.ku.dk

Stephen M. Robinson
University of Wisconsin-Madison
Department of Industrial Engineering
Madison, WI 53706
U.S.A.
smrobins@facstaff.wisc.edu

Sidney I. Resnick
Cornell University
School of Operations Research and Industrial Engineering
Ithaca, NY 14853
U.S.A.
sir1@cornell.edu

*Cover photo:* An F-15E Strike Eagle taken by Staff Sgt. Lee O. Tucker. Courtesy of the United States Air Force.

Mathematics Subject Classification (2000): Primary: 90B05, 90-XX, 93A15, 90B15, 90-01; Secondary: 90Cxx, 60-XX, 60Jxx

Library of Congress Cataloging-in-Publication Data
Muckstadt, J. A., 1940–
    Analysis and algorithms for service parts supply chains / J.A. Muckstadt.
      p.  cm.
    Includes bibliographical references and index.
    ISBN 0-387-22715-6 (alk. paper)
    1. Spare parts.  2. Inventory control.  I. Title.
  TS160.M82   2004
  658.7'87–dc22                   2004056000

ISBN 0-387-22715-6        Printed on acid-free paper.

Printed in the United States of America.    (TXQ/EB)

9 8 7 6 5 4 3 2 1      SPIN 11307594

springeronline.com

# Contents

# Preface

There is an axiom in the military: never volunteer. Nonetheless, many have, and their lives have been altered because of this act. I was one of those who said "Yes sir, I can do that," and my future was changed forever.

This act occurred for me in early 1972. At that time, I was a Captain in the United States Air Force, assigned to Headquarters, Air Force Logistics Command (AFLC). I was an operations research staff officer reporting to Colonel Fred Gluck. Colonel Gluck and I had previously served on the faculty of the Air Force Institute of Technology (AFIT), so we knew each other quite well. Because of this prior relationship, Colonel Gluck gave me a great deal of latitude to select projects on which to work. I chose to assist a pair of former AFIT advisee's, Mr. Thomas Harruff and Captain Michael Pearson, who were both assigned to the propulsion directorate within the Material Management Deputate. This cooperative effort of working on engine management problems with my former advisees had begun while I was still an AFIT faculty member. The three of us were interested in strategic, tactical and operational aspects of procuring, allocating and repairing of the Air Force's jet engines. I knew little about these problems, so I interacted frequently with them while I was an AFIT faculty member to learn more about the variety of problems that were encountered when managing these expensive and important items . This interaction became a full time activity once I was reassigned to the AFLC Headquarters. As I said, Colonel Gluck gave me the go ahead to continue work on this project.

We began our studies largely because the Air Force was experiencing shortages for several types of engines. These shortages arose for a variety of reasons, but one was obvious. The Department of Defense (DOD) methodology for computing engine requirements ignored the operational environment into which engines were placed. The question that I immediately asked was whether or not the use of an alternative methodology for computing spares requirements for engines would result in significantly different estimates of performance – engine availability at operating bases – and therefore in the number of spare engines needed to achieve desired levels of performance. This question directed me to study research papers produced at the RAND Corporation, the unquestioned leader in developing logistics models for the Air Force at that time. I was exposed to METRIC (Multi-Echelon Technique for Recoverable Item Control) and related concepts which were developed by Craig Sherbrooke and his colleagues at RAND. My colleagues and I constructed implementations of METRIC like optimization models and corresponding simulation models to study a broad range of tactical and operational planning problems. The results of these preliminary studies indicated that the DOD methodology needed to be changed.

Another event was occurring at that time. The F-15 weapon system was being developed. The F-15 is a remarkable aircraft in many ways. Its design is focused on maintainability, and, in particular, maintainability of its engines. The engines were designed in a modular way by Pratt and Whitney. The basic idea behind the maintenance concept was to remove a defective engine from an aircraft, and replace it on the aircraft with a serviceable one. Once the defective engine is removed from the aircraft, a defective module is identified and is removed and replaced, thereby returning the engine to a serviceable condition. The goal was to minimize an engine's repair cycle time and thereby minimize the number of serviceable engines and modules that were needed to keep the fleet flying. Keeping the repair cycle time for engines low could be done only if adequate quantities of serviceable spare modules were available at each base. Of course the defective modules had to be repaired, too. Thus their repairs, which were often conducted at the depot in San Antonio, TX, and their repair cycle times had a significant affect on aircraft availability. This maintenance concept and the corresponding engine design strategy were clearly great ideas. However, one question existed. How many engines and modules should be purchased? Even supporters of the simple DOD spare engine requirements methodology recognized that the DOD method was inadequate to address this problem. The propulsion directorate had the responsibility to establish these quantities for initial spares procurement and to estimate future spares requirements for budgeting purposes. Obviously, the leaders of this office were facing a problem which they did not know how to address.

Opportunities for an individual often come in the moment of organizational crisis. By this I mean that large bureaucratic organizations are usually able to embrace new ideas only when it is obvious that they are doomed if they do not. Obviously, the DOD is a very large bureaucratic organization, and it was forced to change because its standard operating procedure clearly would not work well. At this juncture, I committed the ultimate sin. I violated axiom number one. I volunteered to help create a solution to the engine and module mix problem, and to develop it quickly. Colonel Gluck and his counterpart in the propulsion shop never flinched. Colonel Gluck got the authority (from Brigadier General George Rhodes, the head of the Material Management Deputate) to develop the method.

Based on Sherbrooke's ideas, our team created the first tactical planning model for multi-indentured repairable items that was implemented in a multi-echelon environment. We called this model Modified-METRIC or simply MOD-METRIC. This model was used by our team, which had been expanded by this time to include Major Gene Perkins, Captain Jon Reynolds and MSgt Robert Kinsey, to compute requirements for all repairable spare parts for the F-15. In subsequent years, MOD-METRIC, and its successor models, have been used by the U.S. Air Force to compute spares requirements for other weapon systems as well.

As I reflect on this time, some 30 years later, I see that there were so many events that occurred simultaneously that resulted in my having been given the opportunity to volunteer. To have people be supportive of the creation and implementation of our ideas was truly remarkable.

Throughout my active duty years, I had the good fortune to work with many exceptional people. One person was George Babbitt. I met George when he was an AFIT student, where he became my advisee. As I worked with him that year, it was clear to me that George was an extraordinary person. From that time onward, we, and our families, interacted socially and professionally. George and I argued endlessly about logistics system's management. These discussions occurred periodically and with great passion over a thirty year time span. (Louise, his wife, constantly called us two very boring people.) However, by challenging each other, I developed a much clearer understanding of how to design and manage service parts systems. I was not the only one to benefit from George's wisdom. His abilities were widely recognized by his superiors, and he eventually achieved the rank of General and was the Commander of the Air Force Materiel Command (AFLC's successor organization) at the time of his retirement.

When developing our first METRIC based models for managing engines, I also met Mr. Bernie Rosenman, who led the Army Inventory Research Office in Philadelphia. Bernie was responsible, both through his office and personal intellect, for the development of many of the inventory models which are found in standard text books on operations management and inventory control. Some of the models were developed under contract with MIT, where Herb Galliher worked. Herb, in subsequent years, had a profound impact on my thinking about inventory modelling. Specifically, he taught me to look carefully at data before making modelling choices. Bernie's team included Alan Kaplan and Karl Kruse, both of whom, at one time or another, guided me in structuring spares procurement models. In fact, one of Karl's papers on waiting times is the basis for a section in this book.

As time went on and the efforts of our AFLC team expanded, we were also very fortunate to interact with the great logistics thinkers of that period at the RAND Corporation, Murray Geisler, Irv Cohen, Bob Paulson, Mort Berman, John Lu, Hy Shulman, Steve Drezner, among others. Subsequently, the work of others at RAND had profound impact on the Air Force and my thinking. These included Lou Miller, Dick Hillestad, M.J. Carrillio, Jack Abell and Gordon Crawford. Many of the interactions that I had with my RAND colleagues occurred after my joining the Cornell faculty in 1974. I had the privilege of spending several summers in Santa Monica learning from these highly skilled, dedicated and genuinely good people. I am forever in their debt. As such, it is not surprising that a large portion of this book is based directly and indirectly on ideas that originated from work at the RAND Corporation and the interactions I had with the people who worked there.

My interest and work in spare parts inventory management continued upon joining Cornell's faculty. In addition to continuing my work at RAND, I also had immediate opportunities to work with the US Navy, GE, and most importantly, with XEROX. XEROX had a Management Sciences department in Rochester, NY, which was one of the premier industrial management science organizations in the world. Jack Chambers was its leader. XEROX had an enormous investment in parts and its service infrastructure. Hence, it was not surprising that Jack's team was engaged in a number of studies in this area. One member of this team was Ron Hudson, who contacted me to establish my interest in working with this group on this class of problems. Ron was the key intellectual leader in the logistics activities of the Management Science department. Of all the many technical people I have worked with in industry, Ron stands out as one of the most competent. His professionalism and uncompromising commitment to doing the best possible analysis have had a major impact on me over the quarter of a century that I worked with Ron. There, of course, were others at XEROX with whom I interacted who shaped my views on the differences between military systems and high tech service parts replenishment systems. Chuck Mitchell and Ron Nawrocki exposed me, and a generation of Cornell students, to the complexities in strategic, tactical and operational planning and execution in commercial service parts environments. The Management Sciences department at XEROX, and at other companies, was disbanded in the very early 1980s.

Another life changing event occurred in 1980 when I met Thomas P. Latimer, the President, CEO and Chairman of the Chicago Pneumatic Tool Company. At about the same time, I became keenly aware of the demise of the manufacturing might of United States companies. Tom gave me the opportunity to work on several very interesting problems relating to the manufacture and distribution of both finished goods and spare parts. Although most of the activities I was engaged in with Tom's company were focused on evaluating various alternative manufacturing strategies for fabricating components, I did work on one significant service parts study throughout 1982. This study involved the stocking of parts in distribution centers in several European countries.

Although I was aware of the competition that existed throughout western Europe, I quickly learned about how governmental policy and interference impact the flow of material and hence the distribution strategies taken and the costs incurred by companies doing business there. Regulations affected both the movement of both physical product and information among company facilities located in different countries. The social costs and restrictions of doing business that were a consequence of government policy were in evidence everywhere. Observing these hard economic and operational realities greatly affected my thinking about global manufacturing and distribution. Other work I did in Europe later in the 1980s with Bell Atlantic reinforced these observations.

In the 1990s, two more significant opportunities arose related to service parts. My colleague and very good friend, Dennis Severance, was responsible for both of them. Dennis had been on the Cornell faculty in 1974. We and our families became close friends on a personal level at that time. On a professional level, however, Dennis had, and continues to have a tremendous influence on my thinking and teaching. The importance of his presence in my life cannot be overstated, as you will now see.

Late in the 1980s, Dennis introduced me to Howard Selland, who subsequently became the president of the Aeroquip Corporation. After assisting Howard on a manufacturing and distribution problem in the late 80's, he asked Dennis and me to assist him in restructuring Aeroquip's manufacturing and distribution systems. Their operations were global in scope, although much of their market was in the United States. Aeroquip was in the hydraulic hose and fittings business. All fittings contain machined components as well as others. My earlier studies of manufacturing practices, largely supported by Tom Latimer and Chicago Pneumatic Tool Company, made it clear to me what Aeroquip should do. About half of Aeroquip's business was for spare parts. Hence, again the problems were familiar. As was the case for Chicago Pneumatic Tool, Aeroquip's demand patterns for parts were highly erratic. Furthermore, there were many tens of thousands of part numbers. The nature of the problem immediately suggested that the traditional batch manufacturing and MRP based planning strategy used by Aeroquip, and many other companies, was totally inappropriate. At this point, I began working with Mike Hoverman, whose professionalism, dedication, and sheer doggedness, led to the design of an extraordinarily effective manufacturing and distribution system. Mike deployed a software environment that permitted Aeroquip to comprehend the idiosyncracies of various customer's ordering patterns. This analysis formed the basis for a production strategy for producing physical products, for planning of production of the components and products, and for developing the underlying inventory policies. As part of this work, Mike was the force behind the design of a new physical distribution center and all its operational rules, and software. This distribution center and the backbone manufacturing strategy made the cost effective production of parts possible and dramatically improved service to customers.

It is impossible to overstate the lessons that Dennis and I learned from Howard, Mike and literally hundreds of other Aeroquip people. In aggregate, they forced us to take general concepts and to work with them to create concrete proposals for manufacturing and distribution system design and operations. Collaboratively, the group constructed a remarkable environment. They stimulated our thinking and provided us with the opportunity to further develop our ideas and concepts pertaining to the management of service parts.

While the efforts at Aeroquip were underway, Dennis also introduced me to Stu Wagner, the director for strategic planning in General Motor's Service Parts Operations organization. This organization is responsible for the acquisition, processing, storing, and distributing of hundreds of thousands of different types of parts to many thousands of automotive dealers, wholesale distributors and mass merchandisers. Through Stu, and his team, David Sergeant, Mary Shaw and others, we have had the opportunity to learn about this fascinating business, a business that is matched only in scale and complexity found in the US military. Over the past decade we have been exposed to numerous issues relating to this system.

Although he is now retired, Stu Wagner was a guiding force in this organization. He was relentless in his quest for uncovering better ideas and communicating them to others in General Motors. His quick mind and sharp tongue always forced us to think more clearly and to articulate our thoughts more strongly. His no nonsense approach to the analysis of problems and opportunities and reporting the results of these analyses has affected our actions greatly. We deeply appreciate his leadership and that of the many others we have worked with at GM.

Most recently, my colleagues Jim Rappold and Kathy Caggiano, both of the University of Wisconsin, David Murray of the College of William and Mary, and Peter Jackson and I have been exposed to even a broader range of interesting problems facing high tech companies, aircraft manufacturers, and airlines. As often happens, the introduction to these problems has come from an earlier source of stimulation. In this case, from Ron Hudson. Remember, a quarter of a century earlier, Ron gave me my first experiences in commercial spare parts resupply systems. In 2000, Ron was working for XELUS, a company that provides software for planning various aspects of service parts acquisition and distribution. Ron articulated clearly his dissatisfaction with available models. After exploring his reasons for this dissatisfaction, he challenged us to think of new approaches for modelling and solving various types of strategic, tactical and operational service parts problems. And the journey, of course, has not ended. We continue to be stimulated by the opportunities that XELUS provides to us. Although our friend and respected colleague, Ron Hudson, died recently, his spirit of inquiry remains in us, and we devote our energies to improving our modelling methodologies in his memory.

During the past twenty years I have also had the distinct privilege of working with my friend and colleague, Peter Jackson. Peter and I have worked on several projects for GM and XELUS. But these are but two of the many ways that we have collaborated. Peter and I have struggled with many research problems, often related to service parts, during the past decades. We have also spent countless hours preparing related teaching materials. Peter represents all that is best in a teacher and colleague, and I am blessed to have had the privilege of working with him.

I have provided just a glimpse of the events in my life that have led me to examine service parts systems. This sequence of events began with a seemingly harmless act. That I volunteered clearly changed the direction of my life completely. The wonderful people that I have met, and who have inspired me, have made me a very happy and thankful person. Their friendships, in addition to their professional guidance, have made my life a better one. My wish is that all have such good experiences; but, my advice to you is to be careful when you volunteer, since the consequences may be long lasting.

*Jack Muckstadt*
Cornell University
September 2004

# Acknowledgments

I have had the benefit of working with some outstanding people throughout my career, as I have mentioned in the Preface. Besides the ones mentioned there, others have helped form my thinking in a substantial way. These include Richard Wilson, my doctoral advisor and long time friend; William Maxwell and Robin Roundy, my close Cornell colleagues and co-authors on a number of papers; and Andrew Schultz, our mentor and leader in so many ways.

I have also had the distinct privilege of having so many outstanding students with whom to interact. In particular, Jim Rappold, Kathy Caggiano, Ganesh Janakiraman, Amar Sapra, Hui Wang, and Janis Chang have all helped in the preparation and writing of this book. I especially thank Hui Wang for her constant assistance to me during the assembling of the manuscript.

I especially want to thank Jim Rappold, Kathy Caggiano, Ganesh Janakiraman, and Peter Jackson for their significant contributions to this book. They were collaborators in writing research papers that have provided the basis for many of the sections of this book. Their enthusiasm and creativity have been a constant source of inspiration to me.

Finally, I thank Sharon Hobbie for her careful preparation of the manuscript. Deciphering my handwriting was a time consuming and demanding challenge I am sure.

*Analysis and Algorithms for Service Parts Supply Chains*

# 1

# Introduction

When repairing equipment, there is often a requirement to replace defective components or parts. These replacement parts are commonly called service parts. Clearly, service parts play a critical role in the effective operation of commercial and military systems, as well as in the personal lives of individuals. Without adequate stocks of these parts, power plants cannot function, construction equipment must be idled, airliners will be grounded, and computers cannot be repaired. The list of such commercial environments requiring service parts is seemingly unlimited. In other words, our technology-dependent world depends on the careful creation and distribution of service parts. As individuals, we cannot repair our appliances, our furnaces, or our cars, if parts are not readily available. Thus each of our lives is critically dependent on the abundant availability of service parts within short periods of time.

We clearly recognize that there are many different types of service parts and that they perform many different functions. For example, to keep our cars operational, we can buy inexpensive parts, such as filters, and also very expensive parts, such as transmissions or engines. To ensure timely repair of our cars, extensive supply chain systems have been developed by car manufacturers and other companies, such as NAPA. But how should these resupply systems be designed and operated? That is, how many warehouses should there be? Which warehouses should resupply which other warehouses? How should car dealers be resupplied? What parts should be stocked at dealers and at the many warehouses in the system? Clearly, car parts differ in terms of their cost and demand rates. However, they also differ in terms of criticality. We can wait for certain parts but not for others: a car is often operable if it has a damaged interior trim part, but not if it has a faulty transmission. Thus decisions concerning what parts to stock at what locations is of central importance to the car-using public and to the providers of the parts. If too few parts are available in the resupply system, customers will be forced to wait to have use of their vehicles. If too many parts exist in the system, the cost to operate the system – inventory investment, facilities, transportation, and other operating costs – will be too great. Thus the best design of resupply networks and the optimal allocation of inventories within these networks is of unquestionable importance to the economical maintenance of equipment.

The types of decisions that must be made relating to service parts can roughly be divided into three planning categories: strategic planning, tactical planning, and operational planning. Strategic planning is an on-going activity that has two primary functions. First, an organization must determine what customer requirements are and will be over the next several years. In military environments, the time horizon may be decades in length. Customer requirements in this context refer to the range of service parts needed by customers as well as the timeliness of these needs. For example, the permissible time to return an important medical device in a hospital or the principle radar system in an air traffic control system to serviceable condition is different than the permissible time to replace a knob on the dashboard of a car. Thus, the location in which parts are stocked depends on a customer's needs. Even the same parts may have different time requirements for different customers, as is evidenced by the service contracts that customers purchase.

After establishing the need for various types of service by different categories of customers, the second strategic planning activity is to determine how to allocate resources to meet these requirements. This means a company must decide to what extent it will directly provide service to customers, where it will locate parts and in what quantities, how it should structure its operating systems, what its information systems requirements are, what its supply chain partners' and its own business processes should be, and how its proposed operating environment will be executed on a moment-by-moment basis to meet its contractual obligations over the planning horizon.

These strategic decisions must be made recognizing not only customer needs but also the competitive forces present in the marketplace. Simply put, there is a tremendous amount of uncertainty about the future environment. Therefore, the strategic planning activity for different companies in different industries can result in highly different outcomes. For example, all the major car companies in the United States have chosen to create substantial infrastructures to provide parts to their dealers. Parts systems for companies in the computer business, such as Dell, are very different. You are very likely to be able to get a key part for a seven year old Buick directly from General Motors; however, you will not likely get a part for a seven year old computer directly from Dell. As a consequence of its business strategy, General Motors has well over 10 million square feet of service parts storage space in the United States in which many hundreds of thousands of different part types are stocked. It continually struggles with determining the proper range and depth of stock to locate at each facility in its network and with providing guidance to dealers as to which parts they, in turn, should stock. Thus General Motors must consider its supply chain partners – its individual car dealers as well as its suppliers – when making strategic decisions about the nature of its internal service parts supply chain.

Strategic decisions that affect parts acquisition and repair positioning is a key requirement for military planners, too. Since the United States military maintains service (spare) parts inventories with a value exceeding $100 billion, it is imperative that complex systems exist for their physical storage as well as the planning of their disbursement throughout the world. Strategic planning encompasses establishing what the design characteristics of weapons should be, recognizing that the life cycle costs are a consequence of a weapon's design. Deployment of weapon systems and the accompanying costs of maintaining them must be estimated as a function of defense and foreign policy. Infrastructure is expensive to create and maintain. It also takes significant amounts of time to build for political as well as technical reasons. Thus with all the uncertainties that exist, a strategic plan must be created that will provide the flexibility needed to meet a wide range of mission scenarios, while also satisfying these uncertain requirements in a cost effective manner.

While strategic planning is obviously of considerable importance to both commercial and military organizations, we will not discuss this topic further in this book. We assume that these decisions have been made and that operating environments exist.

The second category of planning is called tactical planning. From the perspective of service parts, tactical planning establishes what inventories will be required to meet operational objectives at some future time, given the design and operational characteristics of an existent resupply system infrastructure. Thus we set stockage objectives for each part at each location consistent with procurement and resupply times from location to location inherent in the system's design, the uncertain demands for various types of items at different locations, the strategies for repairing certain parts, and the timeliness of service that is needed. In most real world situations, tactical planning is associated with budget, procurement, and repair decisions. That is, for some planning horizon, we determine how much money should be allocated for investment in which service parts. The length of the planning horizon differs by application. For example, in military applications, the horizon's length is dependent on the Federal budgeting cycle and procurement lead times, and hence, can be years in duration. Procurement decisions involve determining what aggregate inventory levels should be and how much should be purchased in the relatively near future, say the next few months, in some environments. Thus a planner at American Airlines may place an order for some quantity of turbine blades for an engine based on perceived future needs. Procurement lead times are often lengthy for complex parts: it takes months to fabricate many new chips and boards; it can take even longer to acquire them given the business processes that exist in purchasing departments. Finally, repair capacity must be planned to meet forecasted requirements for repair and refurbishment of parts and systems. This capacity exists in many locations, and, in many instances, may be performed by other organizations. For example, the person who repairs your Dell computer is not an employee of Dell. Yet Dell must ensure that there is enough capacity in each geographical region to meet its service commitments.

Much of this book is devoted to constructing mathematical models that can be effectively used to carry out the tactical planning goal of determining system stock levels. The models address this problem in a variety of service parts resupply networks for both repairable and consumable (nonrepairable) item types. As we will see, these models are based on many assumptions, but the objective is always the same. The goal is to answer the following questions: how much do I need of each part type to meet my goals, given the nature of the resupply network?

The third, and final, category of planning models addresses the following simple question. Given the stocks I now have on hand, in serviceable or repairable condition, at each location, what do I repair now and what do I ship now from one location to another via what mode of transport? While tactical planning models are, by nature, normally formulated assuming a stationary operating environment, these real-time execution models are not. That is, they are formulated over short planning horizons, and also contain more details on current operating limitations. They may, for example, recognize that shipping from point A to B is not possible today, or that individuals or equipment are unavailable to conduct repairs today. Basically, they are designed to make the best possible decisions consistent with the current situation while ignoring the longer term potential consequences of making these decisions.

We will devote only one chapter of this book to these real-time planning models. As we will see, these models differ substantially from the types of models employed for making tactical decisions.

## 1.1  Taxonomy of Service Parts Inventory Systems

We now consider in greater detail the different elements that affect the amounts of inventory found in various portions of a service parts inventory system. In general, there are numerous reasons for choosing to stock inventory of an item type within a system, often at multiple locations.

The underlying echelon or network resupply structure will have a substantial impact on the amount of inventory needed. There are clearly many possible structures. However, for each one, there is usually a well defined resupply plan. Consider the system depicted in Figure 1.1. In this system, demands for service parts arise due to the failure of some equipment operated by a customer. To repair that equipment, a repair technician diagnoses the failure type, and, if necessary, removes and replaces the defective unit with a serviceable one. The technician obtains the serviceable parts from a stock room at the service center location that is responsible for resupplying him. That location's stock is subsequently resupplied from the appropriate regional warehouse, according to the ordering policy followed at the service center location. The regional warehouse is in turn resupplied by its supporting central warehouse, as prescribed by the inventory policy employed by the regional warehouse. The central warehouse likewise receives replenishment stocks from the factories that manufacture the particular item. Typically, the parts managed in resupply systems of the type we have described are not repaired. They are called consumable service parts.

There are many variations on this theme; some systems have many more echelons, some have fewer. Nonetheless, they are similar in structure. Note, however,

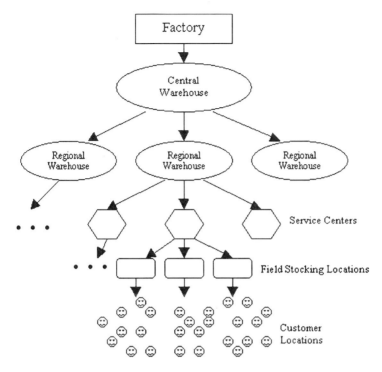

**Fig. 1.1.** A Typical Service Parts Resupply Network

that while the basic structure may be similar to the one represented in Figure 1.1, the detailed characteristics of the actual operating environments can vary dramatically, as we shall discuss shortly.

Let us now examine another system in which failed parts are repaired after they are removed from an assembly. We illustrate such a system in Figure 1.2, which corresponds to one often found in military environments. There are two echelons in this system consisting of a set of bases supported by a depot. Flying activity occurs at the lower echelon, which we call bases. When an item on an aircraft becomes inoperable, it is removed from an aircraft. The failed part is then either entered into the base's repair facility, sent to the upper echelon, which we call the depot, where repairs can also take place, or condemned. The part repair location depends on the nature of the failure. To replenish condemned parts, new ones are purchased from external suppliers. In any case, the aircraft is repaired by removing a serviceable unit of the same type from base supply and placing it in the aircraft. Base supply is resupplied either from the base's repair shop or from depot supply, depending on where the failed part was repaired. We shall study this and related problems extensively in this book.

There are obvious differences in the significance of the items serviced by the systems depicted in Figures 1.1 and 1.2. Different item characteristics create op-

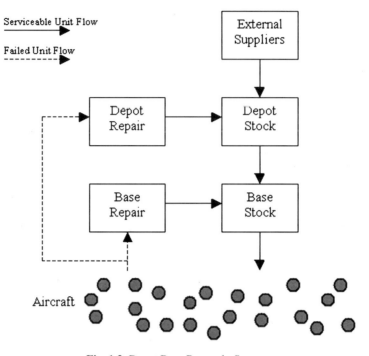

**Fig. 1.2.** Depot Base Resupply System

erational differences between systems, whether the systems are of the same type or not.

### 1.1.1  The Items

Each service parts resupply system is designed to accommodate the items found in it. The systems, and the items within them, can have varying characteristics.

First, systems differ in the number of items that are managed. In some environments, there are just a few hundred or a few thousand items. In other cases, there may by hundreds of thousands of items, or, as is the case for the Department of Defense, there may be a few million items. Because of the differences in the number of items, very different methods must be employed to manage the system's operation. In some cases, rudimentary reporting systems coupled with decision support systems for ordering and shipping parts can easily be used to manage the system using spreadsheet level software run on personal computers. This is roughly the case for General Motor's SATURN division. However, for the remainder of General Motors, the management system is extraordinarily complex because of the number of parts considered and the volumes of parts moved per year.

Second, the demand rate among items can vary substantially. Demand rates of items also differ dramatically by location within a resupply system, as well as between different resupply systems. Figure 1.3 illustrates this difference for one system we have examined. As the graph in this figure shows, a high fraction of total unit demand is concentrated in a small fraction of the items. Over 80% of the demand is concentrated in about 8% of the items. Over 50% of the lowest demand rate items constitute about 1% of the annual unit demand. In automotive

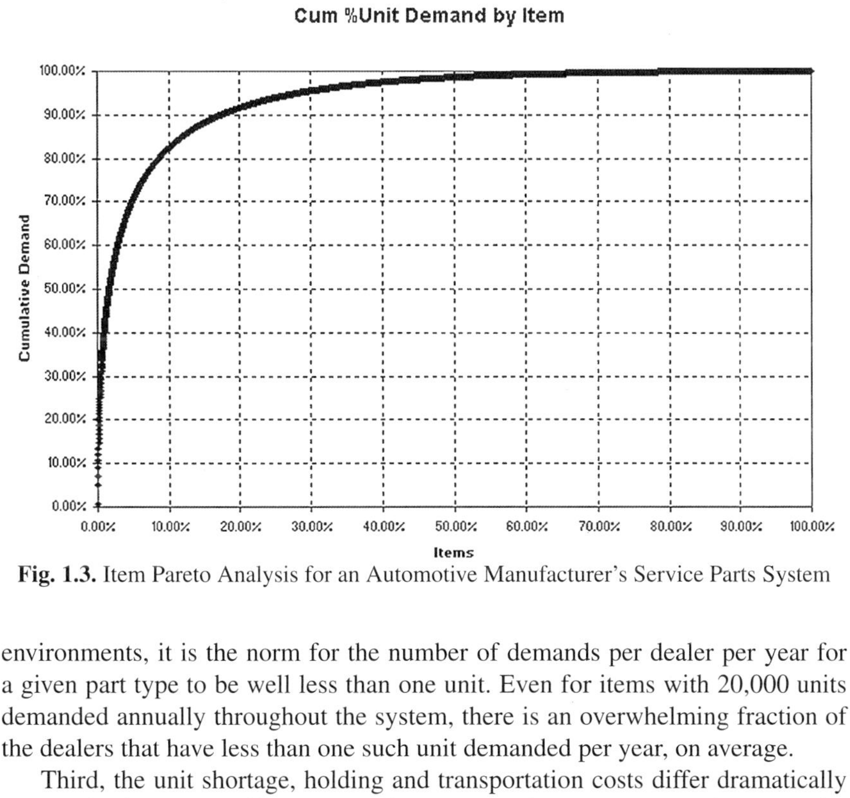

**Cum %Unit Demand by Item**

**Fig. 1.3.** Item Pareto Analysis for an Automotive Manufacturer's Service Parts System

environments, it is the norm for the number of demands per dealer per year for a given part type to be well less than one unit. Even for items with 20,000 units demanded annually throughout the system, there is an overwhelming fraction of the dealers that have less than one such unit demanded per year, on average.

Third, the unit shortage, holding and transportation costs differ dramatically among the items as well. In automotive cases, unit costs can vary from pennies per unit to thousands of dollars per unit. Engines stocked by the US Air Force cost millions of dollars per engine. Thus there is a huge difference in holding and investment costs experienced in different systems. There is also a substantial difference in shortage costs when a system is out of stock. It is clear that having an aircraft wait many hours to be repaired has a far greater cost than waiting for a hood ornament on a car, although these costs are often difficult to estimate. Nonetheless, these backorder costs are estimated either directly or indirectly. For example, in environments in which fill rate constraints are imposed, shortage costs are usually not stated explicitly. But these constraints imply a shortage cost. In some models we will study, we will explicitly consider the costs of holding inventory and backordered units. In others, we will focus on investment costs, while enforcing minimum fill rate performance requirements. We also note that transportation costs can be a substantial component of operating a service parts resupply system. In some instances the total cost of moving material can amount to hundreds of millions of dollars per year. The size and weight of each item along with their demand rates obviously determine the volume, weight, and quantity of material that must be transported. But the mode of transport selected to move this material is an important factor in determining the annual transportation costs.

Fourth, the procurement, transportation, and other components of lead times associated with each item determine the amount of inventory carried in the system in two ways. There is pipeline stock that exists because of the time it takes to receive orders after they are placed, that is, the resupply time. Based on Little's Law, this time results in an average number of units in the resupply system. Thus, the choices of suppliers, transportation modes, and, as we will see, inventory policies, all affect the average resupply lead time and hence the average pipeline stock. Furthermore, lead times are not always constant. For example, the length of time it takes to ship material from a General Motors warehouse in Michigan to another warehouse in Boston varies substantially from shipment to shipment. Another factor that influences the resupply time is the inventory policy followed by the supplying location. When orders are shipped immediately because stock is on hand at the supplier, then resupply lead times are one value. If the supplier does not have stock available to ship, then the resupply action is delayed for some amount of time. This uncertainty in lead times is an important factor when setting stock levels. We note that the average and uncertain length of resupply lead times also affects the second type of stock that is required: safety stock. There will be inherent variability in the demand processes for each item. The degree of difference in this variation of demand can be substantial. The data displayed in Figures 1.4 through 1.10 show demand for several items that we have observed in one environment. Uncertain demand over uncertain replenishment lead times yields a requirement for safety stock. In many real world situations, safety stock is the predominant component of total stock for most items. All the models we will consider are based on the premise that demand over replenishment lead times is governed by a random process.

Fifth, as we mentioned earlier, some service parts are consumable and some are repairable after they fail. We will consider both types of items in our analyses; however, the majority of our discussions will focus on repairable items. This focus is largely a result of both the economic importance of such items as well as personal interest.

There are many other characteristics associated with items that are of importance when setting inventory levels. These include the physical characteristics of the items (physical volume, weight, shape), the special temperature and humidity storage requirements, the possibility of items becoming obsolete, and the substitutability of one item type for another. The models we will construct do not consider these other factors.

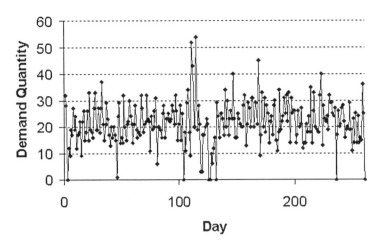

**Fig. 1.4.** Time Series of Demand for an Item

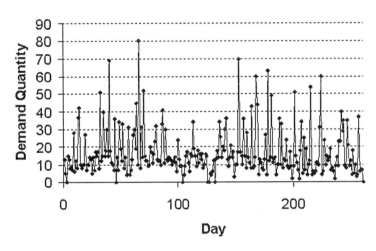

**Fig. 1.5.** Time Series of Demand for an Item

### 1.1.2 Inventory Policies

Many types of inventory policies are found in practice and are discussed in academic literature. These range from policies that are location specific, such as reorder point/reorder quantity or (s,S) policies, to echelon-stock-based inventory policies in which total system stock and system performance across all items at all locations are considered. The inventory position at a location for an item is equal to its on-hand plus on-order minus backordered inventory. Reorder points are normally expressed in terms of inventory position. When following an (s,S) policy, a location places an order when its inventory position falls to s or below

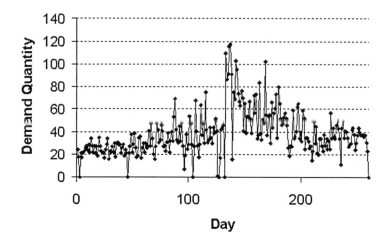

**Fig. 1.6.** Time Series of Demand for an Item

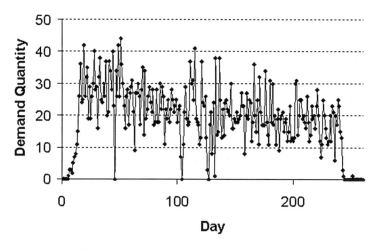

**Fig. 1.7.** Time Series of Demand for an Item

and an order is placed to raise the inventory position to S. Echelon stock in a resupply network refers to the inventory position at that location plus all the inventory found in the resupply system for successor (downstream) locations in the resupply network. In some environments, inventory levels are monitored continuously while in others they are monitored only periodically. Policy implementation obviously depends on whether reviews are continuous or periodic.

One important class of policies are called base stock, order-up-to, or (s–1,s) policies. When employing these policies in a continuous review environment, an order is placed every time a demand arises. The quantity ordered equals the quan-

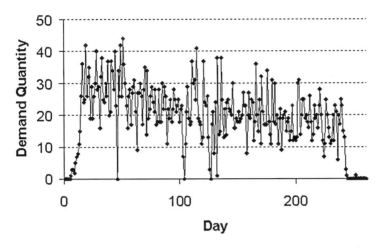

**Fig. 1.8.** Time Series of Demand for an Item

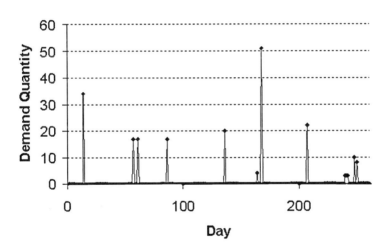

**Fig. 1.9.** Time Series of Demand for an Item

tity demanded. In periodic review situations, an order is placed in a period to raise the inventory position to some specified level. In both cases, some target inventory level, based on either echelon or installation inventory position, is used to trigger an order. Thus, whenever the inventory position is below s when a review occurs, an order is placed immediately to raise the inventory position for the location to s.

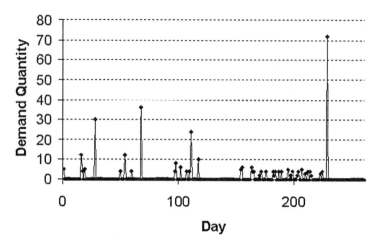

**Fig. 1.10.** Time Series of Demand for an Item

## 1.2  An Overview of the Book

The environments that we will study in detail are a subset of those that we have briefly discussed in the previous section. First, we will limit ourselves to constructing models for expensive items. We assume the costs of these items are high enough so that all items are managed using either continuous review or periodic review (s-1,s) policies. For the most part, we will assume that demand rates for each item are low so that the stochastic processes generating the demands are represented by discrete-valued random variables. Specifically, we assume that the random variables representing the number of units demanded over lead times are integer-valued.

Second, the resupply networks that we will examine are multi-echelon in nature. In some cases, we will study two-echelon systems and in others three-echelon systems.

Third, the items in certain cases are assumed to be consumable while in others they are assumed to be repairable. The models are often constructed in two steps. To begin, we develop probability distributions of the number of units on-order or in repair at each location. Next, we usually construct economic models and algorithms that can be used to calculate stock level values for each location in the multi-echelon, multi-item systems. These economic models are based on the probability model of the number of units on-order or in repair at all locations in the resupply system.

Fourth, we will consider both periodic and continuous review situations. These models are constructed to represent different operating environments and to address different questions.

The contents of this book reflect some of our interests and experiences related to the mathematical analysis of service parts systems. As such our goal is to synthesize only a portion of the literature on this topic. While our focus will be on reviewing many important results, the book is in no sense encyclopedic. However, to make the reader aware of a larger set of materials on this topic of service parts inventories, we have provided an extensive bibliography on the subject.

Let us now summarize the contents of the remainder of the book.

Recall that we will restrict our attention to (s–1,s) policies for controlling inventories. In the next chapter we will prove that such policies are optimal in single location and serial type systems with both constant and random lead times. Two alternative methods of proof are presented, which are due to Karlin and Scarf [147] and Muharremoglu and Tsitsiklis [184].

Once we establish that the class of (s–1,s) policies is optimal for a broad class of problems, we turn our attention in Chapter 3 to proving some key results used throughout the book. We establish mathematical properties associated with important performance measures, develop rudimentary optimization models, and demonstrate how alternative methods can be used to find system stock levels. To demonstrate these properties and techniques we consider single location systems. The material found in Chapter 3 was originally published largely by Feeney and Sherbrooke [89], Fox and Landi [92], and Everett [80].

We present an exact analysis of a two-echelon depot-base system for repairable parts in Chapter 4. This analysis focuses on the development of the exact distribution of the number of units on-order (in resupply) at each location in the two-echelon system. We will see that while these exact distributions can be calculated, the computational effort is too great to be of practical value. A work by Simon [230] provides the basis for much of the material provided in this chapter.

In the fifth chapter, we consider a variety of tactical planning models for repairable items. Most are based on the works of Sherbrooke [223], Graves [99], and Muckstadt [174]. These models contain approximations to the exact distributions developed in the fourth chapter. We develop economic models and provide algorithms for computing stock levels at each echelon and location for a collection of items. These models are formulated to optimize a system performance measure subject to constraints on investment in inventory. The solution methods we examine are of two types; one is a marginal analysis method, and the other employs a Lagrangian relaxation technique. We also introduce models that consider multiple indenture relationships for the repairable items. Multi-indentured repairable items are ones that contain components that are themselves repairable. Both continuous and periodic review models are discussed in this chapter.

A continuous-time model for the management of consumable items is the topic considered in Chapter 6. In most of the models developed in earlier chapters, the objective functions measure the expected number of backorders outstanding at a random point in time subject to a single constraint on system investment. Other models have as their objective the minimization of the average per period cost of carrying inventory and shortages. The objective considered in this model is the minimization of investment in system inventory subject to a collection of constraints associated with time-based fill rate requirements for various contracts. These contract constraints exist at each level of a three echelon system. In addition, we present an algorithm for finding stock levels for each item at each location in the three echelon system. This material is due to Caggiano, Muckstadt, Jackson and Rappold [37].

Several models are presented in Chapter 7 that address the effectiveness of pooling and lateral resupply in multi-echelon settings. The models developed in previous chapters preclude the sharing of inventories among locations in the same echelon. We now consider a few models that permit this sharing of stock among locations. Two of these models are continuous review models and a third is a periodic review model. These models are due to Axsäter [17], Lee [156], and Caggiano, Jackson, Muckstadt, and Rappold [35]. We also show the effect on inventory levels of having a three echelon rather than a two echelon resupply system. A periodic review model, due to Eppen and Schrage [78], and a continuous review model are presented to demonstrate the reasons for operating three echelon systems.

All models developed to this point assume that repair capacity or production capacity is unlimited. Resupply times for an order are assumed to be independent of the resupply times for all other orders. In Chapter 8, we investigate the impact of capacity on setting inventory levels. Both continuous-time and periodic review models are developed. Some basic ideas are presented related to the presence of capacity constrained systems. These ideas are given in Roundy and Muckstadt [206].

The final two chapters address environments in which the demand and resupply processes are not necessarily stationary. A nonstationary generalization of Palm's theorem is discussed first in Chapter 9. A time dependent representation of the probability distribution of the random variables measuring the number of units in resupply at a depot and its bases is given. This model permits us to estimate, for example, the effect on aircraft availability when both the demand and repair processes are nonstationary. In Chapter 10, we present models that address real-time decision-making rather than tactical planning, which is the focus of the models in the first eight chapters. These real-time models are used to determine what to repair in a period and what to allocate to bases from a depot. Furthermore, two models consider both emergency and regular resupply of the bases by the depot. The material in this chapter is due to Caggiano, Muckstadt, and Rappold [39].

Clearly there are many topics related to service parts management that are not discussed in this book. As mentioned earlier, the topics covered are ones with which we have had experiences and interest over the past several decades. For additional summaries of other topics on this subject, we refer the reader to the excellent review articles written by Nahmias [186] and Daniel, Guide and Srivastava [67].

# Background: Analysis of (s–1, s) and Order-Up-To Policies

Throughout our analyses, we limit our attention to (s–1,s) policies or order-up-to policies in the continuous review and periodic review cases, respectively. Since there are many types of policies that can be invoked, the obvious question is why we confine our attention to only these types. While these policies are intuitively appealing and hence are often applied in practice because of this appeal, it is nonetheless important for us to know that they are optimal in many circumstances. In this chapter, we will show the optimality of these policies in several environments via two different methods of proof.

We begin by considering a single location, single item system in which the inventory is measured at the end of each period and an order is placed at that time. In this environment, demand is assumed to be independent and identically distributed from period to period, cost functions are the same in each period, and the planning horizon is infinite. Fixed ordering costs are assumed to be negligible compared to other costs. Resupply lead times are assumed to be constant and known. We show the optimality of the order-up-to policy in this case using a classical dynamic programming approach following a proof by Karlin and Scarf [147].

We next show the optimality of the (s–1,s) policy for managing a single item in both single location and serial systems. Again, ordering decisions are made periodically. Demand in each period is described by a discrete random variable and is independent from period to period. Resupply lead times are assumed to be random variables with the property that lead times of successive orders do not cross. The method of proof is based on novel ideas presented by Muharremoglu and Tsitsiklis [184].

The optimality of order-up-to policies in serial systems was first shown by Clark and Scarf [49]. They introduced the notion of echelon stocks and proved that each stage in the system makes its ordering decision based on its echelon stock level. They employed a dynamic programming approach in their analysis, which we will summarize subsequently.

Finally, we will study the optimality of the (s–1,s) policy when inventory levels are monitored continuously. In this case, we will study in detail the situation where demand is described by a Poisson process. We also discuss the cases in which demand is described by a compound Poisson process or a general renewal process. We will carefully examine the case when resupply lead times are constant; however, we also will comment on the cases where lead times are described by random variables. The method of proof is new, although based on the observations of Muharremoglu and Tsitsiklis [184].

## 2.1  Optimality of Order-Up-To Policies in a Single Location, Periodic Review, Backorder Environment

We examine a situation in which we periodically review the inventory level at a single location and, based on this level, decide how much, if anything, should be ordered. The time between making ordering decisions is fixed. Demand in each period is characterized by a random variable that possesses a positive, continuous density function, which we denote by $g(x)$. While we can extend the ideas to cases with arbitrary demand distributions, we limit the discussion to this case to simplify notation and technical details. Demand is also independent from period to period.

We assume the system operates as follows. At the beginning of each period, inventory arrives that was ordered $\tau$ periods previously. An order is then placed, if required. At the end of the period, demand occurs, and period costs are charged. We assume there are three types of costs: ordering, holding and backorder costs. Ordering costs are incurred proportional to the quantity ordered in a period. Holding and backorder costs are charged proportional to the number of units of stock on-hand or backordered at the period's end.

Let $y$ represent the net inventory at the beginning of a period, after the arrival of stock ordered $\tau$ periods in the past, $q_j$ the amount ordered previously and due to arrive $j$ periods in the future, $c$ the unit purchase cost, $h$ the per unit holding cost, $b$ the per unit backorder cost, $\tau$ the resupply lead time, and $\alpha$ the discount rate with $\alpha < 1$. The quantity $q_1$ will arrive at the beginning of the next period, $q_2$ will arrive in the following period, and $q_{\tau-1}$ will arrive just after the beginning of the $(\tau - 1)^{st}$ period in the future. Let $u$ be the amount ordered in the current period. Finally, let $f(y, q_1, \ldots, q_{\tau-1})$ represent the minimum expected cost when following an optimal policy given $(y, q_1, \ldots, q_{\tau-1})$, which completely describes the state of the system.

In this environment, the dynamic programming functional equation is

$$f(y, q_1, \ldots, q_{\tau-1}) =$$
$$\min_{u \geq 0} \left\{ c \cdot u + L(y) + \alpha \int_0^\infty f(y + q_1 - x, q_2, \ldots, q_{\tau-1}, u) g(x) \, dx \right\} \quad (2.1)$$

where

$$L(y) = \begin{cases} h \int_0^y (y-x)g(x)\,dx + b \int_y^\infty (x-y)g(x)\,dx, & y > 0, \\ b \int_0^\infty (x-y)g(x)\,dx, & y \le 0. \end{cases}$$

We noted that $(y, q_1, \ldots, q_{\tau-1})$ represents the system's state at the period's beginning before an order of size $u$ is placed. Hence $u$ is a function of the system's state. An important question relates to how $u$ depends on this state. This relationship is made clear in the following theorem.

**Theorem 1.** *The optimal policy for managing the system, $u(y, q_1, \ldots, q_{\tau-1})$, is a function of the system's inventory position, which is equal to $y + q_1 + \cdots + q_{\tau-1}$.*

The theorem states that the optimal order policy depends only on the value of the inventory position and not on the specific values of $y$ and $q_j$, $j = 1, \ldots, \tau-1$. While this result may be obvious, it requires proof, which we now present.

*Proof.* Observe that

$$\begin{aligned} f(y, q_1, \ldots, q_{\tau-1}) &= \min_{u \ge 0} \Big\{ c \cdot u + L(y) \\ &\quad + \alpha \int_0^\infty f(y + q_1 - x, q_2, \ldots, q_{\tau-1}, u)g(x)\,dx \Big\} \\ &= L(y) + \min_{u \ge 0} \Big\{ c \cdot u \\ &\quad + \alpha \int_0^\infty f(y + q_1 - x, q_2, \ldots, q_{\tau-1}, u)g(x)\,dx \Big\}. \quad (2.2) \end{aligned}$$

Clearly $u^*$, an optimal value for $u$, is a function of $(y + q_1, q_2, \ldots, q_{\tau-1})$. Thus $f(y, q_1, \ldots, q_{\tau-1}) = l(y) + p(y + q_1, q_2, \ldots, q_{\tau-1})$, where $l(y) = L(y)$. Observe that the optimal cost function depends on $q_1$ only by knowing $y + q_1$. Substituting this function into (2.2), we see that

$$\begin{aligned} f(y, q_1, \ldots, q_{\tau-1}) &= l(y) + \min_{u \ge 0} \Big[ c \cdot u + \alpha \int_0^\infty [l(y + q_1 - x) \quad\quad (2.3) \\ &\quad + p(y + q_1 + q_2 - x, q_3, \ldots, q_{\tau-1}, u)]g(x)\,dx \\ &= l(y) + l_1(y + q_1) \\ &\quad + \min_{u \ge 0} \Big[ c \cdot u + \alpha \int_0^\infty p(y + q_1 + q_2 - x, q_3, \ldots, q_{\tau-1}, u)g(x)\,dx \Big]. \quad (2.4) \end{aligned}$$

Observe that this functional equation implies that $u$ is a function of $y + q_1 + q_2, q_3, \ldots, q_{\tau-1}$. Substituting the resulting function $u$ of these values shows that

$$f(y, q_1, \ldots, q_{\tau-1}) = l(y) + l_1(y + q_1) + p_1(y + q_1 + q_2, q_3, \ldots, q_{\tau-1}).$$

Obviously, this line of reasoning can be repeated and ultimately shows that

$$f(y, q_1, \ldots, q_{\tau-1}) = l(y) + l_1(y + q_1) + \cdots + l_{\tau-1}(y + q_1 + \cdots + q_{\tau-1})$$

and that $u^*$ is of the form $u^* = u^*(y + q_1 + \cdots + q_{\tau-1})$. □

The function $l_j(y + q_1 + \cdots + q_j)$ represents the expected holding and backorder costs incurred in the $jth$ period in the future, $j < \tau - 1$. These costs depend on the cumulative supply, $y + q_1 + \cdots + q_j$, and the cumulative demand through that period. Thus if cumulative supply exceeds cumulative demand, then inventory will be on-hand and carrying costs will be incurred; otherwise, cumulative demand exceeds cumulative supply and backorders will exist and backorder costs incurred.

In summary, we have demonstrated that the optimal ordering policy has the property that the order quantity that minimizes expected future discounted costs is a function of the inventory position at the time the order is placed.

We now turn our attention to the exact structure of the optimal ordering policy. The form of the policy is given in the following theorem.

**Theorem 2.** *Given linear purchase, holding and backorder costs, the optimal policy is of the following form: there exists a value $s^*$ such that the order quantity $u^*$ is*

$$u^* = \max \left\{ 0, s^* - \left( y + \sum_{i=1}^{\tau-1} q_i \right) \right\}.$$

*Proof.* We prove this theorem in the manner presented by Karlin and Scarf [147]. To simplify notation, we assume $\tau = 1$. It is straightforward to show that Theorem 2 holds in 1- and 2-period problems; the proof is left to the reader. We begin by assuming the planning horizon is $n$ periods long, $n \geq 2$, indexing the periods from earliest to latest by $n, n - 1, \ldots, 1$, respectively. We then show that the earliest period's optimal order quantity $u_n^*$ is of the desired form, and then induct on $n$. Later we let $n \to \infty$. Since $n$ is finite, we construct the recursion for $f_n(y)$, the minimum expected discounted cost if the current inventory position is $y$ given that there are $n$ periods remaining in the planning horizon. The recursion is

$$f_n(y) = \min_{u \geq 0} \left\{ c \cdot u + L(y) + \alpha \int_0^\infty f_{n-1}(y + u - x) g(x) \, dx \right\}.$$

Suppose $u_n^*(y)$ is the optimal solution to this $n$ period problem, $n \geq 2$, and suppose we have shown that this optimal policy is

$$u_n^*(y) = \begin{cases} s_n^* - y, & \text{if } y < s_n^*, \\ 0, & \text{otherwise,} \end{cases} \tag{2.5}$$

where the value of $s_n^*(y)$ is the unique solution of

$$c + \alpha \int_0^\infty f_{n-1}'(s_n^* - x) g(x) \, dx = 0. \tag{2.6}$$

We also assume for some $n \geq 2$ that we have shown that

(a) $s_n^* \geq s_{n-1}^*$,

(b) $f_n'(y) = \begin{cases} -c + L'(y), & y < s_n^*, \\ L'(y) + \alpha \int_0^\infty f_{n-1}'(y - x)g(x)\,\mathrm{d}x, & y \ge s_n^*, \end{cases}$

(c) $f_n(y)$ is a convex function, with $f_n''(y)$ existing everywhere except possibly at $y = s_n^*$; at $s_n^*$, both right and left hand derivatives exist,

(d) $f_n'(y) \le f_{n-1}'(y)$.

Given that (a) through (d) hold, and that $u_n^*(y)$ is given by the policy form stated in (2.5) and satisfies equation (2.6), for $j = 1, \ldots, n$, we now show that (a) through (d) are also satisfied by the optimal policy for period $n + 1$. The recursion for the $n + 1$ period horizon is

$$f_{n+1}(y) = \min_{u \ge 0} \left\{ c \cdot u + L(y) + \alpha \int_0^\infty f_n(y + u - x)g(x)\,\mathrm{d}x \right\}.$$

The derivative of the minimand with respect to $u$ is

$$c + \alpha \int_0^\infty f_n'(y + u - x)g(x)\,\mathrm{d}x. \tag{2.7}$$

Suppose $w = y + u$, the total supply available to meet the total demand for periods $n + 1$ and $n$. Then we restate (2.7) as

$$F_n(w) = c + \alpha \int_0^\infty f_n'(w - x)g(x)\,\mathrm{d}x. \tag{2.8}$$

To make the problem interesting, we assume that $b > \frac{1-\alpha}{\alpha}c$ ; i.e., the potential backorder cost outweighs the discount obtained by deferring purchases.

Recall that $f_n(y)$ is a convex function. Hence $F_n(w)$ is a nondecreasing function of $w$ because $f_n''(y) \ge 0$. When $y \le 0$, recall that

$$L(y) = b \int_0^\infty (x - y)g(x)\,\mathrm{d}x \quad \text{and}$$

$$L'(y) = -b \int_0^\infty g(x)\,\mathrm{d}x = -b.$$

From property (b), observe that as $w \to -\infty$,

$$F_n(w) = c + \alpha \int_0^\infty f_n'(w - x)g(x)\,\mathrm{d}x$$

$$= c + \alpha \int_0^\infty [-c - b]g(x)\,\mathrm{d}x = (1 - \alpha)c - b\alpha < 0.$$

Note that $\lim_{w \to \infty} F_n(w) = \lim_{w \to \infty} \left\{ c + \alpha \int_0^\infty f_n'(w - x)g(x)\,\mathrm{d}x \right\}$.

By the dominated convergence theorem,

$$\lim_{w \to \infty} \left\{ c + \alpha \int_0^\infty f_n'(w - x)g(x)\,\mathrm{d}x \right\} = c + \alpha \int_0^\infty \lim_{w \to \infty} f_n'(w - x)g(x)\,\mathrm{d}x,$$

because $f_n'(\cdot)$ is bounded above and below by $h/(1-\alpha)$ and $-(c+b)$, respectively. Since $\lim_{w\to\infty} f_n'(w-x) > 0$, $\lim_{w\to\infty} F_n(w) > 0$, too. Because $F_n(w)$ is a continuous function, $F_n(w)$ must have at least one point $w$ for which $F_n(w) = 0$. We have assumed that $g(x) > 0$ and $x > 0$. Now $F_n(w)$ is constant for an interval beginning at $-\infty$ and then strictly increases as $w$ increases. Therefore, $F_n(w)$ has a unique value $w$ for which $F_n(w) = 0$. This value is $s_{n+1}^*$.

Remember that $f_n'(y) \leq f_{n-1}'(y)$. Then

$$F_n(w) - F_{n-1}(w) = \alpha \int_0^\infty \left[ f_n'(w-x) - f_{n-1}'(w-x) \right] g(x)\,dx$$

$$\leq 0$$

and thus $F_n(w) \leq F_{n-1}(w)$. This in turn implies that $s_{n+1}^* \geq s_n^*$.

Suppose $y < s_{n+1}^*$. Then the minimand of the following problem

$$\min_{u \geq 0} \left\{ cu + L(y) + \alpha \int_0^\infty f_n(y + u - x)g(x)\,dx \right\}$$

occurs for $u_{n+1}^* = s_{n+1}^* - y$, where $s_{n+1}^*$ solves $F_n(w) = 0$.

We must now ensure that (a) through (d) hold for $f_{n+1}(\cdot)$. We have already shown that $s_{n+1}^* \geq s_n^*$.

Now we have

$$f_{n+1}(y) = \begin{cases} c \cdot \left[ s_{n+1}^* - y \right] + L(y) \\ \quad + \alpha \int_0^\infty f_n(s_{n+1}^* - x)g(x)\,dx, & y < s_{n+1}^*, \\ L(y) + \alpha \int_0^\infty f_n(y - x)g(x)\,dx, & y \geq s_{n+1}^*. \end{cases} \tag{2.9}$$

Hence

$$f_{n+1}'(y) = \begin{cases} -c + L'(y), & y < s_{n+1}^*, \\ L'(y) + \alpha \int_0^\infty f_n'(y - x)g(x)\,dx, & y \geq s_{n+1}^*, \end{cases} \tag{2.10}$$

which establishes property (b).

Next, let us show that property (d) holds. When $y > s_{n+1}^* \geq s_n^*$,

$$f_{n+1}'(y) = L'(y) + \alpha \int_0^\infty f_n'(y - x)g(x)\,dx$$

$$\leq L'(y) + \alpha \int_0^\infty f_{n-1}'(y - x)g(x)\,dx = f_n'(y);$$

when $y \leq s_n^*$,

$$f_{n+1}'(y) = -c + L'(y) = f_n'(y);$$

finally, when $s_n^* < y < s_{n+1}^*$,

$$f'_{n+1}(y) = -c + L'(y).$$

But, $-c < \alpha \int_0^\infty f'_{n-1}(y-x)g(x)\,dx$ in this range. Thus $f'_{n+1}(y) = -c + L'(y) < L'(y) + \alpha \int_0^\infty f'_{n-1}(y-x)g(x)\,dx = f'_n(y)$ when $y > s^*_n$. Hence, property (d) is established for $f_{n+1}(y)$.

Given our assumptions, $L(y)$ is a convex function of $y$. Observe also that $f'_n(y)$ is a continuous function of $y$, since $F_{n-1}(s^*_n) = 0$. Furthermore, $f''_{n+1}(y)$ exists everywhere except possibly at $y = s^*_{n+1}$ (at $y = s^*_{n+1}$ both right and left hand derivatives exist), and $f''_{n+1}(y) \geq 0$ because $f''_n(y) \geq 0$. Hence $f_{n+1}(y)$ is convex and possesses the properties expressed in (c).

Thus we conclude the induction step. To complete the proof, we need to verify properties (a) through (d) hold for recursions $f_2(y)$ and $f_1(y)$, where $f_1(y) = L(y)$. The verification of these properties follows using the same method of analysis we have just completed; we leave this as an exercise.  □

To this point, we have assumed that $n$ is finite. As $n \to \infty$, we conjecture that there exists a finite $s^*$ such that

$$u^*(y) = \begin{cases} s^* - y, & y < s^*, \\ 0, & \text{otherwise,} \end{cases}$$

where $s^*$ is the unique solution of

$$F(s^*) = c + \alpha \int_0^\infty f'(s^* - x)g(x)\,dx = 0$$

and $f'_n(y) \to f'(y)$ as $n \to \infty$. It is easy to see that $s^*$ is the unique solution of

$$(1 - \alpha)c + \alpha \int_0^\infty L'(s^* - x)g(x)\,dx = 0.$$

We will not prove this conjecture, however.

The induction proof we presented for the case where $\tau = 1$ can easily be extended to the general resupply lead time case where $\tau$ is an integral number of periods. In this case, we have the following theorem.

**Theorem 3.** *Given our cost model with lead time $\tau$, the optimal policy is*

$$u^* = \begin{cases} s^* - (y + \sum_{j=1}^{\tau-1} q_j), & y + \sum_{j=1}^{\tau-1} q_j < s^*, \\ 0, & \text{otherwise,} \end{cases}$$

*where $s^*$ solves*

$$(1 - \alpha)c + \alpha^\tau \int_0^\infty L'_{\tau-1}(s^* - x)g(x)\,dx = 0,$$

*and $L_j(y) = \int_0^\infty L_{j-1}(y - x)g(x)\,dx$, $L_0(y) = L(y)$.*

## 2.2  Optimality of Order-Up-To Policies in Serial Systems

In the previous section, we demonstrated how dynamic programming can be used to prove the optimality of base-stock policies in single stage (single installation) systems. In their seminal paper, Clark and Scarf [49] proved the optimality of base-stock policies for uncapacitated, periodic review, finite horizon, serial systems using a dynamic programming approach to obtain their results. The system they studied consists of $M$ physical installations, in which installation $j$ procures inventory from $j + 1$, $j = 1, 2, \ldots, M$. Installation $M + 1$ is assumed to have an infinite supply of inventory. In any period, the amount procured by installation $j$ is limited to the inventory on hand at $j + 1$. External demand occurs only at installation 1. The lead time between installations $j + 1$ and $j$ is assumed to be a known and constant number of periods. The cost model consists of the following components: (i) linear purchasing and shipping costs for moving inventory between successive stages, (ii) linear holding and shortage costs at installation 1, (iii) for the higher numbered installations, holding costs charged proportional to *the stock in each echelon*, that is, *echelon stock*. Echelon stock at installation $j$ ($2 \leq j \leq M$) is defined to be inventory on hand at installation $j$ plus inventory in transit to installation $j$ plus echelon stock for installation $j - 1$; echelon stock at installation 1 is defined as the inventory position at installation 1. Thus echelon $j$ stock is the total inventory on hand plus on-order at installation $j$ plus all inventory downstream of $j$ less any backorders at installation 1. We note that linear purchase and shipping costs can be assumed to be zero under very general assumptions, as shown by Janakiraman and Muckstadt [139]. A brief discussion of Clark and Scarf's proof follows.

The state variable in Clark and Scarf's dynamic programming formulation for determining the optimal inventory policy for this system consists of a vector that specifies the amount of inventory in each stage of the system (including each stage of the pipeline between two installations). They prove that the cost of operating this system can be decomposed into a sum of costs, one corresponding to each echelon. The cost at echelon $j$, $j > 1$, is a function of the echelon inventory position at that installation. An additional term is included that measures the expected cost impact on installation $j - 1$ of not meeting the target inventory level. Furthermore, they showed that the cost function associated with each echelon has exactly the same form as that of a single stage system. Consequently, the optimal policy for each installation is an echelon base-stock policy. That is, there is a target level $s_j$ corresponding to every installation $j$. An order is placed in every period to raise the echelon $j$ stock to its inventory position, $s_j$, which is the echelon $j$ stock plus the inventory in transit to installation $j$, if possible. If the inventory on hand at installation $j + 1$ is insufficient, installation $j$ orders all the on-hand inventory at installation $j + 1$.

This dynamic programming approach has been successfully used by several authors in the last forty five years to establish the forms of the optimal policies for several inventory systems. Another proof approach was introduced by Federgruen and Zipkin [84] to prove the optimality of echelon base-stock policies in the infinite horizon case when the performance measure is the infinite horizon average cost per period. Subsequently, the same method was used by Chen and Song [44] to prove the optimality of state dependent base-stock policies for these serial systems when demand is Markov modulated. The arguments are based on a lower bound on cost, an interesting approach which is powerful and worth learning.

A third approach for establishing the forms of optimal policies in inventory systems is the "single-item single-customer" approach introduced recently by Muharremoglu and Tsitsiklis [184]. They proved that state dependent echelon base-stock policies are optimal for uncapacitated multi-echelon serial systems for both the finite and infinite horizon models when lead times and demands are Markov modulated. We will now present and discuss their approach. For ease of exposition, we will first restrict ourselves to analyzing a single stage system with deterministic lead times and a finite planning horizon. We will then discuss how to extend the ideas when the lead times are random and there are multiple stages.

### 2.2.1   The Single-Unit Single-Customer Approach: Single Location Case

As in Section 1, we will analyze a single location inventory system in which we manage a single item. We assume that time is divided into periods of equal length. We also assume the system operates as follows. At the beginning of each period, an order is placed on an outside supplier and arrives exactly $m - 1$ periods in the future. Subsequent to the time the order is placed, the order due in that period is received from the supplier; customer demands are then observed. Demand in each period is governed by an exogenous, stationary Markov Chain. All excess demand is backordered. At the end of each period, both holding and backorder costs are incurred.

#### 2.2.1.1   Notation and Definitions

We begin our analysis by presenting the notation and key definitions. As stated, we initially assume the planning horizon of the system consists of $N$ periods, numbered $n = 1, 2, \ldots, N$, in that order. We assume that there is an exogenous finite-state, ergodic Markov Chain $s_n$ that governs the demand process, where $s_n$ is observed at the beginning of each period $n$. The transition probabilities for the Markov Chain $s_n$ are assumed to be known. Furthermore, given $s_n$, the probability distribution of $D_n$, the demand in period $n$, is known.

We consider each unit of demand as an individual customer. Suppose at the beginning of period 1 there are $v_0$ customers waiting to have their demand satisfied. We index these customers $1, 2, \ldots, v_0$ in any order. All subsequent customers are indexed $v_0 + 1, v_0 + 2, \ldots$ in the order of the period of their arrivals, arbitrarily breaking ties among customers that arrive in the same period.

Next, we define the concept of *the distance of a customer* at the beginning of any period. See Figure 2.1. Every customer who has been served is at distance 0; every customer who has arrived, placed an actual order, but who has not yet received inventory, is at distance 1; all customers arriving in subsequent periods are said to be at distances 2, 3, ... corresponding to the sequence in which they will arrive. Distances are assigned to customers that arrive in the same period in the same order as their indices. This ensures that customers with higher indices are always at "higher" distances.

Next, we define the concept of a *location for a unit*. Again see Figure 2.1. There are $m + 2$ possible locations at which a unit can exist. If the unit has been used to satisfy a customer's order, the location of this unit is 0. If it is part of the inventory on hand, it is in location 1. If the unit has not been ordered from the supplier, it is in location $m + 1$.

At the beginning of period 1, we assign an index to all units in a serial manner, starting with units at location 1, then location 2, ..., location $m+1$, and arbitrarily assign an order to units present at the same location. We assume a countably infinite number of units is available at the supplier, that is, location $m + 1$, at all times.

We will use indices $j$ and $k$ to denote the indices of units as well as customers. We define $y_{jn}$ to be the distance of customer $j$ at the beginning of period $n$ and $z_{jn}$ to be the location of unit $j$ at the beginning of period $n$.

Number units and customers 1,2,3... based on their order of availability/arrival (present and future)

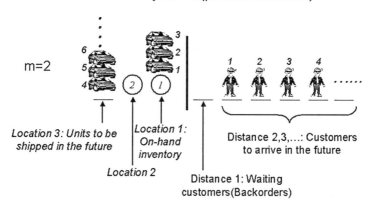

**Fig. 2.1.** Locations of Units and Distances of Customers

We define the state of the system at the beginning of period $n$ to be the vector $x_n = (s_n, (z_{1n}, y_{1n}), (z_{2n}, y_{2n}), \ldots)$.

Next, we explain the sequence of events in period $n$ in detail.

1. $s_n$ is observed. $(z_{jn}, y_{jn})$ is known for all $j$, $j = 0, 1, 2, \ldots$ .
2. An order is placed, which we denote by $q_n$, where $q_n$ is a nonnegative integer. All units in locations $j = 2, 3, \ldots, m$ move to the next location prior to placing the order, that is, location $j - 1$. The $q_n$ units move from location

$m + 1$ to location $m$. If it has been ordered from the supplier $\ell$ periods ago ($1 \leq \ell < m$), it is in location $m - \ell$.

3. Demand $D_n$ is realized and these new customers arrive and are at distance 1. That is, customers at distances $2, 3, \ldots, 2 + D_n - 1$ all arrive and are, by definition, now at distance 1. All customers at distances $2 + D_n, 3 + D_n, \ldots$ at the beginning of the period move $D_n$ steps towards distance 1.
4. Units on-hand and waiting customers are matched to the extent possible. That is, as many waiting customers are satisfied as possible and as many units on hand are consumed as possible.
5. $h$ dollars are charged per unit of inventory remaining on hand (at location 1) and $b$ dollars are charged per waiting customer (at distance 1). Clearly, only one of these costs will be incurred in any period. We assume $b > h$. This ensures that if the inventory position is negative in some period, then the optimal policy will be to increase the inventory position to some nonnegative level.

The performance measure under consideration in our initial discussion is the expected sum (discounted or undiscounted) of costs over the $N$ period planning horizon.

Next, we define a *policy*. Let $u_{jn} \in \{Release, Hold\}$ denote the decision made in period $n$ for unit $j$. By *Release*, we mean that an order is placed for a unit and that it enters the supplier's production/distribution system. However, note that the only units over which we have control are the units at the supplier, that is, at location $m + 1$. The movement of all other units is governed by the lead time and demand processes as defined previously. A *policy* is a function that maps every possible realization of $x_n$ to a vector of Release/Hold actions for each unit at location $m + 1$.

Observe, however, that when a decision is made to release $q$ units from location $m + 1$, it does not matter which $q$ units are released. Consequently, we can consider a class of policies that always releases the units with the smallest indices from $m + 1$. We call such a policy a *monotone policy*. *A class of policies is said to be optimal if it contains at least one optimal policy.* The class of monotone policies is clearly optimal. A monotone policy releases unit $j$ before or at the same time as unit $j + 1$ but never in any period following the one in which unit $j + 1$ is released. Similarly, we define a *monotone state* to be one where lower indexed units are in the same or lower indexed locations. That is, $z_{kn} \leq z_{jn}$ if $k \leq j$. Since lead times are the same for all units and we started period 1 in a monotone state, the system is always in a monotone state when a monotone policy is followed. Furthermore, by definition, the customers also arrive in the order of their indices. We define a policy as a *committed policy* if it ensures that the only customer that the $j^{\text{th}}$ unit can satisfy is customer $j$'s demand and that the only unit that customer $j$ can receive is the $j^{\text{th}}$ unit. Assume that the units at location 1 that are picked to satisfy customers at distance 1, as well as the customers at distance 1 picked to consume units at location 1, are those with the lowest indices. Hence, every monotone policy is also a committed policy. Consequently, the class of committed policies is also optimal. Since this is an important fact, we state this as a lemma.

**Lemma 1.** *The class of monotone policies is optimal. Furthermore, every monotone policy is a committed policy and hence the class of committed policies is also optimal.*

In the next section, we develop a proof of the optimality of base-stock policies for periodic review, single stage uncapacitated systems of the type we have described.

### 2.2.1.2  Optimality of Base-stock Policies

In this section, we first show that the system can be decomposed into a collection of countably infinite subsystems, each having a single unit and a single customer. Subsequently, we prove that each subsystem can be managed optimally by using a policy we call a "critical distance" policy. We prove that when the same "critical distance" policy is used to manage each subsystem, the system follows a base-stock policy.

#### 2.2.1.2.1  Decomposition of the System into Subsystems

Let us first outline the proof technique. First, we observe that the cost of the system is the sum of the costs incurred for each unit-customer pair. Second, we show that each of these pairs can be controlled independently and optimally and that the resulting policy is optimal for the entire system. Third, we examine the individual unit-customer problem and show that the optimal policy is a "critical distance" policy: Release a unit if and only if the corresponding customer is closer than a critical distance. Last, we observe that operating each unit-customer pair using a critical distance policy produces an echelon base-stock policy in the original system.

Let us now precisely define the concepts of the system, the subsystems, and the sets of constraints that govern these systems and subsystems.

**Definition 1.** *Let $S$ refer to the entire system with all the units and all the customers. Subsystem $w$, represented by $S_w$, $1 \leq w$, refers to the unit-customer pair with index $w$.*

**Definition 2.** *Constraints on Monotone and Committed Policies in $S$:*
   **Monotonicity:** *Unit $j$ ( $j = 1, 2, \ldots$ ) can not be released before unit $j - 1$.*
   **Commitment:** *Unit $j$ ( $j = 1, 2, \ldots$ ) serves customer $j$.*

**Definition 3.** *Constraint on Committed Policies in $S_w$:*
   **Commitment:** *Unit $w$ serves customer $w$.*

We will now show that the optimal cost for the system $S$ is equal to the sum of the optimal costs for the subsystems $S_w$. We will prove this fact by demonstrating that every monotone and committed policy for system $S$ corresponds to a set of monotone and committed policies for the subsystems, $S_w$, and that any set of monotone and committed policies for the subsystems yields a feasible policy for the system $S$. We will also show that when the individual subsystems are managed "independently and optimally", the resulting policy for the system $S$ is optimal.

From now on, we will use $\tilde{S}$ to denote the group of all subsystems, that is, $\tilde{S} = (S_1, S_2, \dots)$. When only monotone and committed policies are considered, the constraints in definition 2 apply to $S$ while the constraint in definition 3 applies to $\tilde{S}$. When we say "the (optimal) expected cost for $\tilde{S}$", we mean the sum of the (optimal) expected subsystem costs.

We have assumed so far that $x_n$ is the state information available to us while managing the entire system $S$ or any subsystem $S_w$. *However, observe that the subsystems are "operationally independent" in the sense that each subsystem can be managed independently without being affected by the policies used to manage the other subsystems.* Consequently, we can find an optimal policy for managing $S_w$ that uses only those parts of the state vector $x_n$ that pertain to unit $w$ and customer $w$. We define $x_n^w =_{def} (s_n, z_{wn}, y_{wn})$. Thus, $x_n^w$ is a sufficient state descriptor for $S_w$. This means that an optimal policy for $\tilde{S}$ can be found by managing the subsystems independently. A subtle point to be noted here is that the subsystems, though operationally independent, are stochastically dependent through the demand process.

We are now ready to state and prove the results relating the optimal costs and policies for the system $S$ and the subsystems $S_w$, $w = 1, 2, \dots$.

**Theorem 4.** *For any starting state $x_1$ in period 1, the optimal expected discounted (undiscounted) cost in periods 1, 2, ..., N for system $S$ equals the optimal expected discounted (undiscounted) cost in periods 1, 2, ..., N for the group of subsystems $\tilde{S}$. Furthermore, when each subsystem $w$ is managed independently and optimally using the state vector $x_n^w$ in every period n, the resulting policy is optimal for the entire system, $S$.*

*Proof.* First, observe that the cost incurred by $S$ is the sum of the costs incurred by every unit and every customer, since the holding and backorder costs are linear.

Second, observe that every monotone and committed policy for $S$ produces a set of committed policies, one for each subsystem $S_w$. Consequently, the optimal expected cost for $\tilde{S}$ is a lower bound on the optimal expected cost for $S$ over any number of periods, since a monotone and committed policy is optimal for $S$.

Third, observe that operating each subsystem independently using any committed policy is a feasible policy for $S$. Consequently, the optimal expected cost for $S$ is a lower bound on the optimal expected cost for $\tilde{S}$.

Combining the two lower-bound arguments above proves that the optimal expected costs for $S$ and $\tilde{S}$ are equal. The earlier discussion about the "operational independence" of the subsystems and this equality result show that when each subsystem $w$ is managed independently and optimally using the state vector $x_n^w$, the resulting policy for the entire system, $S$, is optimal.                                $\square$

Next, we show the existence of an optimal policy with a very special structure for every subsystem.

### 2.2.1.2.2   Optimal Policy Structure for a Subsystem

Before examining an individual subsystem, we first observe that all subsystems are identical in the sense that (i) they have identical cost structures and (ii) given a state $(x_n^w)$ and a fixed operating policy for a subsystem, the stochastic evolution of the subsystem is independent of the index $w$. Consequently, the optimal policies are identical across all subsystems.

We define $R_n^*(s_n, y) \subseteq \{Release, Hold\}$ to be the set of optimal decisions for subsystem $w$ at time $n$ if the state of the exogenous Markov Chain is $s_n$ and if $y_{wn}$ is $y$ and if $z_{wn}$ is $m + 1$.

Next, we show that there is a "critical distance" policy that is optimal for a subsystem. We need the following Lemma to prove this fact. The lemma states that if it is *uniquely* optimal for subsystem $w$ to release unit $w$ (if it is at location $m + 1$) in period $n$ when the system is in the Markovian-state $s_n$ and customer $w$ is at a distance $y + 1$, then it would be optimal to release it if the customer were any closer and the unit were at location $m + 1$.

**Lemma 2.** $R_n^*(s_n, y + 1) = \{Release\}$ *implies that* $R_n^*(s_n, y) \supseteq \{Release\}$.

*Proof.* The proof is by contradiction. Assume the statement is not true. That is, there exists $n$, $s_n$ and $y$ such that $R_n^*(s_n, y + 1) = \{Release\}$ and $R_n^*(s_n, y) = \{Hold\}$. Another way of saying this is as follows: it is suboptimal for a subsystem to hold unit $w$ if customer $w$ is at a distance $y + 1$ while it is suboptimal for a subsystem to release unit $w$ if customer $w$ were at a distance $y$.

Consider some monotone and committed policy for $S$. Assume the exogenous Markov Chain is at state $s_n$ in period $n$ and that we can find subsystems $w$ and $w + 1$ such that $y_{wn}$ is $y$ and $y_{(w+1)n}$ is $y + 1$. Monotonicity implies that this policy would choose one of the following three pairs of actions for units $w$ and $w + 1$: (a) release both $w$ and $w + 1$, (b) hold both $w$ and $w + 1$ and (c) release $w$ and hold $w + 1$.

Cases (a) and (c) are suboptimal for subsystem $w$, while cases (b) and (c) are suboptimal for subsystem $w + 1$ due to our initial assumption. This implies that any monotone and committed policy for $S$ is suboptimal for at least one of subsystems $w$ and $w + 1$. So, any monotone and committed policy for $S$ has a higher expected cost than the optimal cost for $\tilde{S}$ from period $n$ onwards, which is the same as the optimal cost for $S$. This implies that no monotone and committed policy can be optimal for $S$, which contradicts our earlier assertion about the optimality of some such policy. Therefore, our assumption about $R_n^*(s_n, y)$ and $R_n^*(s_n, y + 1)$ is invalid.                                $\square$

We use this Lemma to develop the notion of a "critical distance" policy. Let us define

$$y^*(n, s_n) \stackrel{\text{def}}{=} \max\{ \, y \, : \, R_n^*(s_n, y) \supseteq \{Release\} \, \} \, .$$

$y^*(n, s_n)$ is defined in such a way that it is optimal to release unit $w$ if and only if customer $w$ is at a distance of $y^*(n, s_n)$ or closer. This distance $y^*(n, s_n)$ is the "critical distance" in period $n$ and Markovian state $s_n$ for every subsystem.

Consider the policy

$$R_n(s_n, y) \; = \; \{Release\} \;\; \text{if and only if} \;\; y \; \leq \; y^*(n, s_n) \, .$$

Policy $R_n$ is an optimal policy for every subsystem. The next observation we make is that when policy $R_n$ is used in period $n$ for every subsystem, the resulting policy for the original system $S$ is an order-up-to policy. This can be shown either using an algebraic proof or using a more intuitive argument, which we now provide.

**Theorem 5.** *The optimal policy for $S$ is to release as many units as necessary to raise the inventory position to $y^*(n, s_n) - 1$ in period $n$ when in Markovian state $s_n$ and the planning horizon consists of $N$ periods. That is, a state dependent order-up-to or base-stock policy is optimal for the entire system when the planning horizon is finite.*

*Proof.* We know that policy $R_n$ is optimal for every subsystem. It can be seen that if $R_1, \ldots, R_{n-1}$ are the policies used on each of the subsystems in periods $1, \ldots, n - 1$, we will start period $n$ in a state where the units that are in location $m + 1$ bear consecutive labels. Consequently, the corresponding customers who have not arrived are in consecutive distances. Among these customers, those in locations $2, 3, \ldots, y^*(n, s_n)$ are all within the critical distance $y^*(n, s_n)$. All backordered customers are also within the critical distance. The policy $R_n$ dictated that we should release the waiting unit in just the right number of subsystems in period $n$ so that all waiting customers and all future customers within the critical distance can be satisfied with the units on-hand or on-order. That is equivalent to saying we would raise the inventory position to $y^*(n, s_n) - 1$.                                 □

This concludes the proof of the finite horizon, optimality result for uncapacitated single stage systems with constant lead times. Muharremoglu and Tsitsiklis [184] present the analysis of the infinite horizon problem. In the next two sections, we will discuss how this approach can be extended to more general situations.

### 2.2.2   Stochastic Lead Times

So far, we have assumed that the lead time is exactly $m - 1$ periods. Let us now relax this assumption by allowing stochastic lead times subject to the restriction that orders can not cross, that is, the sequence in which orders are received from the supplier corresponds to the sequence in which orders were placed on the supplier. We permit the lead time distribution to be governed by the Markov Chain $s_n$. This lead time model is described below.

The lead time process evolves as follows. There is a random variable $\rho_n$, whose distribution is determined completely by $s_n$, that specifies the least "age" of orders that will be delivered in period $n$ to location 1. This means all outstanding orders placed in period $n - \rho_n$ or earlier are delivered in period $n$. We assume that the sample space of the random variable $\rho_n$ is $\{0, 1, 2, \ldots, m - 1\}$ and consequently, the maximum lead time of an order is $m - 1$ periods.

The sequence of events in a period as described in Section 2.2.1.1 is now modified slightly. Due to the possibility of more than one period's orders arriving at location 1 in a period, we include the following event just prior to observing the demand in period $n$.

• $\rho_n$ *is realized; if $\rho_n \leq m - 2$, all units in locations 2 through $m - \rho_n$ arrive from the supplier and are at location 1. If $\rho_n = m - 1$, then no units arrive at location 1.*

It is easy to verify that all the analysis and results that we presented for the "deterministic lead time model" hold for the stochastic lead time model described above. Therefore, state-dependent echelon base-stock policies are optimal for the single stage system even when lead times are stochastic and noncrossing.

### 2.2.3  The Serial Systems Case

The next extension to our analysis is the case of serial systems. In the single echelon case, the only location from which a unit could be released using a control policy was location $m + 1$. In the multiple echelon case, there are more "physical locations" from which a unit can be released using the control policy. These physical locations correspond to stages in the production/distribution system. In addition, there are as many "artificial locations" between successive stages. The number of these artificial locations corresponds to the maximum possible lead time between these stages. We still assume that orders do not cross. The cost model is the same as discussed in the beginning of Section 2.2. The optimality of state dependent echelon base-stock policies can be verified by repeating the following arguments, which we used in the analysis of the single stage system.

First, the cost for system $S$ is still the sum of the costs for the subsystems because of the linear cost structure. Second, monotone and committed policies are still optimal. Third, each subsystem can be operated independently and optimally, which results in an optimal policy for system $S$. Fourth, the optimal policy for a subsystem should be such that if it is optimal to release a unit from a stage or physical location when the corresponding customer is at a distance $y$, then it would also be optimal to release the unit from that stage if the customer were any closer. Consequently, an appropriately defined critical distance policy is optimal for every subsystem. Now there is a critical distance corresponding to each stage. When this policy is used for every subsystem, the resulting policy is a state-dependent echelon base-stock policy for the system $\mathcal{S}$. The proofs of all these results are identical to the proofs for the corresponding results for the single stage system. Consequently, it is clear that state-dependent echelon base-stock policies are optimal for serial systems with Markov modulated demands and noncrossing lead times.

In the next section, we discuss how the single-unit, single-customer approach can be used for continuous review systems.

### 2.2.4   Continuous Review Systems

We now examine single stage, continuous review systems and present some modelling assumptions and arguments to extend the optimality results to these systems. Let us assume the planning horizon is infinite and the goal is to find the policy that minimizes the average cost per unit time. The following discussion is meant to be intuitive rather than technically rigorous.

The following are the main differences between continuous-time and periodic review systems. First, the echelon holding costs at all echelons as well as the backorder costs at stage 1 are now charged continuously. That is, the holding and backorder cost parameters have the units of dollars/unit/year. Second, customers and orders can arrive at any time and orders can be placed at any time, not just at pre-specified points in time.

The concept of customer distance stays the same while the location of a unit is now a continuously changing process. In fact, if $L$ is the maximum lead time, the location of unit $j$ at time $t$ is defined as (i) $L + 1$ if it is at the supplier, (ii) $(1 + L - (t - t_j))$ if it was released at time $t_j$ but has not arrived by time $t$, (iii) 0 if it has satisfied the demand of a customer and (iv) 1 if it is part of inventory on hand.

This section has several purposes. The first purpose is to argue intuitively that base-stock policies are optimal when the demands are modelled as a compound Poisson process. More will be said about these processes in the next chapter. The next purpose is to extend the argument to the more general case of compound renewal demand processes. We conclude the section with a procedure to compute the order-up-to level for the special case of Poisson demand processes.

#### 2.2.4.1   The Optimality Proof for Compound Poisson Demand Processes

Let us now examine the case where demands are modelled as a compound Poisson process; i.e., the size of a customer order is a random variable with an arbitrary distribution; the arrival of customers follows a stationary Poisson process. Lead times are assumed to be constant. It is intuitively clear that the following propositions that we proved in the periodic review case hold for the continuous review case as well, where the objective is to minimize the average cost per unit time. First, monotone and committed policies are optimal. Second, managing each unit-customer pair independently and optimally produces an optimal policy for the entire system.

Consider any unit-customer pair $j$ at any time $t$ such that the unit is at the supplier. Intuitively, the only quantity that the optimal *Release/Hold* decision should depend on is the amount of time left for customer $j$ to arrive. This "time to arrive" is a random variable, the distribution of which can be completely determined using $y_{jt}$, the distance of customer $j$ at time $t$ because of the memoryless property of the Poisson process. Furthermore, the class of policies that is restricted to *Release* item $j$ only at customer arrival epochs is optimal because of the memoryless property.

The system is now identical to a periodic review system except that the length of a period is now the time between the arrival of two consecutive customer orders. Recall, that the proof that a critical distance policy is optimal for a single-unit single-customer subsystem, in the periodic review case, does not use the fact that all periods are of equal length. Consequently, this proof holds for this continuous review system as well. Similarly, the proof of the claim that using a critical distance policy for every subsystem produces a base-stock policy for the original system does not depend on the length of a period. Therefore, order-up-to or base-stock policies are optimal for the continuous review system with compound Poisson demands.

### 2.2.4.2  The Optimality Proof for Compound Renewal Demand Processes

Next, suppose that the time between the receipt of two consecutive customer arrivals is a random variable. These inter-arrival times are independent and identically distributed; that is, the customer arrival process is a renewal process. The number of units ordered by each customer is described by a discrete random variable whose distribution function is arbitrary. The compound Poisson model is a special case of this model. Lead times are assumed to be constant.

In the compound Poisson case, it was sufficient to restrict *Release/Hold* decisions to customer arrival epochs because of the memoryless property of the Poisson process. However, in this more general renewal process environment, it might be optimal to release a unit from the supplier at a point in time between two arrivals since the distribution of the time until the next arrival changes through time continuously.

The arguments about a critical distance policy being optimal for the single-unit single-customer problem and about using an optimal policy for each pair to produce an optimal policy for the entire system are still valid. The only difference between the compound Poisson case and the more general compound renewal case is that the critical distance is now a function of the time since the time of the arrival of the previous customer. Consequently, state dependent base-stock policies are optimal for these systems, where the state includes the time since the previous arrival.

In fact, the interarrival times, the demand size at each arrival epoch, and the lead times could be dependent on a continuous-time Markov Chain $\{s_t\}$. The preceding discussion holds for these generalities as well. Note, however, that our discussion pertains only to the proof of the optimality of state dependent base-stock policies and not the computation of the optimal base-stock level, which could be a computational challenge when a Markov modulated demand environment is considered.

### 2.2.4.3   A Computational Procedure for Minimizing the Expected Steady State Cost when Demands are Poisson

We now consider a single stage system that is reviewed continuously. The performance measure of interest is the expected steady state cost per unit time. The demand process is assumed to be a Poisson process with arrival rate $\lambda$. We will first present a set of conjectures for this system, without presenting rigorous proofs. Finally, we use these conjectures to develop a computational procedure to determine the optimal base-stock level for this system.

First, we present a conjecture about the expected long run inflow of units.

*Conjecture 1.* Let $IN(t)$ be the expected number of units released into the system by time $t$. Then, $\lim_{t \to \infty} [IN(t)/t] = \lambda$. That is, the expected long run rate of inflow of units equals the expected long run demand rate, $\lambda$.

Next, we present a conjecture that relates the expected steady state cost per unit time to the expected total cost incurred by every unit-customer pair.

*Conjecture 2.* Let $C^*$ be the optimal expected steady state cost per unit time for $S$. Let $\mu^*$ be the expected cost incurred by any unit-customer pair when it is managed by the policy that minimizes the total expected cost incurred by this pair during the time interval $[0, \infty)$. Then, $C^* = \lambda \cdot \mu^*$. Furthermore, each subsystem (unit-customer pair) can be managed optimally with the expected total cost performance measure and the resulting policy is optimal for $S$ with the steady state expected cost per unit time performance measure.

This conjecture implies that it is sufficient to find the policy that minimizes the total expected cost associated with each unit-customer subsystem.

Next, we present a computational procedure to determine such a policy for every subsystem. First, we develop some necessary notation.

Let $\mu(y)$ be the expected cost associated with a subsystem during the time interval $(t, \infty)$ if the distance of the customer at time $t$ is $y$ and the unit is just released. Let $t + t(y)$ be the arrival time of this customer. That is, $t(y)$ is the length of time for $y - 1$ arrivals to take place. Given that the arrival process is Poisson, $t(y)$ is gamma distributed with parameters $(y - 1, 1/\lambda)$. Therefore, $\mu(y)$ can be calculated as

$$\mu(y) \;=\; E_{t(y)} \left( h \cdot (t(y) - L)^+ \;+\; b \cdot (L - t(y))^+ \right) .$$

Since monotone policies are optimal, it has to be true that if it is optimal to release a unit when the corresponding customer is at a distance $y$, then it would be optimal to release the unit if the customer were at distance $y - 1$. In other words, $\mu(y)$ is such that it first decreases in $y$ and then increases in $y$ with the minimum occurring at the smallest value of $y$ such that $\mu(y + 1) \geq \mu(y)$. This value of $y$, which we denote by $y^*$, is that distance that triggers a release.

The optimal policy for the original system is therefore an order-up-to $(y^* - 1)$ policy.

## 2.3 Problem Set, Chapter 2

**2.1.** In Section 2.1, we proved the optimality of an order-up-to policy for a particular single location environment (Theorem 2). The proof was by induction. We assumed that properties (a) through (d) held for recursions $f_2(y)$ and $f_1(y)$ as defined in that section. Complete the proof of this theorem by verifying properties (a) through (d) do hold when there are but two periods in the planning horizon.

**2.1** Suppose there are three periods in the planning horizon and demand in each period has an exponential distribution with a mean of 5 units. Suppose the lead time is one period in length and there are currently 5 units on order. The unit cost is \$20, the holding cost is \$1 per unit per period, the backorder cost is \$10 per unit per period. The discount factor, $\alpha$, is equal to 1. Find the optimal policy for the planning horizon.

**2.2.** Prove Theorem 3.

**2.3.** In Section 2.2.1, we proved the optimality of order-up-to policies for a single location system by employing a "single-unit, single-customer" approach. We assumed that an exogenous finite state, ergodic Markov Chain governs the demand process. Suppose this chain is trivial, that is, it has only a single state $s$. Determine the states that the system attains in the first five periods given the following information. The maximum lead time is 4 periods. At the beginning of period 1, there are 3 backorders and there is no inventory in the system. Furthermore, $q_1 = 10, q_2 = 7, q_3 = 12, q_4 = 3, q_5 = 8; \rho_1 = 0, \rho_2 = 2, \rho_3 = 2, \rho_4 = 1, \rho_5 = 0; d_1 = 5, d_2 = 8, d_3 = 14, d_4 = 5, d_5 = 7$.

**2.4.** Suppose the operating environment described in Section 2.1 and Section 2.2 of this chapter is modified slightly. Suppose all inventory left over at the end of the planning horizon can be returned at the purchase cost $c$, and suppose all backorders that exist at the end of the horizon can be cleared by purchasing an equal amount of inventory. This purchased quantity arrives instantaneously and at a cost of $c$ per unit. In this modified environment, show that the costs $(c, h, b)$ can be transformed into a new set of costs $(\tilde{c}, \tilde{h}, \tilde{b})$ where $\tilde{c}$ is zero. Furthermore, show that a myopic policy is optimal for the finite horizon and infinite horizon cases where the cost model is either to minimize the expected discounted cost or the average cost per period.

**2.5.** Assume the system described in Section 2.1 and Section 2.2 is altered as follows: demand in period $n + \tau$ is observed in period $n$. That is, advance demand information is available for $\tau$ periods into the future. How do the results change when (a) the order lead time is longer than $\tau$, and (b) when the lead time is shorter than $\tau$?

**2.6.** Consider a single stage inventory system with the following characteristics: (a) it is reviewed continuously, (b) ordering costs are linear, that is, there is no fixed ordering cost, (c) there are linear holding and backorder costs, (d) the demand process is Poisson, and (e) the order lead time is $\tau$. For this system, conjecture the structure of the optimal ordering policy. Outline a proof based on the single-unit decomposition approach discussed in Section 2.2. Can you extend this to a stationary demand process with an arbitrary interarrival distribution and an arbitrary distribution of order sizes?

# 3

# Background Concepts: An Introduction to the (s–1, s) Policy under Poisson and Compound Poisson Demand

We will now discuss the implications of following a (s–1,s) inventory policy when inventories are reviewed continuously in time. Recall that the stock level, $s$, measures the amount of inventory on-hand plus on-order minus backorders, that is, the stock level represents the inventory position for a particular location. In certain situations, we will refer to the on-order quantity as the "in resupply" quantity. This "in resupply" terminology is often used in military and aviation applications in which items fail and are repaired or are procured from an external source. When a (s–1,s) policy is followed, an order is placed immediately whenever a demand occurs for one or more units of an item. The order quantity matches exactly the size of the demand. Hence, the inventory position is constant in this case.

Our specific objective in this chapter is to show how to compute the stationary probability distribution of the quantity of units in resupply. The amount in resupply at a random point in time is a key random variable in the study of the behavior of systems managed using a (s–1,s) policy. Once its stationary distribution is known, we can easily determine the stationary distribution for on-hand and backordered inventory. We will focus primarily on the case where backorders are allowed, since the analysis is simpler. As a special case, however, we will also analyze a situation where excess demand over supply is lost.

We first show how to compute the distribution for the quantity in resupply in the backorder case when the replenishment lead times or equivalently the resupply times are independent and identically distributed. We do this by first assuming the demand process is a Poisson process, then generalize this result to the case where the demand process is a compound Poisson process.

After we show how to calculate the stationary distributions, we show how to determine key statistical measures of supply system performance. Lastly, we present optimization models and algorithms for computing stock levels when items are managed using an (s–1,s) policy.

## 3.1  Steady State Distribution of the Number of Units in Resupply

The construction of the steady state distribution of the number of units in resupply in either the Poisson or compound Poisson demand cases follows from the properties of the underlying Poisson processes generating the orders. Let us begin by reviewing some of these properties.

Let $\lambda$ represent the demand rate of the underlying Poisson customer order process. Now, first suppose exactly one order occurs during the time interval $[0, t]$. Given that this order has occurred, let us establish the distribution of the time at which the order was placed. Intuitively, this distribution should be uniform since a Poisson process has stationary and independent increments. Let $T$ be the time at which this event occurs, and let $N(t)$ represent the number of customer orders received in $[0, t]$. Then, for $s < t$,

$$
\begin{aligned}
P[T < s \mid N(t) = 1] &= \frac{P[T < s; N(t) = 1]}{P[N(t) = 1]} \\
&= \frac{P[N(s) = 1; N(t - s) = 0]}{P[N(t) = 1]} \\
&= \frac{P[N(s) = 1] \cdot P[N(t - s) = 0]}{P[N(t) = 1]} \\
&= \frac{\lambda s\, e^{-\lambda s}\, e^{-\lambda(t-s)}}{\lambda t\, e^{-\lambda t}} \\
&= \frac{s}{t}.
\end{aligned}
$$

Hence the time at which the customer arrival occurs is uniformly distributed over the interval $[0, t]$.

This result can be generalized as follows. Suppose $X_1, \ldots, X_n$ are $n$ independent and identically distributed random variables. The random variables $X_{(1)}, \ldots, X_{(n)}$ are order statistics corresponding to $X_1, \ldots, X_n$ if $X_{(k)}$ corresponds to the $k^{\text{th}}$ smallest value among the random variables, $X_1, \ldots, X_n$. Let $f(x_i)$ represent the common density function for the $X_i$. Then the joint density function for the $X_{(i)}$ is

$$
f_{X_{(1)}, \ldots, X_{(n)}}(x_1, \ldots, x_n) = n! \Pi_{i=1}^{n} f(x_i), \quad x_1 < \cdots < x_n. \tag{3.1}
$$

The term $n!$ appears because there are that many permutations of $X_1, \ldots, X_n$ that lead to the same order statistic.

Now suppose that $N(t) = n$ and suppose $X_1, \ldots, X_n$ are the arrival times of the $1st, 2nd, \ldots, nth$ customer orders, respectively. Then $X_1, \ldots, X_n$ have the same distribution as do the order statistics corresponding to $n$ independent random variables that have uniform distributions over the interval $[0, t]$. We can prove this fact in the following manner.

Suppose we have times $t_1, \ldots, t_n$, where $0 < t_1 < t_2 < \cdots < t_n < t$, and $\Delta_i$ small enough in value so that

$$t_i + \Delta_i < t_{i+1} \text{ and } t_n + \Delta_n < t. \tag{3.2}$$

Then

$$P[t_1 \leq X_1 \leq t_1 + \Delta_1, \ldots, t_n \leq X_n \leq t_n + \Delta_n | N(t) = n]$$

$$= \frac{P[1 \text{ cust order is placed in } [t_i, t_i + \Delta_i], i = 1, \ldots, n, \text{ and no cust orders are placed elsewhere in } [0, t]]}{P[N(t) = n]}$$

$$= \frac{(\lambda \Delta_1 e^{-\lambda \Delta_1}) \ldots (\lambda \Delta_n) e^{-\lambda \Delta_n} \cdot (e^{-\lambda(t - \sum_{i=1}^{n} \Delta_i)})}{e^{-\lambda t} \frac{(\lambda t)^n}{n!}} = \frac{n!}{t^n} \Pi_{i=1}^{n} \Delta_i,$$

and therefore $\dfrac{P[t_1 \leq X_1 \leq t_1 + \Delta_1, \ldots, t_n \leq X_n \leq t_n + \Delta_n | N(t) = n]}{\Delta_1 \ldots \Delta_n} = \dfrac{n!}{t^n}.$

Taking the limit of the left hand side as $\Delta_i \to 0$ for all $i$, we obtain

$$f_{X_1, \ldots, X_n}(t_1, \ldots, t_n) = \frac{n!}{t^n}, \quad 0 < t_1 < t_2 < \cdots < t_n < t, \tag{3.3}$$

which is the desired result. Thus, we may conclude that if $n$ customer orders are placed in $[0, t]$, then the times at which these orders are placed, considered as unordered times, are independent and uniformly distributed over the interval $[0, t]$.

### 3.1.1  Backorder Case

We are now ready to establish a remarkable result, which is a restatement of a theorem attributed to Palm [191].

**Theorem 6.** *Suppose s is the stock level for an item whose demands are generated by a Poisson process with rate $\lambda$. Suppose further that the resupply time random variables have density functions $g(\tau)$ with mean $\bar{\tau}$, and have distribution functions $G(\tau)$. Suppose further that the resupply times are independent and identically distributed from customer order to customer order. Then the steady state probability that x units are in resupply is given by*

$$e^{-\lambda \bar{\tau}} \frac{(\lambda \bar{\tau})^x}{x!}. \tag{3.4}$$

*Proof.* Suppose $N(t) = n$ customer orders have been placed in $[0, t]$. We know that

$$P[N(t) = n] = e^{-\lambda t} \frac{(\lambda t)^n}{n!}. \tag{3.5}$$

Since a (s–1,s) policy is employed to manage the inventory, each customer order generates a corresponding request on the resupply system. Next, let

$$q_t(x|n) = P[x \text{ units are in resupply at time } t | N(t) = n]. \tag{3.6}$$

Consider any one of the $n$ orders. As we just demonstrated, the time of its placement is uniformly distributed over the interval $[0, t]$. Suppose this order was
placed at time $s \epsilon [0, t]$. Then the probability that the corresponding unit remains
in the resupply system at time $t$ is $1 - G(t - s)$.

Let $p$ be the common probability that any unit that arrives during $[0, t]$ remains in the resupply system at time $t$. Since $1 - G(t - s)$ measures the conditional probability that the unit entering the resupply system at time $s$ remains
unsatisfied at time $t$, the unconditional probability is given by

$$
\begin{aligned}
p &= \int_0^t [1 - G(t - s)] \frac{ds}{t} \\
&= \frac{1}{t} \int_0^t [1 - G(t - s)] \, ds \\
&= -\frac{1}{t} \int_t^0 [1 - G(u)] \, du \\
&= \frac{1}{t} \int_0^t [1 - G(u)] \, du.
\end{aligned}
$$

Since each arriving order in $[0, t]$ has a probability $p$ that its corresponding
resupply request is not satisfied by time $t$, the probability that $x$ of the $n$ arriving
units in the resupply system remain in it at time $t$ is given by

$$
q_t(x|n) = \binom{n}{x} p^x (1 - p)^{(n-x)}. \tag{3.7}
$$

Now the unconditional probability that $x$ units remain in the resupply system
at time $t$ is

$$
\begin{aligned}
q_t(x) &= \sum_{n=x}^{\infty} q_t(x|n) \cdot P[N(t) = n] \\
&= \sum_{n=x}^{\infty} \binom{n}{x} p^x (1 - p)^{n-x} e^{-\lambda t} \frac{(\lambda t)^n}{n!} \\
&= \sum_{n=x}^{\infty} \frac{n!}{(n-x)! x!} p^x (1 - p)^{n-x} e^{-\lambda t} \frac{(\lambda t)^n}{n!} \\
&= \frac{e^{-\lambda t} (p\lambda t)^x}{x!} \sum_{n=0}^{\infty} \frac{[\lambda t (1 - p)]^n}{n!} \\
&= \frac{e^{-\lambda t} e^{\lambda t - \lambda t p} (p\lambda t)^x}{x!} \\
&= e^{-\lambda t p} \frac{(\lambda t p)^x}{x!}.
\end{aligned}
$$

But $p = \frac{1}{t} \int_0^t [1 - G(u)] \, du$, so

$$q_t(x) = e^{-\lambda \int_0^t [1 - G(u)] du} \frac{[\lambda \int_0^t [1 - G(u)] du]^x}{x!}. \tag{3.8}$$

Let

$$q(x) = \lim_{t \to \infty} q_t(x). \tag{3.9}$$

Recall that

$$\lim_{t \to \infty} \int_0^t [1 - G(u)] du = \int_0^\infty [1 - G(u)] du = \bar{\tau}, \tag{3.10}$$

and therefore

$$q(x) = e^{-\lambda \bar{\tau}} \frac{(\lambda \bar{\tau})^x}{x!}. \tag{3.11}$$

Hence, the probability that there are $n$ units in the resupply system is Poisson distributed with mean $\lambda \bar{\tau}$; i.e., we do not need to know the density function for the resupply time, but only the mean of the resupply time, $\bar{\tau}$.                    □

Let us now turn to the case where the demand process is a compound Poisson process. In this case, customer orders arrive according to a Poisson process, but the order quantity is not necessarily for one unit. We assume the order quantities for arriving customers are independent and identically distributed where the probability that an order is of size $j$ is represented by $u_j$.

There are two important and commonly used choices for the values of $u_j$. One is where the compounding distribution is a geometric distribution, that is, where

$$u_j = (1 - p)p^{j-1}, \; j \geq 1, \; 0 \leq p \leq 1; \tag{3.12}$$

the other is where the compounding distribution is a logarithmic distribution with

$$u_j = -\frac{(1 - p)^j}{j} (\ln p)^{-1}, \; 0 < p < 1, j \geq 1. \tag{3.13}$$

We will discuss the implications of using each of these two distributions. But first let us consider some general properties of compound Poisson distributions.

Let $\{X_k\}$ be the set of mutually independent and identically distributed random variables corresponding to customer order sizes. Then

$$P[X_k = j] = u_j \text{ for all } k. \tag{3.14}$$

Furthermore, the generating function for $X_k$ for all $k$ is given by

$$G(v) = \sum_j v^j u_j. \tag{3.15}$$

We are interested in the distribution of the total number of units ordered by $N$ customers, that is, the distribution of $S_N$, where

$$S_N = X_1 + \cdots + X_N. \tag{3.16}$$

The random variable $N$ is independent of the random variables $X_k$. Suppose we let

$$g_n = P[N = n] = e^{-\lambda t} \frac{(\lambda t)^n}{n!}. \tag{3.17}$$

Let $w_m$ be the probability that the total number of units ordered by the random number of customers is equal to $m$. Then

$$w_m = \sum_{n=0}^{\infty} P[N = n] \cdot P[X_1 + \cdots + X_n = m]. \tag{3.18}$$

For a given value of $n$, the distribution of $X_1 + \cdots + X_n$ is simply the $n$-fold convolution of the distribution $u_j$ with itself. The generating function for this convolution is $[G(v)]^n$. Also, the generating function of the sum $S_N$ is

$$H(v) = \sum_m v^m w_m$$

$$= \sum_m v^m \sum_{n=0}^{\infty} P[N = n] \cdot P[S_N = m | N = n]$$

$$= \sum_{n=0}^{\infty} \left[ \sum_m v^m P[S_N = m | N = n] \right] \cdot g_n$$

$$= \sum_{n=0}^{\infty} [G(v)]^n \cdot g_n.$$

Since $g_n = e^{-\lambda t} \frac{(\lambda t)^n}{n!}$,

$$H(v) = e^{-\lambda t + \lambda t G(v)} \tag{3.19}$$

because the generating function $f(y)$ for a Poisson distributed random variable is

$$f(y) = e^{-\lambda t + \lambda t y}. \tag{3.20}$$

Let us now return to the two example cases for the compounding distributions. When the order size distribution is a geometric distribution, the generating function is given by

$$G(v) = \sum_{j \geq 1} v^j (1 - p) p^{j-1}$$

$$= \sum_{j \geq 1} (1 - p) v (pv)^{j-1}$$

$$= (1 - p) v \sum_{j \geq 0} (pv)^j$$

$$= \frac{(1 - p) v}{1 - pv}, \quad \text{when } pv < 1.$$

Therefore,

$$H(v) = e^{-\lambda t + \lambda t \cdot \left( \frac{(1-p)v}{1-pv} \right)}.$$

Also note that

$$[G(v)]^n = \left[ \frac{(1-p)v}{1-pv} \right]^n, \tag{3.21}$$

which is the generating function of a negative binomially distributed random variable that is translated to the right. This random variable has mean $\frac{n}{1-p}$ and variance $\frac{np}{(1-p)^2}$.

When the compounding distribution is geometric, the resulting compound Poisson distribution is called a stuttering Poisson distribution.

Now suppose the compounding distribution has a logarithmic distribution. Then

$$G(v) = \sum_{n=1}^{\infty} v^n u_n$$

$$= -\sum_{n=1}^{\infty} v^n \frac{(1-p)^n}{n \ln p}$$

$$= -\frac{1}{\ln p} \sum_{n=1}^{\infty} \frac{[(1-p)v]^n}{n}$$

$$= \frac{1}{\ln p} \ln[1 - (1-p)v], \quad \text{when } |(1-p)v| < 1.$$

Suppose we let $\lambda = -\ln p$. Then

$$H(v) = e^{-\lambda t + \lambda t G(v)}$$

$$= (e^{\ln p})^t (e^{-\ln p \frac{1}{\ln p} \ln[1-(1-p)v]})^t$$

$$= \left[ \frac{p}{1-(1-p)v} \right]^t.$$

But this is the generating function for a negative binomial distribution. Hence, when the compounding distribution is logarithmic, the compound Poisson distribution describing total demand is a negative binomial distribution when $\lambda = -\ln p$.

Let us now state the generalization of Palm's theorem for the case of a compound Poisson process for the backorder case.

**Theorem 7.** *Suppose demands occur according to a compound Poisson process where $\lambda$ is the customer order arrival rate. Suppose also that the resupply times are independent and identically distributed with density $g(\tau)$ with mean $\bar{\tau}$. Assume when a customer order is received, the resupply time for all units in the* *order is the same and is drawn from the resupply time distribution. The steady state probability of x units in resupply is given by the compound Poisson distribution with mean $\lambda\overline{\tau}\overline{u}$, where $\overline{u}$ is the average customer order size.*

*Proof.* The proof of this theorem is straightforward. Again let $X$ represent the random variable describing the number of units in resupply in steady state. From Palm's theorem we know that the probability distribution for the number of customer orders in resupply has a Poisson distribution with parameter $\lambda\overline{\tau}$. If $u_n^{(j)}$ represents the probability that $j$ customers have a total demand of $n$ units, then

$$P[X = n] = \sum_{j=1}^{\infty} u_n^{(j)} e^{-\lambda\overline{\tau}} \frac{(\lambda\overline{\tau})^j}{j!}, n \geq 1, \text{ and} \qquad (3.22)$$

$$P[X = 0] = 1 - \sum_{n\geq 1} P[X = n] = e^{-\lambda\overline{\tau}}. \qquad (3.23)$$

Hence $X$ has a compound Poisson distribution with mean $\lambda\overline{\tau}\overline{u}$.  □

### 3.1.2  Lost Sales Case

To this point we have assumed that all customer orders in excess of the supply $s$ are back-ordered. Let us now assume that this is not the case; that is, when a customer order is placed and there is no on-hand inventory, then the order is lost. We will prove a version of Palm's theorem for a special case of the lost order situation. A general and complicated proof of the lost order case when demand is compound Poisson distributed is given by Feeney and Sherbrooke [89] and discussed by Baganha [23]. We will focus on a relatively simple situation where the order lead times are exponentially distributed. Specifically, the theorem that we will prove is as follows.

**Theorem 8.** *Suppose customer orders arrive according to a Poisson process with arrival rate $\lambda$. Furthermore, suppose the stock level is $s$. Assume resupply times for accepted customer orders are independent and identically distributed with common density $g(\tau) = \beta e^{-\beta\tau}$, with mean $\overline{\tau} = 1/\beta$. Then the steady state probability that $x$ units are in resupply in the lost order case is given by*

$$\frac{e^{-\frac{\lambda}{\beta}}(\lambda/\beta)^x/x!}{\sum_{n=0}^{s} e^{-\lambda/\beta}(\lambda/\beta)^n/n!} = \frac{e^{-\lambda\overline{\tau}}(\lambda\overline{\tau})^x/x!}{\sum_{n=0}^{s} e^{-\lambda\overline{\tau}}(\lambda\overline{\tau})^n/n!}.$$

*Proof.* When $g(\tau) = \beta e^{-\beta\tau}$, we can derive the desired result based on an argument used when analyzing queuing systems. Let $P_j(t)$ represent the probability that $j$ units are in resupply at time $t$. Note that if $j < 0$ or $j > s$, then $P_j(t) = 0$.

Since the order arrival process is a Poisson process and the resupply time distribution is exponential, for $0 \leq j \leq s$,

$$P_j(t + \Delta t) = [1 - (\lambda + j\beta)\Delta t]P_j(t) + \lambda \Delta t \cdot P_{j-1}(t)$$
$$+ (j + 1)\beta \Delta t \cdot P_{j+1}(t) + o(\Delta t). \quad (3.24)$$

Then

$$P_j'(t) = \lim_{\Delta t \to 0} \frac{P_j(t + \Delta t) - P_j(t)}{\Delta t}$$
$$= -(\lambda + j\beta)P_j(t) + \lambda P_{j-1}(t) + (j + 1)\beta P_{j+1}(t).$$

Passing to the limit $(t \to \infty)$, $P_j'(t) \to 0$.

Let $\pi_j$ represent the steady state probability that $j$ units are in the resupply system. Then

$$0 = -(\lambda + j\beta)\pi_j + \lambda \pi_{j-1} + (j + 1)\beta \pi_{j+1}. \quad (3.25)$$

For $j = 0$, we have

$$\lambda \pi_0 = \beta \pi_1 \quad \text{or} \quad (3.26)$$

$$\pi_1 = \frac{\lambda}{\beta} \pi_0. \quad (3.27)$$

For $j = 1$, we have

$$\lambda \pi_0 + 2\beta \pi_2 = (\lambda + \beta)\pi_1 \quad (3.28)$$

or

$$2\beta \pi_2 = (\lambda + \beta)\frac{\lambda}{\beta}\pi_0 - \lambda \pi_0 \quad (3.29)$$

$$\pi_2 = \frac{\lambda^2}{2\beta^2}\pi_0, \quad (3.30)$$

and, as is easily shown, for $0 < j < s$,

$$\pi_j = \frac{1}{j!}\left(\frac{\lambda}{\beta}\right)^j \pi_0. \quad (3.31)$$

For the case where $j = s$,

$$s\beta \pi_s = \lambda \pi_{s-1} \quad (3.32)$$

or

$$\pi_s = \frac{\lambda}{s\beta}\left[\frac{1}{(s-1)!}\left(\frac{\lambda}{\beta}\right)^{s-1}\right]\pi_0$$
$$= \frac{1}{s!}\left(\frac{\lambda}{\beta}\right)^s \pi_0.$$

Since $\sum_{j=0}^{s} \pi_j = 1$,

$$\pi_0 \sum_{j=0}^{s} \frac{1}{j!}\left(\frac{\lambda}{\beta}\right)^j = 1 \quad \text{or} \tag{3.33}$$

$$\pi_0 - \left[\sum_{j=0}^{s} \frac{1}{j!}\left(\frac{\lambda}{\beta}\right)^j\right]^{-1}$$

$$= \frac{e^{-\lambda/\beta}}{\sum_{j=0}^{s} e^{-\lambda/\beta} \frac{(\lambda/\beta)^j}{j!}}$$

$$= \frac{e^{-\lambda\bar{\tau}}}{\sum_{j=0}^{s} e^{-\lambda\bar{\tau}} \frac{(\lambda\bar{\tau})^j}{j!}}.$$

Thus,

$$\pi_j = \frac{e^{-\lambda\bar{\tau}}(\lambda\bar{\tau})^j/j!}{\sum_{i=0}^{s} e^{-\lambda\bar{\tau}}(\lambda\bar{\tau})^i/i!}. \tag{3.34}$$

□

In the general case, that is, where $g(\tau)$ is an arbitrary density with mean value $\bar{\tau}$, the steady state probability that $x$ units are in the resupply system is also given by the above expressions.

### 3.1.3  Another Backorder Case: Delayed Customer Due Dates

Let us conclude this section by returning to the backorder case. Suppose that when a customer order is received the system does not have to ship the required quantity immediately but rather must meet the demand within a time of length $T$ after the order is received. Of course, if $T = 0$, we are back to the original case we examined. When $T$ is positive or equal to 0 we have the following modification of Palm's theorem.

**Theorem 9.** *Suppose the demand process is a compound Poisson process with customer order rate $\lambda$. Suppose further that the customer order resupply times are independent and identically distributed with density function $g(\tau)$ and with mean resupply time $\bar{\tau}$. In the backorder case, the steady state probability of n units in resupply, each of which has been in the resupply system for at least T time units, is given by*

$$P[X = n] = p(n|\lambda\bar{\tau}\alpha) = \sum_{y=0}^{n} \frac{(\lambda\bar{\tau}\alpha)^y e^{-\lambda\bar{\tau}\alpha}}{y!} u_n^{(y)},$$

*where*

$$\alpha = \frac{1}{\tau} \int_T^\infty [1 - G(t)] dt, \tag{3.35}$$

$$G(t) = \int_0^t g(\tau) d\tau, \text{ and} \tag{3.36}$$

$u_n^{(y)}$ is the probability that y customers generate a total demand of n units. As we did earlier, we assume the entire customer order quantity shares a common resupply time.

We will prove this theorem for the case where the demand process is a Poisson process.

*Proof.* Suppose there are y customer orders in the resupply system at time $t + T$ that were also in the resupply system at time $t$. Then one of the following must have occurred: there were y orders in resupply at time $t$, none of which completed resupply by time $t + T$; there were $y + 1$ orders in resupply at time $t$ and one completed the resupply process by time $t + T$, and so on. Hence

$$P\Big\{ y \text{ orders are in resupply at time } t + T$$

$$\text{each of which was in the resupply system at time } t\Big\}$$

$$= \frac{e^{-\lambda\bar{\tau}}(\lambda\bar{\tau})^y}{y!} \binom{y}{0} \alpha^y (1-\alpha)^0 + \frac{e^{-\lambda\bar{\tau}}(\lambda\bar{\tau})^{y+1}}{(y+1)!} \binom{y+1}{1} \alpha^y (1-\alpha)$$

$$+ \frac{e^{-\lambda\bar{\tau}}(\lambda\bar{\tau})^{y+2}}{(y+2)!} \binom{y+2}{2} \alpha^y (1-\alpha)^2 + \cdots$$

$$= \frac{e^{-\lambda\bar{\tau}}(\lambda\bar{\tau})^y \alpha^y}{y!} \left[ 1 + \frac{(\lambda\bar{\tau})(1-\alpha)}{1!} + \frac{(\lambda\bar{\tau})^2(1-\alpha)^2}{2!} \right.$$

$$\left. + \cdots + \frac{(\lambda\tau)^n(1-\alpha)^n}{n!} + \cdots \right]$$

$$= e^{-\lambda\bar{\tau}} \frac{(\lambda\bar{\tau}\alpha)^y}{y!} e^{\lambda\bar{\tau}(1-\alpha)}$$

$$= e^{-\alpha\lambda\bar{\tau}} \frac{(\alpha\lambda\bar{\tau})^y}{y!} \qquad \qquad \Box$$

The proof when the demand process is a compound Poisson process follows due to our assumption that resupply times are the same for all units in a customer order.

## 3.2 Performance Measures

To this point, we have developed the steady state probabilities for the number of units that are in the resupply system at a random point in time when the demand process is either a Poisson or compound Poisson process. Based on these probabilities, we can calculate different measures of system performance. These measures relate to performance at a single location. In subsequent chapters, we will see how steady state probabilities and performance measures are computed in multi-echelon situations.

We will begin by considering several measures that are single-item measures, confining our discussion to the backorder case. The first performance measure we consider, the fill rate, is the most commonly used measure in practice, and is defined as follows. Given a stock level of $s$, the fill rate, $F(s)$, is the expected fraction of demands that can be satisfied immediately from on-hand stock. As is intuitively clear, as $s$ increases the fill rate will increase. We will develop an explicit expression for $F(s)$ in this section and will discuss its properties in the next section of this chapter.

A second performance measure is called the ready rate corresponding to stock level $s$. The ready rate measures the probability that an item observed at a random point in time has no backorders, that is, its net inventory is nonnegative. We denote the ready rate by $R(s)$. This is an all or nothing measure. Either there are backorders or there are no backorders at a random point in time.

Observe that when computing either a fill rate or ready rate we are not concerned with the duration of backorders when they occur. Thus, for example, a fill rate of say 95% implies that, on average, 95 of every 100 units that are ordered have that request satisfied immediately. But we are not measuring how long it takes to satisfy the other 5% of the units requested. Thus it is not always clear that a firm that maintains a high fill rate is truly satisfying its customers needs. This is particularly true when fill rates are calculated for a large number of item types. In this case, the fill rate would measure the fraction of demands satisfied immediately over all items. Thus some items could have nearly 100% of the demands satisfied immediately while others could have a 0% fill rate.

Note also that the ready rate is always at least as high as the fill rate. For example, when $s = 0$, $F(s) = 0$. But $R(s)$ could approach 1 if the demand rate is low and the lead time is very short. It is not unusual that the measures $F(s)$ and $R(s)$ are confused in practice.

A third single-item performance criterion measures the expected number of backorders outstanding at a random point in time, and is denoted by $B(s)$. This measure accounts for the length of time backorders exist. Hence, it is a response-time focused measure. Observe that $B(s)$ is equal to the demand rate times the average "waiting time" of a demand. This is a consequence of Little's law, $L = \lambda W$, where $B(s)$ is $L$, $\lambda$ the demand rate, and $W$ the average waiting time. We could also compute the conditional value of $W$, given that backorders exist.

Let us now see how these performance measures can be computed. Recall that, in the backorder case when the demand is a compound Poisson process, the steady state probability that $x$ units are in resupply is given by

$$P\{X = x\} = p(x|\lambda\overline{\tau}) = \sum_{j=1}^{\infty} e^{-\lambda\overline{\tau}} \frac{(\lambda\overline{\tau})^j}{j!} u_x^{(j)}, \ x \geq 1,$$

$$P\{X = 0\} = e^{-\lambda\overline{\tau}}$$

where $\lambda$ is the demand rate, $\overline{\tau}$ the average resupply time, and $u_x^{(j)}$ the probability that $j$ customer orders generate a total demand of $x$ units.

The ready rate is the probability that there are no backorders existing at a random point in time. This is the probability that the number of units in resupply is $s$ or less. That is,

$$R(s) = \sum_{x=0}^{s} p(x|\lambda\overline{\tau}).$$

The computation of the fill rate is more difficult, but it is obtained from the steady state probabilities, $p(x|\lambda\overline{\tau})$. Suppose a customer order is received. There will be one unit of the order satisfied if there are $s - 1$ or fewer units in resupply. A second unit will be sent to the customer if the order is for two or more units and there are $s - 2$ or fewer units in resupply. Remember that the timing and size of a customer order are independent of all past orders and resupply times. Hence, the expected number of units filled per customer order is given by

$$F_1(s) = \sum_{x \leq s-1} p(x|\lambda\overline{\tau}) + (1 - u_1) \sum_{x \leq s-2} p(x|\lambda\overline{\tau})$$

$$+ (1 - u_1 - u_2) \sum_{x \leq s-3} p(x|\lambda\overline{\tau})$$

$$+ (1 - \sum_{j \leq s-1} u_j) p(0|\lambda\overline{\tau}), \ \text{where, as before,}$$

$u_j$ measures that probability that a customer order is for exactly $j$ units. In the case of a simple Poisson demand process (that is, when $u_1 = 1$),

$$F(s) = F_1(s) = \sum_{x \leq s-1} p(x|\lambda\overline{\tau}). \ \text{Hence, in this case,}$$

$$F_1(s) = F(s) = R(s) - p(s|\lambda\overline{\tau}) \ \text{and} \ F(s) < R(s).$$

When the demand process is a compound Poisson process, $\lambda F_1(s)$ measures the expected number of units that can be shipped on time per day, when $\lambda$ is the expected daily rate at which customers place orders. Furthermore, $\lambda\overline{u}$ measures the expected number of units demanded per day, where $\overline{u}$ is the expected number of units demanded per order. Thus

$$\frac{\lambda F_1(s)}{\lambda\overline{u}} = \frac{F_1(s)}{\overline{u}}$$

measures the fraction of the units ordered that are sent to customers on time. We let this quantity be defined as the fill rate, or

$$F(s) = \frac{F_1(s)}{\overline{u}}.$$

Next, we see that the expected number of units in a backorder status in steady state is

$$B(s) = \sum_{x>s} (x - s) p(x | \lambda \overline{\tau}).$$

That is, there are $x - s$ units backordered if and only if there are $x$ units in resupply, $x > s$.

Suppose there are $n$ item types in a system rather than just a single item. Then performance measures are computed somewhat differently.

First, the system fill rate is calculated by computing the conditional fill rate for the item type, multiplying by the probability that a demand was for a specific item type, and summing over item types. Let $\overline{F}(\overline{s})$ measure the system fill rate, where $\overline{s} = (s_1, \ldots, s_n)$ is a vector of item stock levels. If $F_i(s_i)$ measures the fill rate for item type $i$, then

$$\overline{F}(\overline{s}) = \sum_{i=1}^{n} \frac{\lambda_i}{\sum_{j=1}^{n} \lambda_j} \cdot F_i(s_i)$$

because $\frac{\lambda_i}{\sum_{j=1}^{n} \lambda_j}$ is the probability that a customer demand is for item type $i$ when the customer order process is a Poisson process. This calculation is based on the assumption that an order is for a single item type.

The expected number of backorders at a random point in time for $n$ items is simply

$$\sum_{i=1}^{n} B_i(s_i) = \sum_{i=1}^{n} \sum_{x>s_i} (x - s_i) p(x | \lambda_i \overline{\tau}_i).$$

When there is more than one item, the ready rate measure must be modified. The new measure is called the operational rate. We assume a system is operational if and only if all item types are operational. A particular item type will not be available, and hence the system will not be operational, if the number of units in resupply for that item exceeds its stock level. Assuming demand and resupply times are independent from item type to item type, the operational rate is given by

$$OR(\overline{s}) = \Pi_{i=1}^{n} R_i(s_i).$$

Suppose there are many operating systems, say a fleet of aircraft. Furthermore, suppose part shortages can be consolidated into as few aircraft as possible. This process is often called "cannibalization" of the aircraft. Suppose there may be more than one unit of a particular type on an aircraft, say $q_i$ units of type $i$. Then, assuming independence and cannibalization,

$$\Pi_{i=1}^{n} R_i(s_i + q_i) = \text{probability that all aircraft}$$
$$\text{or one less than all aircraft}$$
$$\text{are operational.}$$
$$\text{and } \Pi_{i=1}^{n} R_i(s_i + kq_i) = \text{probability that } k \text{ or}$$
$$\text{fewer aircraft are nonoperational.}$$

Let $Y$ be a random variable that measures the number of nonoperational aircraft. Assuming cannibalization and independence,

$$P\{Y = 0\} = \Pi_{i=1}^{n} R_i(s_j)$$
$$P\{Y \leq 1\} = \Pi_{i=1}^{n} R_i(s_i + q_i)$$
$$P\{Y \leq k\} = \Pi_{i=1}^{n} R_i(s_i + kq_i).$$

Hence, the expected number of nonoperational aircraft at a random point in time given cannibalization and independence is

$$E[Y] = \sum_{k \geq 1} k \cdot P\{Y = k\}$$
$$= \sum_{k \geq 1} P\{Y \geq k\}$$
$$= \sum_{k \geq 1} (1 - P\{Y \leq k - 1\})$$
$$= \sum_{k \geq 0} (1 - P\{Y \leq k\}).$$

Hence, if there are $N$ aircraft, an approximation to the expected number of operational aircraft is

$$N - E[Y].$$

Since the demand process assumes an infinite population, this is a conservative estimate.

Another approximation for the expected number of operational aircraft can be developed as follows. Let us consider item type $i$. Recall that $B_i(s_i)$ measures the expected number of backorders for item type $i$ at a random point in time. Suppose there is one unit of this item type per aircraft and there are $N$ aircraft in the system. The probability that a random aircraft at a random point in time is missing a unit of item type $i$ is $\frac{B_i(s_i)}{N}$, or $1 - B_i(s_i)/N$ is the probability that the aircraft is not missing a unit of type $i$. Assuming independence, the probability that a random aircraft is operational at a random point in time is

$$p = \Pi_{i=1}^{n}(1 - B_i(s_i)/N).$$

The expected number of operational aircraft at a random point in time is given by

$$Np = N\Pi_{i=1}^{n}(1 - B_i(s_i)/N)$$

$$= N\left(1 - \sum_{i=1}^{n} B_i(s_i)/N + \sum_{k \neq j} \frac{B_j(s_j)B_k(s_k)}{N^2}\right.$$

$$\left. - \sum_{i \neq j \neq k} \frac{B_i(s_i)B_j(s_j)B_k(s_k)}{N^3} + \cdots\right)$$

$$\cong N - \sum_{i=1}^{n} B_i(s_i),$$

when $\frac{B_i(s_i)}{N}$ is small for all item types.

Thus there is a simple approximate correspondence between the expected number of backorders outstanding at a random point in time and the expected number of operational aircraft. The latter approximation of the expected number of operational aircraft is particularly useful from a computational viewpoint because of the mathematical properties of the functions $B_i(s_i)$, as we will now see.

## 3.3  Properties of the Performance Measures

Now that we have defined several key performance measures and have shown how to compute them, let us examine them more closely. We begin by studying the fill rate measure.

Let us assume, for simplicity, that the demand process is a simple Poisson process with rate $\lambda$. Furthermore, assume that resupply times for each order are independent and identically distributed with mean $\bar{\tau}$. As we have shown, the probability that $x$ units are in the resupply system in steady state is given by

$$p(x|\lambda\bar{\tau}) = e^{-\lambda\bar{\tau}}\frac{(\lambda\bar{\tau})^x}{x!}.$$

Since the demand process is a simple Poisson process, the fill rate, given a stock level of $s$, is given by

$$F(s) = 1 - \sum_{x \geq s} p(x|\lambda\bar{\tau}) = \sum_{x < s} p(x|\lambda\bar{\tau}).$$

Perhaps our goal might be to choose stock levels for many items so that the average fill rate across items is maximized given some target investment level in inventory. This type of optimization problem would be easy to solve if $F(s)$ were a discretely concave function. Unfortunately, as we will now observe, it is not.

We know that if $F(s)$ were a discretely concave function in $s$, then its second difference must be nonpositive for all $s \geq 0$. Let us now define both the first and second differences of $F(s)$. The first difference, $\Delta F(s)$, is given by

$$\Delta F(s) = F(s+1) - F(s),$$

and the second difference, $\Delta^2 F(s)$, is given by

$$\Delta^2 F(s) = \Delta F(s+1) - \Delta F(s).$$

Hence

$$\Delta F(s) = \sum_{x \leq s} p(x|\lambda\bar{\tau}) - \sum_{x \leq s-1} p(x|\lambda\bar{\tau})$$

$$= e^{-\lambda\bar{\tau}} \frac{(\lambda\bar{\tau})^s}{s!}$$

and

$$\Delta^2 F(s) = e^{-\lambda\bar{\tau}} \frac{(\lambda\bar{\tau})^{s+1}}{(s+1)!} - e^{-\lambda\bar{\tau}} \frac{(\lambda\bar{\tau})^s}{s!}$$

$$= e^{-\lambda\bar{\tau}} \frac{(\lambda\bar{\tau})^s}{s!} \left\{ \frac{\lambda\bar{\tau}}{s+1} - 1 \right\}.$$

When $\lambda\bar{\tau} > s + 1$, then $\Delta^2 F(s) > 0$ and $F(s)$ is not concave in that region. In fact, when $s < \lambda\bar{\tau} - 1$, $F(s)$ is discretely convex. Hence $F(s)$ is discretely concave only when $s \geq \lfloor \lambda\bar{\tau} \rfloor$, when $\lambda\bar{\tau}$ is noninteger, and $s \geq \lambda\bar{\tau} - 1$, when $\lambda\bar{\tau}$ is an integer.

Graphs of $F(s)$ for two cases are given in Figures 3.1 and 3.2. In the first case $\lambda\bar{\tau} = 3.2$ and in the second case $\lambda\bar{\tau} = 3$. The graphs illustrate what we have proven. Tables 3.1 and 3.2 contain values for $F(s)$, $\Delta F(s)$ and $\Delta^2 F(s)$ for the two cases. The table values show that the concavity property holds when $\lfloor \lambda\bar{\tau} \rfloor \leq s$. Note that when $\lfloor \lambda\bar{\tau} \rfloor = \lambda\bar{\tau}$, that is, when $\lambda\bar{\tau}$ is an integer, $\Delta F(\lambda\bar{\tau}) = \Delta F(\lambda\bar{\tau}-1)$.

Next, we observe immediately that the ready rate function, $R(s)$, is also not a concave function of $s$ for all values of $s$.

Thus, neither $F(s)$ nor $R(s)$ possesses the mathematical property of concavity that is desirable when formulating and solving an optimization problem whenever we consider all values of $s \geq 0$. Hence, in practical cases, $s$ is constrained to assume values that are greater than or equal to $\lfloor \lambda\bar{\tau} \rfloor$ to ensure that the fill rate or ready rate functions are concave over the feasible region. Note that the operational rate and the first approximation for the expected number of operational systems are stated in terms of the ready rate. Hence, these performance measures are not easy to work with in optimization models unless $s \geq \lfloor \lambda\bar{\tau} \rfloor$.

The backorder function $B(s)$ does have very desirable mathematical properties, however. Recall that

$$B(s) = \sum_{x > s} (x - s) p(x|\lambda\bar{\tau}).$$

For $B(s)$ to be strictly discretely convex and strictly decreasing requires that

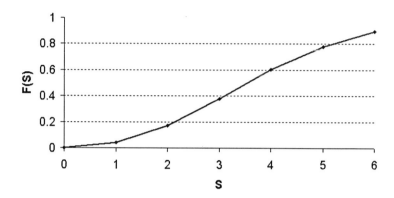

**Fig. 3.1.** Graph of Fill Rate vs Inventory (Case 1)

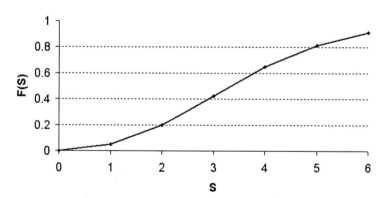

**Fig. 3.2.** Graph of Fill Rate vs Inventory (Case 2)

**Table 3.1. Fill Rate vs Inventory Tradeoff with Poisson Demand**

| | **Mean Demand** | **3.2** | |
| $s$ | $F(s) = P(\Delta < s)$ | $\Delta F(s) = F(s+1) - F(s)$ | $\Delta 2F(s) = \Delta F(s+1) - \Delta F(s)$ |
|---|---|---|---|
| 0 | 0 | 0.040762204 | 0.089676849 |
| 1 | 0.040762004 | 0.130439053 | 0.078263432 |
| 2 | 0.171201257 | 0.208702484 | 0.013913499 |
| 3 | 0.379903741 | 0.222615983 | −0.044523197 |
| 4 | 0.602519724 | 0.178092787 | −0.064113403 |
| 5 | 0.780612511 | 0.113979383 | −0.053190379 |
| 6 | 0.894591895 | 0.060789005 | −0.032999745 |
| 7 | 0.955380899 | 0.027789259 | −0.016673556 |
| 8 | 0.983170158 | 0.011115704 | −0.007163453 |
| 9 | 0.994285862 | 0.00395225 | −0.00268753 |
| 10 | 0.998238112 | 0.00126472 | −0.000896801 |
| 11 | 0.999502832 | 0.000367919 | −0.000269807 |
| 12 | 0.999870751 | 9.81116E−05 | −7.39611E−05 |
| 13 | 0.999968862 | 2.41506E−05 | −1.86304E−05 |
| 14 | 0.999993013 | 5.52013E−06 | −4.3425E−06 |
| 15 | 0.999998533 | 1.17763E−06 | −9.42102E−07 |
| 16 | 0.999999711 | 2.35525E−07 | −1.91191E−07 |
| 17 | 0.999999946 | 4.43342E−08 | −3.64526E−08 |
| 18 | 0.999999991 | 7.88163E−09 | −6.5542E−09 |
| 19 | 0.999999998 | 1.32743E−09 | −1.11504E−09 |

$$\Delta B(s) = B(s+1) - B(s) < 0$$

and

$$\Delta^2 B(s) = \Delta B(s+1) - \Delta B(s) > 0.$$

We see that

$$\Delta B(s) = \sum_{x \geq s+1} (x - (s+1)) p(x|\lambda\overline{\tau})$$
$$- \sum_{x \geq s+1} (x - s) p(x|\lambda\overline{\tau})$$
$$= - \sum_{x \geq s+1} p(x|\lambda\overline{\tau}) = -\left(1 - \sum_{x \leq s} p(x|\lambda\overline{\tau})\right)$$

and

$$\Delta^2 B(s) = - \sum_{x \geq s+2} p(x|\lambda\overline{\tau}) + \sum_{x \geq s+1} p(x|\lambda\overline{\tau})$$
$$= p(s+1|\lambda\overline{\tau}) > 0$$

and hence $B(s)$ is a strictly (discretely) convex function of $s$ for all $s \geq 0$.

**Table 3.2.** Fill Rate vs Inventory Tradeoff with Poisson Demand

| Mean Demand | | 3 | |
|---|---|---|---|
| $s$ | $F(s) = P(\Delta < s)$ | $\Delta F(s) = F(s+1) - F(s)$ | $\Delta^2 F(s) = \Delta F(s+1) - \Delta F(s)$ |
| 0 | 0 | 0.049787068 | 0.099574137 |
| 1 | 0.049787068 | 0.149361205 | 0.074680603 |
| 2 | 0.199148273 | 0.224041808 | 0 |
| 3 | 0.423190081 | 0.224041808 | −0.056010452 |
| 4 | 0.647231889 | 0.168031356 | −0.067212542 |
| 5 | 0.815263245 | 0.100818813 | −0.050409407 |
| 6 | 0.916082058 | 0.050409407 | −0.028805375 |
| 7 | 0.966491465 | 0.021604031 | −0.01350252 |
| 8 | 0.988095496 | 0.008101512 | −0.005401008 |
| 9 | 0.996197008 | 0.002700504 | −0.001890353 |
| 10 | 0.998897512 | 0.000810151 | −0.000589201 |
| 11 | 0.999707663 | 0.00022095 | −0.000165713 |
| 12 | 0.999928613 | 5.52376E−05 | −4.24904E−05 |
| 13 | 0.999983851 | 1.27471E−05 | −1.00156E−05 |
| 14 | 0.999996598 | 2.73153E−06 | −2.18522E−06 |
| 15 | 0.99999933 | 5.46306E−07 | −4.43873E−07 |
| 16 | 0.999999876 | 1.02432E−07 | −8.4356E−08 |
| 17 | 0.999999978 | 1.80763E−08 | −1.50636E−08 |
| 18 | 0.999999996 | 3.01272E−09 | −2.53702E−09 |
| 19 | 0.999999999 | 4.75692E−10 | −4.04338E−10 |

## 3.4  Finding Stock Levels in (s–1, s) Policy Managed Systems: Optimization Problem Formulations and Solution Algorithms

Setting stock levels for items managed using an (s–1,s) policy will depend on the objectives and constraints that are stipulated. For example, we could choose to minimize the average number of outstanding backorders across $n$ item types subject to a constraint on investment in inventory. We could also select stock levels that minimize investment cost subject to an average fill rate constraint across items. Other optimization models could be formulated as well for complex resupply networks. We will study several such problems in later chapters. In this chapter we will examine solution methods that will be employed subsequently for more general problems. The problem that we will study now is concerned with setting stock levels for many items at a single location.

   One solution approach that we could use is to construct a Lagrangian relaxation of a particular optimization problem. We begin by solving the resulting relaxed problem for a given set of Lagrange multiplier values. We then adjust these multiplier values, and re-solve the relaxed problem. We continue in this manner until a stopping criterion of some sort is satisfied.

Before we construct and solve an example problem using the Lagrangian relaxation technique, let us make some important observations.

### 3.4.1  Everett's Theorem

Suppose we have a general optimization problem

$$\min f(x)$$

subject to

$$g(x) \leq b, \tag{3.37}$$

$$x \in S,$$

where $x$ is a vector and $S$ is a set of vectors that constrains the choice of an optimal solution. The single constraint $g(x) \leq b$ is the one that will be relaxed. We assume both $f(x)$ and $g(x)$ are convex functions. We call the above problem, Problem 1. The following is a relaxation of Problem 1

$$\min_{x \in S}[f(x) + \theta(g(x) - b)] \tag{3.38}$$

for a given scalar $\theta \geq 0$. $\theta$ is called the Lagrange multiplier associated with the constraint $g(x) \leq b$. We call this relaxation, Problem 2. The question we will address is: What is the relationship between the solutions to Problem 2 and Problem 1?

The answer to this question is found in the following theorem, which is due to Everett [80].

**Theorem 10.** *Suppose $x^0(\theta)$ is an optimal solution to Problem 2 with the Lagrange multiplier set to $\theta$. Let $b' = g(x^0(\theta))$. Then $x^0(\theta)$ also solves*

$$\min_{x \in S} f(x)$$

$$g(x) \leq b', \text{ which we call Problem 3.} \tag{3.39}$$

*Proof.* To see why this is the case, we first observe that $x^0(\theta)$ is a feasible solution to Problem 3. Let $\hat{x}$ be an optimal solution to Problem 3. Hence $g(\hat{x}) \leq b'$ and $f(\hat{x}) \leq f(x^0(\theta))$. We also know that $\hat{x} \in S$, and hence $\hat{x}$ is a feasible solution to Problem 2. Thus

$$f(x^0(\theta)) + \theta(g(x^0(\theta)) - b) \leq f(\hat{x}) + \theta(g(\hat{x}) - b)$$

or

$$f(x^0(\theta)) + \theta g(x^0(\theta)) \leq f(\hat{x}) + \theta g(\hat{x})$$

Since $g(x^0(\theta)) = b'$,

$$f(\hat{x}) \leq f(x^0(\theta)) \leq f(\hat{x}) + \theta(g(\hat{x}) - b') \leq f(\hat{x})$$

because $g(\hat{x}) \leq b'$. Therefore $x^0(\theta)$ is an optimal solution to Problem 3.   □

Thus, by varying the value of $\theta$, we can find optimal solutions to problems of the form of Problem 3. If $b' = b$ for some choice of the Lagrange multiplier $\theta$, then we have also solved Problem 1.

For many of the problems we will study, we can construct graphs that show how the minimum value of $f(x)$ relates to the value of $b$. For example, suppose $f(x)$ measures the expected number of outstanding backorders at a random point in time, $g(x)$ measures the required investment corresponding to the vector of stock levels, $x$, and $b$ represents the budget limitation on the investment in inventory. Then we may want to construct the relationship between the minimum expected number of outstanding backorders at a random point in time and the investment in inventory. Let $h(b) = \min_{x \in S} \{f(x) : g(x) \le b\}$.

Thus we would want to construct the graph depicted in Figure 3.3. Often times, we are not interested in solving Problem 1 for a single value of $b$ but rather for a range of values of $b$.

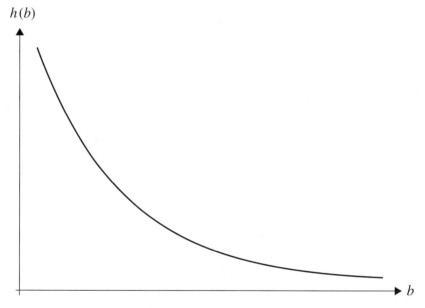

**Fig. 3.3.** The graph of $h(b)$ versus $b$

Recall that every choice of $\theta$ determines an $x^0(\theta)$ when solving Problem 2. In turn, $x^0(\theta)$ yields $g(x^0(\theta))$, which is a value of $b$. Thus for each choice of $\theta$ there exists a corresponding value of $b$, as illustrated in Figure 3.4. We will use this fact in our subsequent analysis. Let us now apply this idea to a specific problem.

$b(\theta)$

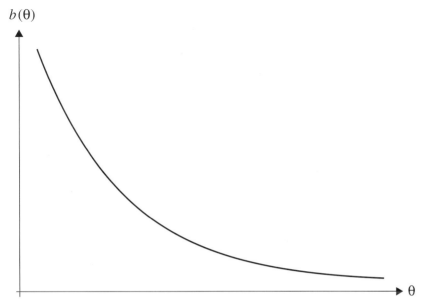

**Fig. 3.4.** The relationship of $b$ as a function of $\theta$

### 3.4.2 First Example: Minimize Expected Backorders Subject to an Inventory Investment Constraint

Suppose a firm manages a group of item types at a single location. The inventory policy followed for all items is an (s–1,s) policy. Thus a replenishment order is placed on an external supplier whenever a unit is withdrawn from the firm's stock to satisfy a customer demand. The goal is to select the stock levels so that the average number of outstanding backorders is minimized subject to a constraint on the average investment in inventory. The demand process is assumed to be a stationary compound Poisson process for each of the $n$ item types being managed. Order lead times on the supplier for replenishment stock are assumed to be independent, identically distributed random variables for each item type and across item types.

Let

$b$ represent the budget limit on the average value of on-hand inventory;

$c_i$ is the unit cost for item type $i$,

$s_i$ is the stock level for item type $i$,

$\lambda_i \bar{\tau}_i \bar{u}_i$ is the expected demand over a lead time for item type $i$, and

$B_i(s_i)$ is the expected number of backorders outstanding at a random point in time for item type $i$.

From Palm's theorem, we know that the steady state probability distribution for the number of units on order with the supplier has a compound Poisson distribution for each item type. Let us denote the probability that $x$ units of item type $i$ are on-order with the supplier by $p(x|\lambda_i\bar{\tau}_i\bar{u}_i)$.

The inventory position when following an (s–1,s) policy is a constant, $s$. In general, the inventory position is defined as

$$\text{Inventory Position} = \text{On-Hand} + \text{On-Order} - \text{Backorders}$$

In this case

$$s = E\,[\text{Inventory Position}] = E[\text{On-Hand}] + E\,[\text{On-Order}] - B(s).$$

The expected number of units on-order for item type $i$ is $\lambda_i\bar{\tau}_i\bar{u}_i$, from Little's law. Hence, for item type $i$,

$$E[\text{On-hand}] = s_i - \lambda_i\bar{\tau}_i\bar{u}_i + B_i(s_i).$$

Let $\mu_i = \lambda_i\bar{\tau}_i\bar{u}_i$. Then the average investment in on-hand inventory for item $i$ is $c_i[s_i - \mu_i + B_i(s_i)]$.

We are now in a position to state the optimization problem as

$$\text{minimize} \sum_{i=1}^{n} B_i(s_i)$$

subject to

$$\sum_{i=1}^{n} c_i[s_i - \mu_i + B_i(s_i)] \le b, \quad s_i = 0, 1, \ldots. \tag{3.40}$$

To solve this problem, which we will call Problem 4, we will use the Lagrangian relaxation method discussed earlier. Let $\theta$ represent the multiplier associated with the budget constraint that links the item stock level decisions. The relaxation is

$$\min \sum_{i=1}^{n} B_i(s_i) + \theta\left[\sum_{i=1}^{n} c_i(s_i - \mu_i + B_i(s_i)) - b\right]$$

subject to $s_i = 0, 1, \ldots$

$$= \min_{s_i=0,1,\ldots} \sum_{i=1}^{n}\left[(1 + \theta c_i)B_i(s_i) + \theta c_i s_i\right] - \left[\theta\sum_{i=1}^{n} c_i\mu_i + \theta b\right]$$

$$= -\theta\left[\sum_{i=1}^{n} c_i\mu_i + b\right] + \sum_{i=1}^{n}\min_{s_i=0,1,\ldots}[(1 + \theta c_i)B_i(s_i) + \theta c_i s_i].$$

Thus, given a value of $\theta$, the resulting relaxed optimization problem is separable by item type. The problem that must be solved for each item is of the same form so we will temporarily drop the item subscript.

Let $f(s) = (1 + \theta c)B(s) + \theta c s$. Since $B(s)$ is discretely strictly convex in $s$, $f(s)$ is convex, too. Define

$$\Delta f(s) = f(s + 1) - f(s)$$
$$= (1 + \theta c)\{B(s + 1) - B(s)\} + \theta c.$$

Since we previously showed that

$$B(s + 1) - B(s) = -\left(1 - \sum_{x \leq s} p(x|\mu)\right),$$

$$\Delta f(s) = -(1 + \theta c)\left(1 - \sum_{x \leq s} p(x|\mu)\right) + \theta c.$$

Due to the convexity of $f(s)$, the optimal stock level, given $\theta$, is the smallest nonnegative integer, $s^*$, for which

$$\Delta f(s) \geq 0,$$

that is, the smallest value for which

$$(1 + \theta c)\left(1 - \sum_{x \leq s} p(x|\mu)\right) \leq \theta c$$

or

$$\sum_{x \leq s} p(x|\mu) \geq \frac{1}{1 + \theta c}.$$

Clearly the value of $s^*$ depends on the value of $\theta$. Observe that as $\theta$ increases, $s^*$ is nonincreasing, and, similarly, as $\theta$ decreases, $s^*$ is nondecreasing. Let

$$C(\theta) = \sum_{i=1}^{n} c_i \left[s_i(\theta) - \mu_i + B_i(s_i(\theta))\right].$$

It is clear that $C(\theta)$ is also nonincreasing as $\theta$ increases and nondecreasing as $\theta$ decreases. A graph of this relationship is shown in Figure 3.5. The goal is to find a value of $\theta$ such that $C(\theta)$ is approximately equal to $b$. It is generally not possible to find a value of $\theta$ that yields $C(\theta) = b$, as illustrated in Figure 3.5. Hence, the goal is to construct the graph of the minimum expected backorders as a function of the average investment in on-hand inventory. Each value of $\theta$ yields a set of stock levels, a corresponding inventory investment, and a minimum number of average outstanding backorders. Thus it is obvious that by solving the relaxation corresponding to a set of multiplier values, $\theta_1 > \theta_2 > \ldots > \theta_M$, we can construct a graph of minimum expected backorders as a function of $\theta$, as illustrated in Figure 3.6.

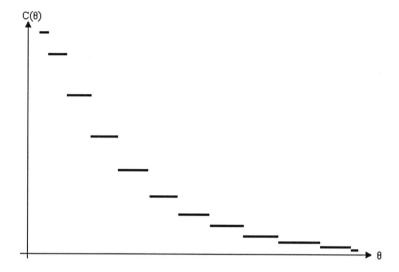

**Fig. 3.5.** Graph of $C(\theta)$ as a function of $\theta$

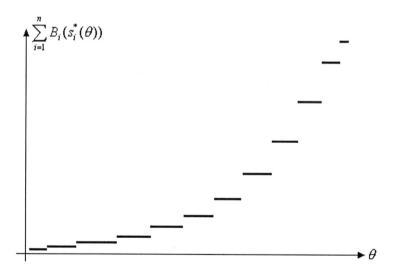

**Fig. 3.6.** Graph of Minimum Expected Backorders as a Function of $\theta$

Suppose we are given values $\theta_1 > \theta_2 > \ldots > \theta_M$, $M$ Lagrange multiplier values. As stated earlier, since $\frac{1}{1+\theta_1 c} < \frac{1}{1+\theta_2 c} < \cdots < \frac{1}{1+\theta_M c}$, $s^*(\theta_1) \leq s^*(\theta_2) \leq \ldots \leq s^*(\theta_M)$. To find $s^*(\theta)$ recall we find the smallest nonnegative integer value of $s$ for which

$$\sum_{x \leq s} p(x|\mu) \geq \frac{1}{1+\theta c}.$$

Thus to find $s^*(\theta_i)$ we know that

$$\sum_{x \leq s^*(\theta_{i-1})} p(x|\mu)$$

is a starting point for our calculation. Since we have already computed this value to determine $s^*(\theta_{i-1})$, the amount of computational effort required to find $s^*(\theta_i)$ may be reduced significantly.

Observe that there exists a $\theta > 0$ such that

$$p(0|\mu_i) = \frac{1}{1+\theta c_i}, \text{ or}$$

$$\theta = \frac{1}{c_i} \left\{ \frac{1}{p(0|\mu_i)} - 1 \right\}.$$

Let $\theta_{max} = \max_i \frac{1}{c_i} \left\{ \frac{1}{p(0|\mu_i)} - 1 \right\}$. If $\theta = \theta_{max}$, then $s_i^*(\theta_{max}) = 0$ for all $i$.

Let us now state an algorithm that can be used to solve the original problem approximately using the Lagrangian method we have discussed.

### Algorithm for Solving Problem 4

Step 0: Set $\theta_{min} = 0$; $\theta_{max} = \max_i \frac{1}{c_i} \left\{ \frac{1}{p(0|\mu_i)} - 1 \right\}$ and $N = 0$.

Step 1: Compute $\theta = \frac{\theta_{min} + \theta_{max}}{2}$; $N = N + 1$.

Step 2: For each item $i$, find the smallest value of $s_i$ such that

$$\sum_{x \leq s_i} p(x|\mu_i) \geq \frac{1}{1+\theta c_i} \text{ and call it } s_i^*(\theta).$$

Step 3: Calculate $A = \sum_{i=1}^{n} c_i \left[ s_i^*(\theta) - \mu_i + B_i(s_i^*(\theta)) \right]$.
If $|A - b| < \epsilon$, or if $N >$ max iterations, stop; otherwise, if $A > b$, set $\theta_{min} = \theta$ and if $A < b$, set $\theta_{max} = \theta$. Return to Step 1.

Some of the ideas discussed in this section were described first in Fox and Landi [92] and later reviewed in Muckstadt [177].

### 3.4.3  Second Example: Maximize Expected System Average Fill Rate Subject to an Inventory Investment Constraint

As in our first example, suppose a firm manages $n$ item types at a single location. We assume a (s–1,s) policy is used for each of these items. We assume requests for these items are placed by customers on the firm. Each request that is made corresponds to a failure of a single unit of a particular part type. The failed parts are repaired at a repair facility. Repair times for an item type are independent and identically distributed; repair times across item types are also independent. We assume requests for serviceable parts for each item type $i$ occur according to a Poisson process with rate $\lambda_i$.

In this case our goal is to find the stock levels that maximize the average expected system fill rate subject to an investment constraint. Here, units do not leave the system so that the investment corresponding to stock levels $s_i$ is $\sum_{i=1}^{n} c_i s_i$.

We use the same notation as in the preceding example where appropriate. $F_i(s_i)$ measures the fill rate for item $i$ given a stock level $s_i$.

In this situation, the probability that $x$ units are in the repair process is given by

$$p(x|\lambda_i \bar{\tau}_i) = e^{-\lambda_i \bar{\tau}_i} \frac{(\lambda_i \bar{\tau}_i)^x}{x!},$$

by Palm's theorem.

The optimization problem can be stated as

$$\text{maximize} \sum_{i=1}^{n} \frac{\lambda_i}{\sum_{j=1}^{n} \lambda_j} F_i(s_i) \qquad (3.41)$$

subject to

$$\sum_{i=1}^{n} c_i s_i \leq b,$$

$s_i \geq \lfloor \lambda_i \bar{\tau}_i \rfloor \geq 0$ and integral, which we call Problem 5.

Recall from our earlier discussion that

$$F_i(s_i) = \sum_{x < s_i} e^{-\lambda_i \bar{\tau}_i} \frac{(\lambda_i \bar{\tau}_i)^x}{x!}.$$

Recall also that $F_i(s_i)$ is concave in the region $s_i \geq \lfloor \lambda_i \bar{\tau}_i \rfloor$, and hence we have placed this constraint on $s_i$ in our formulation of the inventory stocking problem.

We could obtain an answer to Problem (5) using the Lagrangian relaxation method described earlier. However, we will use a simpler approach, marginal analysis. This greedy approach will produce an optimal solution for certain values of $b$ and an approximately optimal solution for all other values of $b$, as we will see.

Define

$$\Delta_i(s_i) = \frac{\lambda_i}{\sum_{j=1}^n \lambda_j} \left\{ \frac{F_i(s_i + 1) - F_i(s_i)}{c_i} \right\},$$

which measures the increase in average expected system fill rate per incremental dollar invested in item $i$ given the current stock level is $s_i$.

Suppose we have stock levels $s_i \geq \lfloor \lambda_i \overline{\tau}_i \rfloor$ and want to determine which item's stock level should be increased from $s_i$ to $s_i + 1$. Since $\Delta_i(s_i)$ measures the change in performance per incremental dollar invested, we would choose to increment the stock level of item $i^*$ if

$$i^* = \arg \max_i \Delta_i(s_i).$$

Initially set $s_i = \lfloor \lambda_i \overline{\tau}_i \rfloor$, and compute

$$\sum_{i=1}^n \frac{\lambda_i}{\sum_{j=1}^n \lambda_j} \cdot F_i(\lfloor \lambda_i \overline{\tau}_i \rfloor) \text{ and } \sum_{i=1}^n c_i \lfloor \lambda_i \overline{\tau}_i \rfloor.$$

Next, compute $\Delta_i(\lfloor \lambda_i \overline{\tau}_i \rfloor)$ for all $i$ and increment the stock level for the item having the maximum value of $\Delta_i(s_i)$, say $i^*$. The solution

$$s_i = \lfloor \lambda_i \overline{\tau}_i \rfloor \quad i \neq i^*$$
$$s_{i^*} = \lfloor \lambda_{i^*} \overline{\tau}_{i^*} \rfloor + 1$$

is the optimal solution to Problem (5) when

$$b = \sum_{i \neq i^*} c_i \lfloor \lambda_i \overline{\tau}_i \rfloor + c_{i^*} \{ \lfloor \lambda_{i^*} \overline{\tau}_{i^*} \rfloor + 1 \}.$$

Continuing in this manner it is clear how we would construct a graph of the maximum average expected fill rate as a function of system investment in inventory. Thus the proposed greedy algorithm will find the optimal solution for values of $b$ that would be generated sequentially as a consequence of constructing the solution as outlined.

## 3.5 Problem Set, Chapter 3

**3.1.** Suppose that time is divided into periods of equal length. Demand in each period is Poisson distributed with a mean of $\lambda$ units. An order-up-to policy is followed for managing inventories. Replenishment lead times are assumed to be independent and identically distributed with a mean of $D$ periods. Prove that the number of units on order (in resupply) in steady state has a Poisson distribution with mean $\lambda D$. That is, prove that the discrete time analogue of Palm's theorem.

**3.2.** Plot the logarithmic distribution (3.13) for the following value of $p$ : .05, .1, .3, .5, .7, .9 and .95. What do you observe?

**3.3.** Find the expected value of a random variable having a logarithmic distribution, which we will denote by $\bar{u}$. Suppose the demand process is a compound Poisson process in which $\lambda = -\ln(p)$, where $p$ is the parameter of the logarithmic distribution. What is the mean and variance of this compound Poisson distribution? What is the form of its probability distribution? Suppose we let $\bar{\lambda} = \lambda \bar{u}$ be the mean of a Poisson distribution. How well does a Poisson distribution with this mean match the exact compound Poisson distribution? Make this comparison when $p = .1, .25, .5, .75, .9$. What do you observe?

**3.4.** Plot the probabilities for a random variable that has a Negative Binomial distribution where its mean assumes value 1, 5, 25. For each mean, construct these plots when the variance-to-mean ratios are 1.01, 3, 10. How do these probabilities compare with the corresponding probabilities for a Poisson distributed random variable with the same means?

**3.5.** Prove the extension to Palm's theorem when the arrival process is a compound Poisson process and where every customer is willing to wait $\tau$ time units for delivery of its order.

**3.6.** Suppose a single stage inventory system is managed using a (s–1,s) continuous review policy. Suppose the lead time is known to be two weeks in length. All demand in excess of supply is backordered. Demand occurs according to a Poisson process with arrival rate $\lambda$. Plot the fill rate for this system when $\lambda = .5, 5$, and 10 units per week as a function of the stock level. Suppose next that the arrival process is a compound Poisson process in which the average order size is 2 units. Assume the distribution of demand in this case is a Negative Binomial distribution (the order size distribution is logarithmic with a mean of 2). Assuming the customer order arrival rate is now either .25, 2.5 and 5.65 units per week, again plot the fill rate as a function of the stock level $s$.

**3.7.** Plot the backorder, ready rate, and fill rate functions $(B(s), R(s), F(s))$ for a single stage inventory system managed using a continuous review (s–1,s) policy. Suppose the lead time demand is either Poisson distributed or Negative Binomially distributed with expected lead time demand being either 1, 5 or 10 units. In the case where lead time demand follows a Negative Binomial distribution, construct the plots for the variance-to-mean ratios of 1.01, 2 and 5. What do you observe?

**3.8.** Suppose there are ten critical items on an aircraft. Compute the expected number of nonoperational aircraft for several combinations of stock levels for these items assuming the demand process is Poisson. Expected demands over the lead time for these ten items are 10, 7, 2, 1, 0.7, 0.5, 0.3, 0.1, 0.04, 0.01, respectively.

**3.9.** Prove the Splitting Property of Poisson processes: Let $N, Y_n$ be independent random variables, where the distribution of $N$ is Poisson($\lambda$) and $P(Y_n = j) = p_j$ for $j \in \{1, 2, \ldots, k\}$ and all $n$. Set $N_j = \sum_{n=1}^{N} I(Y_n = j)$, where $I$ is the indicator function. Then, $N_1, N_2, \ldots, N_k$ are independent random variables and the distribution of $N_j$ is Poisson($\lambda p_j$) for all $j$. Prove the converse as well.

**3.10.** An owner of a fleet of aircraft wants to determine how many spares of each of $N$ components to buy. Aircraft failures, which occur according to a Poisson process, are caused by the failure of a component. Each failure is due to the failure of a single component and the probability that it is due to component $i$ is $\lambda_i / \sum_j \lambda_j$, where $\lambda_i$ is the failure rate for component $i$. The average repair time for component $i$ is $t_i$. The owner wants to set the spare parts' stock levels such that the average delay in completing the repair of an aircraft due to parts availability is minimized. There is a constraint on total investment in spare parts of $C$ dollars. Assume each component of type $i$ costs $c_i$ dollars. An (s–1,s) policy is used to manage the system.

(a) Develop a mathematical formulation of this problem. Construct a Lagrangian-based algorithm for finding the optimal stock levels.
(b) Use your algorithm to find the optimal solution when $N = 2, C = 6$,

$$
\begin{array}{ll}
c_1 = 1 & c_2 = 2 \\
t_1 = 1/10 & t_2 = 1/7\, . \\
\lambda_1 = 5 & \lambda_2 = 7
\end{array}
$$

Plot the total investment in spare parts as a function of the Lagrange multiplier and use this plot to obtain a solution for the problem.

**3.11.** Suppose a firm manages $n$ item types at a single location. Assume an (s–1,s) policy is used for each of these items. Requests for these items are placed by customers on the firm. Each request corresponds to a failure of a single unit of a particular part type. The failed parts are repaired at a repair facility. Repair times for an item type are independent and identically distributed; repair times across item types are also independent. Assume requests for serviceable parts for each item type $i$ occur according to a Poisson process with rate $\lambda_i$. Construct a marginal analysis algorithm to maximize the average expected system fill rate subject to an investment constraint. Use this algorithm to determine the optimal stock levels for the data given in the previous problem.

# 4

# An Exact Model for a Depot-Base Two Echelon Inventory and Repair System

We will study a variety of multi-echelon service parts systems in subsequent chapters. For the most part, the analyses that we will present in these chapters are based on approximations of the probability distributions for key random variables. We create these approximations so that we can construct and solve optimization models that will yield stock levels for each location in the system. However, before we develop these approximations and these optimization models, we will develop the exact representations of the probability distributions of the important random variables that relate the interactions present in multi-echelon systems. We will see that the exact calculation of the required probability distributions is too computationally burdensome to be of practical value for large scale systems. Let us now develop the exact probability distributions for a specific two-echelon service parts system.

## 4.1 Introduction

The two-echelon system we will analyze in detail, depicted in the following figure, consists of a depot, at which inventory is stocked and items are repaired, and a set of $n$ bases. We assume this system supports the flying operations of aircraft, which occur at the bases. Inventories of parts are also maintained at the bases. We will focus on a single item in this system since we also assume there are no constraints among the various items.

We further assume that this item fails according to a Poisson process at base $j$ at the rate of $\lambda_j$ units per day. When a removal of a defective part occurs at a base, three events occur simultaneously and instantaneously; first, a unit of stock is withdrawn from base supply to repair the aircraft (if one is available); second, the failed unit is shipped to the depot for repair; and third, the depot resupplies the base (ships a replacement unit to the base) if there is a unit on hand at the depot stocking location. When units are not available at either the base or depot, a backorder occurs and lasts until a unit of stock is available to meet the request for the item.

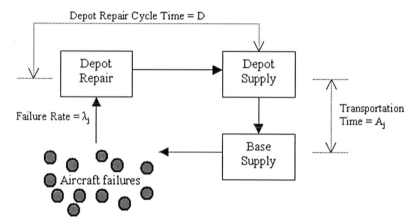

**Fig. 4.1.** Depot-Base Two-Echelon System

## 4.2 Model Development

Our goal is to develop an exact probabilistic representation of this system's operation in steady state. That is, we will find the stationary distribution of certain key random variables. The ideas presented in this section follow those given by Simon[230].

We, of course, are assuming that each base follows an (s–1,s) policy and that the depot does as well. We also assume that all failed items can be repaired. Furthermore, we initially assume that the depot to base $j$ transportation time, $A_j$, is a constant and the depot repair cycle time is the same for all failed items, and is represented by the symbol $D$. The depot repair cycle time measures the total time it takes to ship the failed unit to the depot from a base and to repair the unit at the depot.

Let $t$ be a random point in time. Furthermore, let

$\quad I_j(t) =$ the net inventory random variable for base $j$ at time $t$,
$Z_0(t_a, t_b) =$ a random variable describing total base level demand during the interval of time $(t_a, t_b]$, where $t_a = t - A_j - D$ and $t_b = t - A_j$,
$\quad s_j =$ the target base $j$ stock level (inventory position),
$\quad s_0 =$ the target depot stock level (inventory position), and
$\quad X_j(t) =$ random variable for the number of units on-order (or in resupply) from the depot at time $t$ attributable to base $j$ requests for replenishment of its stock.

Observe that if $I_j(t) = k$, then $s_j - k$ units must be on order from the depot. Thus we see that $I_j(t) = k$ if and only if $X_j(t) = s_j - k$. Our objective is to find $\lim_{t \to \infty} P\{I_j(t) = k\}$.

Since we know that $I_j(t) = k$ if and only if $X_j(t) = s_j - k$,

$$P\{I_j(t) = k\} = P\{X_j(t) = s_j - k\}$$
$$= \sum_{d_0 \geq 0} P\{X_j(t) = s_j - k | Z_0(t_a, t_b) = d_0\} \cdot P\{Z_0(t_a, t_b) = d_0\}.$$

To calculate $P\{I_j(t) = k\}$, we partition the above expression into two regions. In the first region, $d_0 \leq s_0$ and in the second region, $d_0 > s_0$. We will examine these two cases separately.

When $d_0 \leq s_0$, all orders placed on the depot prior to time $t_b$ by base $j$ will be satisfied by time $t$. Furthermore, any demand placed subsequent to time $t_b$ by base $j$ cannot be satisfied by time $t$. Thus when $d_0 \leq s_0$

$$\sum_{d_0 \leq s_0} P\{X_j(t) = s_j - k | Z_0(t_a, t_b) = d_0\} \cdot P\{Z_0(t_a, t_b) = d_0\}$$

$$= \sum_{d_0 \leq s_0} e^{-\lambda_j A_j} \frac{(\lambda_j A_j)^{s_j - k}}{(s_j - k)!} \cdot e^{-\lambda_0 D} \frac{(\lambda_0 D)^{d_0}}{d_0!}$$

$$= e^{-\lambda_j A_j} \frac{(\lambda_j A_j)^{s_j - k}}{(s_j - k)!} \cdot \sum_{d_0 \leq s_0} e^{-\lambda_0 D} \frac{(\lambda_0 D)^{d_0}}{d_0!},$$

where $\lambda_0 = \sum_j \lambda_j$. Thus in this case the probability that $X_j(t) = s_j - k$ is the probability that demand at base $j$ during the interval $(t_b, t]$, whose length is $A_j$, is equal to $s_j - k$, weighted by the probability that this case will occur, that is, that $d_0 \leq s_0$.

Next, suppose $d_0 > s_0$. Since the total base demand during $(t_a, t_b]$ exceeds depot supply, $s_0$, not all requests placed on the depot during that interval will be satisfied by time $t_b$. Hence, the number of units on order at time $t$ at base $j$ consists of those orders placed during $(t_a, t_b]$ that were not shipped to the base by time $t_b$ plus all demands for the item that occurred during $(t_b, t]$.

Now, let

$V_1(t) = $ the random variable for the number of base $j$ units on order at time $t$ for which an order was placed (item failure occurred) by base $j$ on the depot prior to time $t_b$.

$V_2(t) = $ the random variable for the number of units that fail at base $j$ (and are ordered from the depot) during $(t_b, t]$.

Then

$$X_j(t) = V_1(t) + V_2(t).$$

By assumption, the demand process at each base is a Poisson process. Therefore, demands occurring before and after time $t_b$ are independent. Thus

$$P\{X_j(t) = s_j - k | Z_0(t_a, t_b) = d_0\}$$
$$= \sum_{m=0}^{s_j - k} P\{V_1(t) = m | Z_0(t_a, t_b) = d_0\} \cdot P\{V_2(t) = s_j - k - m\}.$$

We know that

$$P\{V_2(t) = s_j - k - m\} = e^{-\lambda_j A_j} \frac{(\lambda_j A_j)^{s_j-k-m}}{(s_j - k - m)!}.$$

Additionally,

$$P\{V_1(t) = m | Z_0(t_a, t_b) = d_0\}$$

$$= \sum_{d_j=m}^{s_0+m} P\{V_1(t) = m | Z_0(t_a, t_b) = d_0, D_j(t_a, t_b) = d_j\}$$

$$\cdot P\{D_j(t_a, t_b) = d_j | Z_0(t_a, t_b) = d_0\},$$

where $D_j(t_a, t_b)$ represents the number of base $j$ demands placed on the depot during $(t_a, t_b]$. Note that

$$P\{V_1(t) = m | Z_0(t_a, t_b) = d_0, D_j(t_a, t_b) = d_j\} = 0$$

when $d_j > d_0$ and when $d_0 < s_0 + m$. Observe when $d_j \le d_0$ and $d_0 \ge s_0 + m$ that

$$P\{V_1(t) = m | Z_0(t_a, t_b) = d_0, D_j(t_a, t_b) = d_j\}$$

$$= \frac{\binom{d_j}{d_j - m}\binom{d_0 - d_j}{s_0 - (d_j - m)}}{\binom{d_0}{s_0}},$$

and that

$$P\{D_j(t_a, t_b) = d_j | Z_0(t_a, t_b) = d_0\} = \binom{d_0}{d_j}\left(1 - \frac{\lambda_j}{\lambda_0}\right)^{d_0-d_j}\left(\frac{\lambda_j}{\lambda_0}\right)^{d_j}.$$

Combining these above results we see in this second case that

$$P\{X_j(t) = s_j - k | Z_0(t_a, t_b) = d_0\}$$

$$= \sum_{m=0}^{s_j-k}\left\{ \sum_{d_j=m}^{s_0+m} \frac{\binom{d_j}{d_j - m}\binom{d_0 - d_j}{s_0 - (d_j - m)}}{\binom{d_0}{s_0}} \right.$$

$$\cdot \binom{d_0}{d_j}\left(1 - \frac{\lambda_j}{\lambda_0}\right)^{d_0-d_j}\left(\frac{\lambda_j}{\lambda_0}\right)^{d_j} \Bigg\}$$

$$\cdot e^{-\lambda_j A_j} \frac{(\lambda_j A_j)^{s_j-k-m}}{(s_j - k - m)!}.$$

Thus

$$
P\{I_j(t) = k\} = \sum_{d_0 \le s_0} P\{X_j(t) = s_j - k | Z_0(t_a, t_b) = d_0\} \cdot P\{Z_0(t_a, t_b) = d_0\}
$$

$$
+ \sum_{d_0 > s_0} P\{X_j(t) = s_j - k | Z_0(t_a, t_b) = d_0\} \cdot P\{Z_0(t_a, t_b) = d_0\}
$$

$$
= e^{-\lambda_j A_j} \frac{(\lambda_j A_j)^{s_j - k}}{(s_j - k)!} \cdot \sum_{d_0 \le s_0} e^{-\lambda_0 D} \frac{(\lambda_0 D)^{d_0}}{d_0!}
$$

$$
+ \sum_{d_0 > s_0} \left\{ \sum_{m=0}^{\min(s_j - k, d_0 - s_0)} \left\{ \sum_{d_j = m}^{s_0 + m} \frac{\binom{d_j}{d_j - m} \binom{d_0 - d_j}{s_0 - (d_j - m)}}{\binom{d_0}{s_0}} \right.\right.
$$

$$
\left. \cdot \binom{d_0}{d_j} \left(1 - \frac{\lambda_j}{\lambda_0}\right)^{d_0 - d_j} \left(\frac{\lambda_j}{\lambda_0}\right)^{d_j} \right\}
$$

$$
\left. \cdot e^{-\lambda_j A_j} \frac{(\lambda_j A_j)^{s_j - k - m}}{(s_j - k - m)!} \right\} \cdot e^{-\lambda_0 D} \frac{(\lambda_0 D)^{d_0}}{d_0!} .
$$

Observe that none of the terms depends on $t$. Thus this expression is also the limiting probability that the net inventory random variable is equal to $k$. Also observe that only under the condition that $s_0$ is large enough so that

$$
\sum_{d_0 \le s_0} e^{-\lambda_0 D} \frac{(\lambda_0 D)^{d_0}}{d_0!} \quad \text{is very close to 1}
$$

does $P\{I_j = k\}$ or $P\{X_j = x\}$ have an approximately Poisson distribution.

Finally, we observe that this probability distribution is computationally intractable for the large scale problems found in practice. Hence approximations will be developed for this distribution as we construct optimization models.

However, before we do so, let us examine some additional exact models.

## 4.3 Some Extensions

In the preceding analysis, we assumed that the depot to base $j$ transportation time, $A_j$, and the depot repair cycle time, $D$, were constants. Let us observe that we can relax these assumptions without affecting the analysis we have performed.

Let us now suppose that each failed item entering the depot repair process has a repair cycle time drawn from a distribution having mean $D$ and a density function $\psi$. Assume repair cycle times are independent from failed part to failed part as well. Define the random variable $\overline{Z}_0$ to be the number of units remaining in the depot repair cycle at a random point in time.

We know from Palm's theorem that the random variable $\overline{Z}_0$ has a Poisson distribution with mean $\lambda_0 D$. The random variable $V_1(t)$ will now measure the number of units back-ordered at the depot that are due to be shipped to base $j$; that is, $V_1(t)$ represents the number of base $j$ backorders at the depot at a random point in time.

Let $N_0$ represent the number of depot backorders at a random point in time. Then

$$P(V_1(t) = m) = \sum_{n_0 \geq m} P\{V_1(t) = m | N_0 = n_0\} P\{N_0 = n_0\}$$

But

$$P\{V_1(t) = m | N_0 = n_0\} = \binom{n_0}{m}\left(\frac{\lambda_j}{\lambda_0}\right)^m \left(1 - \frac{\lambda_j}{\lambda_0}\right)^{n_0-m},$$

$$P\{N_0 = n_0\} = e^{-\lambda_0 D}\frac{(\lambda_0 D)^{s_0+n_0}}{(s_0+n_0)!} = P\{\overline{Z}_0 = s_0 + n_0\}, n_0 > 0,$$

and

$$P\{N_0 = 0\} = \sum_{k=0}^{s_0} e^{-\lambda_0 D}\frac{(\lambda_0 D)^k}{k!}.$$

Also, as in our previous analysis,

$$P\{V_2(t) = k\} = e^{-\lambda_j A_j}\frac{(\lambda_j A_j)^k}{k!}.$$

Since $V_1(t)$ and $V_2(t)$ are independent random variables,

$$P\{I_j(t) = k\} = P\{X_j(t) = s_j - k\}$$

$$= e^{-\lambda_j A_j}\frac{(\lambda_j A_j)^{s_j-k}}{(s_j-k)!} \cdot P\{N_0 = 0\}$$

$$+ \sum_{n_0>0}\left(1 - \frac{\lambda_j}{\lambda_0}\right)^{n_0} e^{-\lambda_0 D}\frac{(\lambda_0 D)^{n_0+s_0}}{(s_0+n_0)!} e^{-\lambda_j A_j}\frac{(\lambda_j A_j)^{s_j-k}}{(s_j-k)!}$$

$$+ \sum_{m=1}^{s_j-k}\left\{\sum_{n_0\geq m}\binom{n_0}{m}\left(\frac{\lambda_j}{\lambda_0}\right)^m\left(1 - \frac{\lambda_j}{\lambda_0}\right)^{n_0-m} e^{-\lambda_0 D}\frac{(\lambda_0 D)^{n_0+s_0}}{(n_0+s_0)!}\right\}$$

$$\cdot e^{-\lambda_j A_j}\frac{(\lambda_j A_j)^{s_j-k-m}}{(s_j-k-m)!}.$$

Again this is the stationary distribution of the net inventory random variable and the distribution of the number of units in resupply for a base. Furthermore, when $P\{N_0 = 0\} \approx 1$, these distributions are approximately Poisson distributions. Specifically

$$P\{X_j = k\} \approx e^{-\lambda_j A_j} \frac{(\lambda_j A_j)^k}{(k)!} = P\{V_2 = k\} \text{ when } P\{N_0 = 0\} \approx 1.$$

Next, let us make a different assumption concerning the manner in which the depot maintenance center operates. To this point, we have assumed repair times are independent and identically distributed random variables. We now assume the repairs can not cross. That is, a unit that enters the repair process can not complete repair prior to one that enters the repair at an earlier time.

Suppose that we know the probability distribution for the length of the repair cycle time for a unit when repair cycle times do not cross. Let $\psi(\cdot)$ be the density function of this random variable.

Now consider a random point in time. Because we have assumed repairs of items are completed in the order in which the failures occur, that is, the resupply process satisfies our no crossing assumption, the number of items in the repair cycle process at a random point in time is equal to the number of failures occurring during a resupply time. That this is the case is due to Svoronos and Zipkin [241, 242].

Let

$$G(k) = P\{Z_0 = k\},$$

where $Z_0$ is a random variable that measures the number of demands (failures) placed on the depot during the repair cycle of a randomly failed item. Then

$$G(k) = \int P\{Z_0 = k|t\}\psi(t)\,dt$$

$$= \int e^{-\lambda_0 t} \frac{(\lambda_0 t)^k}{k!} \psi(t)\,dt.$$

As before, let us compute the probability distribution of $V_1$, the number of units back-ordered at the depot that correspond to base $j$ orders. Again we let $N_0$ be the random variable describing the number of depot backorders at a random point in time. As we observed earlier,

$$P(V_1 = m) = \sum_{n_0 \geq m} P(V_1 = m|N_0 = n_0) \cdot P(N_0 = n_0).$$

But in this case,

$$P(N_0 = 0) = \sum_{k \leq s_0} G(k) = \sum_{k \leq s_0} \int P(Z_0 = k|t)\psi(t)\,dt$$

$$= \sum_{k \leq s_0} \int e^{-\lambda_0 t} \frac{(\lambda_0 t)^k}{k!} \psi(t)\,dt.$$

and $P(N_0 = n_0) = G(s_0 + n_0)$, $n_0 \geq 1$ and integer.

Given $P(N_0 = n_0)$ and, for $n_0 \geq m$,

$$P(V_1 = m | N_0 = n_0) = \binom{n_0}{m} \left(\frac{\lambda_j}{\lambda_0}\right)^m \left(1 - \frac{\lambda_j}{\lambda_0}\right)^{n_0-m},$$

we can compute $P(V_1 = m)$.

Suppose $\psi(t)$ is gamma distributed, that is,

$$\psi(t) = \frac{e^{-t/\beta} t^{\alpha-1}}{\beta^\alpha \Gamma(\alpha)}, t \geq 0.$$

Then the expected repair cycle time is $E(t) = \alpha\beta$ and Var $t = \alpha\beta^2$. Also

$$
\begin{aligned}
G(k) &= \int_0^\infty e^{-\lambda_0 t} \frac{(\lambda_0 t)^k}{k!} \frac{e^{-t/\beta} t^{\alpha-1}}{\beta^\alpha \Gamma(\alpha)} \, dt \\
&= \frac{\lambda_0^k}{k! \beta^\alpha \Gamma(\alpha)} \int_0^\infty e^{-\frac{(\beta\lambda_0+1)t}{\beta}} t^{k+\alpha-1} \, dt \\
&= \frac{\lambda_0^k}{k! \beta^\alpha \Gamma(\alpha)} \Gamma(\alpha+k) \frac{\beta^{\alpha+k}}{(\beta\lambda_0 + 1)^{\alpha+k}} \\
&= \frac{\Gamma(\alpha+k)}{k! \Gamma(\alpha)} \left[\frac{\beta\lambda_0}{\beta\lambda_0 + 1}\right]^k \left[\frac{1}{\beta\lambda_0 + 1}\right]^\alpha.
\end{aligned}
$$

Thus $G(k)$ has a negative binomial distribution with $p = \frac{1}{\beta\lambda_0+1}$, and mean $= \alpha\frac{(1-p)}{p}$ and variance $= \frac{\alpha(1-p)}{p^2}$ when $\psi(t)$, the lead time density, is a gamma density.

Next, suppose the transportation times from the depot to base $j$ do not cross. Let the transportation lead time random variable have a density function denoted by $\gamma(t)$. Let $V_2$ be the stationary distribution of the number of units of demand at base $j$ during a lead time. By employing the results of Svoronos and Zipkin[241, 242], we may compute $P(V_2 = k)$ as follows

$$
\begin{aligned}
P(V_2 = k) &= \int P(V_2 = k | t) \gamma(t) \, dt \\
&= \int e^{-\lambda_j t} \frac{(\lambda_j t)^t}{k!} \gamma(t) \, dt.
\end{aligned}
$$

But $V_1$ and $V_2$ are independent since the demands contributing to their calculation occur in nonoverlapping intervals of time. Thus we use these distributions to determine the probability distribution for the random variable $X_j$, the number of units in the resupply system for base $j$.

$$P(X_j = s_j - k) = P(V_2 = s_j - k) \cdot P[N_0 = 0]$$

$$+ \sum_{n_0 > 0} \left(1 - \frac{\lambda_j}{\lambda_0}\right)^{n_0} \cdot P[N_0 = n_0 + s_0] \cdot P(V_2 = s_j - k)$$

$$+ \sum_{m=1}^{s_j - k} \left\{ \sum_{n_0 \geq m} \binom{n_0}{m} \left(\frac{\lambda_j}{\lambda_0}\right)^m \left(1 - \frac{\lambda_j}{\lambda_0}\right)^{n_0 - m} \right.$$

$$\left. \cdot P[N_0 = n_0 + s_0] \right\} \cdot P(V_2 = s_j - k - m).$$

## 4.4  Problem Set, Chapter 4

**4.1.** Consider the following inventory control policy for managing a single item at a single location. The inventory position is monitored continuously and an order for $Q$ units is placed whenever the inventory position hits the level $r$. This is called an $(r, Q)$ policy. Prove that the inventory position is uniformly distributed over the set $\{r+1, r+2, \ldots, r+Q\}$ when an $(r, Q)$ policy is used and the demand process is Poisson. Suppose the demand process is a renewal process. What is the distribution of the inventory position random variable in this case?

**4.2.** Suppose a single item is managed in a two-echelon system of the type depicted in the following figure.

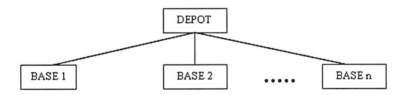

**Fig. 4.2.** Depot-Base System

The item is a repairable item, that is, it normally can be repaired after it fails. We assume all failures occur at the lower echelon - the base echelon. When an item fails at base $j$ it is either repaired there with probability $r_j$ or it is sent to the depot to be repaired. Each failure generates a demand for a spare serviceable part at the base at which the failure occurred. When the broken part is repaired at base $j$, the part is returned to a serviceable condition $B_j$ time units later. When the part is sent to the depot to be repaired, the depot is immediately informed and sends a replacement part to the base as soon as possible. If the depot has a serviceable part in stock, it is sent immediately to base $j$. The transportation time from the depot to base $j$ is $A_j$ time units. If the depot does not have a serviceable part on hand, a replacement part is sent as soon as one becomes available given that the depot follows a first-come-first-serve policy for satisfying demands. Thus resupply of a base's serviceable inventory comes from the base's maintenance organization when the failed item is repaired there and from the depot otherwise. In either case there is a one for one exchange of a broken part for a serviceable part. Thus the base follows an (s–1,s) inventory policy.

If a unit is sent to the depot for repair, $p$ is the probability that it cannot be repaired but must be scrapped. Thus periodically the depot places an order with an outside source to replenish system stock. The procurement lead time is $E$ time units.

We assume that the failure process at base $j$ is a Poisson process with rate $\lambda_j$, the failure processes are independent, the depot repair cycle time is a constant $D$ time units, there is no lateral resupply among the bases, all excess demand is back-ordered, and the depot follows a continuous review $(r, Q)$ policy for replenishing scrapped units. We also assume that $B_j < A_j < D < E$.

Develop the exact probability distribution of the net inventory random variable for base $j$.

**4.3.** In Section 4.2 we developed the probability distribution for net inventory at each base in a continuous review two-echelon inventory system, where the bases and the depot all followed an (s–1,s) policy. Recall that the depot to base transportation time is a constant ($A_j$ days for base $j$) and the depot procurement lead time is a constant $D$ days. We also assumed that the demand for the product is Poisson distributed with rate $\lambda_j$ at base $j$ and that the demand processes are independent among bases.

Assume now that the depot follows a $(r, Q)$ policy. First, extend the results of Section 4.2 to represent this change in the depot's operating policy. Then, carefully develop an algorithm for finding the optimal (or possibly near optimal) values for $Q, r$, and $s_j, j = 1, \ldots, n$. The objective is to select these values so that the average annual costs of holding, procuring, and backordering are minimized. Backorder costs are charged proportional to the expected number of units in a backorder status at any point in time.

**4.4.** For the environment presented in Section 4.2, let us assume that there are 10 bases. Demand at each base arises according to a Poisson distribution. For base $j$, assume that $\lambda_j A_j = .2$. Furthermore, $\lambda_j/\lambda_0 = .5$ and $\lambda_0 D = 2$. Let $X_j(t)$ represent the random variable that measures the number of units in resupply for base $j$. Plot $P\{X_j = k\}, k = 0, 1, \ldots$, for depot stock levels $s_0 = 2, 4$ and 6. Let $B(s_0)$ measure the expected backorders at the depot given the depot stock level is $s_0$, where $B(s_0) = \sum_{x > s_0}(x - s_0)p(x|\lambda_0 d)$ and $p(x|\lambda_0 d) = e^{-\lambda_0 D}(\lambda_0 D)^x/x!$. Then $E[X_j] = \lambda_j A_j + B(s_0)\lambda_j/\lambda_0$.

Suppose we approximate the distribution of $X_j$ with a Poisson distribution with mean $E[X_j]$. For the specified values of $s_0$ given earlier, plot the approximating distribution of $X_j$. How do they compare with the corresponding exact distributions?

# 5

## Tactical Planning Models for Managing Recoverable Items

In 1968, Sherbrooke [223] published a landmark paper in which he described a mathematical model for the management of recoverable or repairable items called METRIC (Multi-Echelon Technique for Recoverable Item Control). Since that time, many extensions and modifications to his model have been proposed, some of which are discussed in this and subsequent chapters. Recall that in the previous chapter we showed how to calculate the exact distribution of the number of units in the resupply system at each base in a two-echelon depot base system. We observed there that the exact expressions are too computationally burdensome to be of practical use. The METRIC model is based on an approximation to this distribution that is easy to compute, and hence has been widely used in many applications.

In this chapter we summarize the key elements of Sherbrooke's ideas and important contributions and improvements to METRIC due to Graves [99] , Sherbrooke [226] and O'Malley [189]. We will also show how METRIC was extended to represent more complicated environments in which there are both repairable assemblies, which are termed LRUs, or line replaceable units, and subassemblies, which are shop replaceable units, or SRUs. This model was originally developed by Muckstadt [174]. Included in the discussion in this chapter is a waiting time analysis that is due to Kruse [152].

### 5.1 The METRIC System

The system modelled by Sherbrooke [223] corresponds to one that was operated by the US Air Force, which consists of a set of bases, at which flying activity occurs, and a depot. Both the depot and bases stock inventory and repair defective parts. The parts removed from the aircraft requiring repair are the LRUs, or Line Replaceable Units. The system is assumed to operate as follows.

When an LRU fails at a base (removed from an aircraft), the following events occur. First, an LRU is withdrawn from base stock and placed on the aircraft, thereby returning the aircraft to an operational status. If serviceable stock is not available in base supply, then a backorder occurs. The failed unit is either repaired at the base or the depot.

The decision as to where the repair occurs depends only on the nature of the failure. Some types of LRU failures can only be diagnosed and repaired at the depot. Others can be diagnosed and repaired at the base level. We assume that whenever a failed unit can be repaired at a base, from a technical viewpoint, then it will be repaired there. Thus the choice of the repair location does not depend on the current on-hand stock at the base or the current workload in the base repair shop.

When the unit is shipped to the depot for repair, a request is made to have an LRU of the same type shipped to that base. If such a unit is on-hand, it will be shipped immediately; otherwise, a backorder will occur at the depot. We assume two things. First, the depot meets demands on a first-come, first-serve basis. That is, no prioritization among bases occurs when making shipping decisions even though such prioritization may often be desirable. Thus the tactical planning model is conservative in that base level performance will be enhanced if proper prioritization practices are put into operation. Second, bases are not resupplied by other bases, that is, lateral resupply is not permitted in the model. Again this is a conservative assumption since lateral resupply will enhance base performance if executed appropriately. We discuss lateral resupply models in Chapter 7.

Because the cost of each unit of an LRU is normally high and demand rates are usually low, the (s–1,s) inventory policy is followed in practice at both the bases and the depot. We assume that all failed units can be repaired; however, if there are condemnations, then orders will be placed on an external supplier. In this case the orders may be for more than a single unit. In these cases, the (s–1,s) policy is not followed and our subsequent discussion must be modified.

We will assume that failures of each LRU type occur according to a Poisson process. This assumption is necessary for some analytic reasons as will become apparent as we proceed. However, we will indicate subsequently how this assumption is relaxed in practice.

### 5.1.1  System Operation and Definitions

As we discussed, the system operates as follows. Removals of LRU $i$ at base $j$ occur according to a Poisson process with rate $\lambda_{ij}$. With probability $r_{ij}$ the unit will be repaired at the base, and with probability $(1 - r_{ij})$ will be repaired at the depot. Hence the arrival process to base maintenance for LRU $i$ is a Poisson process with rate $r_{ij}\lambda_{ij}$. The arrival process to the depot maintenance activity for LRU type $i$ is also a Poisson process with rate $\lambda_{i0} = \sum_j (1 - r_{ij})\lambda_{ij}$. That this process is a Poisson process is a consequence of the fact that failures occur at bases according to Poisson processes for LRU type $i$ and that each such failure has a probability of $r_{ij}$ of being repaired at the base and $1 - r_{ij}$ of being repaired at the depot. Furthermore, the superposition of the independent Poisson arrival processes from the bases corresponding to failures requiring depot repair for LRU type $i$ is also a Poisson process.

We assume the depot repair cycle time for LRU $i$ is denoted by $D_i$ and is not dependent on the base from which the LRU is sent. The repair cycle time includes the packing, transportation, and actual repair times.

We let $B_{ij}$ be the base repair cycle time for LRU $i$ at base $j$ and $A_{ij}$ be the order, shipping and receiving time of LRU $i$ at base $j$ for shipments of LRU $i$ received from the depot.

Let $T_{ij}$ represent the average number of days that it takes to resupply base $j$'s stock for LRU $i$ once a unit enters the resupply system, that is, either depot or base repair cycles.

We assume $\lambda_{ij}$ is measured in units per day and $D_i$, $B_{ij}$ and $A_{ij}$ are measured in days. The flows of LRUs in this system are shown in Figure 5.1.

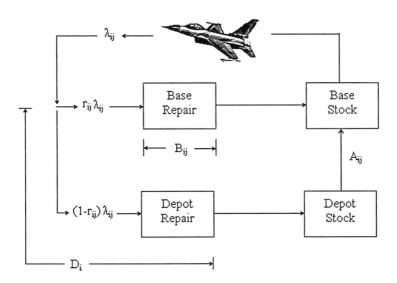

**Fig. 5.1.** The METRIC System

### 5.1.2   The Optimization Problem

Our goal is to develop a model that can be used to determine stock levels $s_{ij}$ for LRU $i$ at location $j$ and to show how to compute these stock levels. The objective of the model is to minimize the total number of average outstanding backorders at the bases at a random point in time for a given level of investment. But why choose the backorder criterion as the performance measure of interest? The goal should be to maximize the expected number of operational aircraft at the bases. Recall that in Chapter 3, we showed, to a first order approximation, that minimizing backorders at bases is equivalent to maximizing the expected number of operational aircraft at the base level. Thus maximizing the average number of available aircraft is approximately equal to minimizing the average number of outstanding base level LRU backorders at a random point in time.

As before,

$$s_{ij} = \text{LRU } i \text{ stock level at base } j \text{ (or depot if } j = 0).$$

Before we state the optimization problem, we first develop some additional relationships.

The key equation that links the base and depot stock levels is the average resupply time equation for LRU $i$ at base $j$. Let

$$T_{ij} = \text{average LRU } i \text{ resupply time at base } j$$
$$= r_{ij}B_{ij} + (1 - r_{ij})(A_{ij} + \text{depot delay } (s_{i0})).$$

Thus the average resupply time is $B_{ij}$ when the repair occurs at base $j$ for LRU $i$ and $A_{ij}$ plus an expected waiting time (depot delay) due to the depot stock level when resupply comes from the depot. These average times are weighted by the probability that resupply occurs either from base maintenance or from depot stock. But how do we measure this expected waiting time?

Let

$$\delta(s_{i0}) = \text{expected depot delay, or waiting time, given the depot}$$
$$\text{stock level is } s_{i0} \text{ for LRU type } i.$$

From Little's Law

$$\delta(s_{i0}) = \frac{\text{Average outstanding depot backorders } (s_{i0})}{\text{Depot demand rate } (\lambda_{i0})},$$

as discussed in Chapter 3.

Let $\mathcal{B}_D(s_{i0}) = $ expected outstanding depot backorders for LRU $i$ given $s_{i0}$. Hence

$$\delta(s_{i0}) = \frac{\mathcal{B}_D(s_{i0})}{\lambda_{i0}}.$$

Let us explore how $\delta(s_{i0})$ behaves. First, we note that $\delta(s_{i0}) \to 0$ quite rapidly as $s_{i0}$ exceeds $\lambda_{i0}D_i$. To illustrate this observation, consider the cases where $\lambda_{i0}D_i$ is equal to 1, 5, 10, 50 and 100. We display the values of $\delta(s_{i0})$ for these cases in Table 5.1 through Table 5.5, where we assume $D_i = 1$.

As can be seen from the data in these tables, the expected depot delay becomes small quite quickly when $D_i = 1$. When $A_{ij}$ is several days in length, the depot delay becomes a relatively inconsequential portion of the expected resupply time as $s_{i0}$ exceeds $\lambda_{i0}D_i$.

As an example, suppose $r_{ij} = 0$, $A_{ij} = 5$, $\lambda_{ij} = 5$ and $\lambda_{i0} = 50$, with $D_i = 1$. Then the expected number of units in resupply for the base is $5 \times 5 + 5 \times \delta(s_{i0})$. If $s_{i0} = 50$, then $\delta(s_{i0}) = .0563$ and the expected number of units in

**Table 5.1.** $\delta(s_{i0})$ when $\lambda_{i0} = 1$, $D_i = 1$

| $s_{i0}$ | $\delta_i(s_{i0})$ |
|---|---|
| 0 | 1 |
| 1 | .3679 |
| 2 | .1036 |
| 3 | .0233 |
| 4 | .0043 |
| 5 | .0007 |
| 6 | .0001 |

**Table 5.2.** $\delta(s_{i0})$ when $\lambda_{i0} = 5$, $D_i = 1$

| $s_{i0}$ | $\delta_i(s_{i0})$ |
|---|---|
| 4 | .2874 |
| 5 | .1755 |
| 6 | .0987 |
| 7 | .0511 |
| 8 | .0244 |
| 9 | .0108 |
| 10 | .0044 |
| 11 | .0017 |
| 12 | .0006 |

**Table 5.3.** $\delta(s_{i0})$ when $\lambda_{i0} = 10$, $D_i = 1$

| $s_{i0}$ | $\delta(s_{i0})$ |
|---|---|
| 9 | .1793 |
| 10 | .1251 |
| 11 | .0834 |
| 12 | .0531 |
| 13 | .0322 |
| 14 | .0187 |
| 15 | .0103 |
| 16 | .0055 |
| 17 | .0028 |
| 18 | .0013 |

**Table 5.4.** $\delta(s_{i0})$ when $\lambda_{i0} = 50$, $D_i = 1$

| $s_{i0}$ | $\delta(s_{i0})$ |
|---|---|
| 49 | .0667 |
| 50 | .0563 |
| 51 | .0471 |
| 52 | .0389 |
| 53 | .0318 |
| 54 | .0258 |
| 55 | .0206 |
| 56 | .0163 |
| 57 | .0127 |
| 58 | .0098 |

**Table 5.5.** $\delta(s_{i0})$ when $\lambda_{i0} = 100$, $D_i = 1$

| $s_{i0}$ | $\delta(s_{i0})$ |
|---|---|
| 100 | .0398 |
| 101 | .0351 |
| 102 | .0308 |
| 103 | .0268 |
| 104 | .0233 |
| 105 | .0200 |
| 106 | .0172 |
| 107 | .0146 |
| 108 | .0124 |
| 109 | .0104 |
| 110 | .0087 |
| 115 | .0032 |

**Table 5.6.** $\delta(s_{i0})$ when $\lambda_{i0} = 5$ and $D_i = 10$

| $s_{i0}$ | $\delta(s_{i0})$ |
|---|---|
| 49 | .667 |
| 50 | .563 |
| 51 | .471 |
| 52 | .389 |
| 53 | .318 |
| 54 | .258 |
| 55 | .206 |
| 56 | .163 |
| 57 | .127 |
| 58 | .098 |

the base's resupply system is 25.2815 units of which only .2815 are attributable to the depot delay in resupply of the base. In this example, we see that this minimal contribution to the expected number of units in the base resupply system occurred when the depot carried no safety stock. Thus as the depot demand rate increases and $A_{ij}$ increases while $r_{ij}$ decreases, there is little to be gained by having much depot safety stock.

Now let us consider another example. Suppose $r_{ij} = 0, A_{ij} = 5, \lambda_{ij} = .5, \lambda_{i0} = 5$ but $D_i = 10$. Thus $\lambda_{i0} D_i = 50$, as was the case in the previous example. To find $\delta(s_{i0})$ in this case we can again use the data in Table 5.4. The data in that table must be multiplied by the value of $D_i$ to obtain the new values of $\delta(s_{i0})$, which are given in Table 5.6. As was the case in the first example, the average resupply time is dominated by the value of $A_{ij}$. Hence if a small amount of safety stock is carried at the depot, say 5 units so that $s_{i0} = 55$, then the average resupply time is 5.206 days. The expected number of units in resupply at the base is now 2.6030 units of which .103 units are attributable to depot delay.

Observe that as $\lambda_{i0} D_i$ remains constant but $D_i$ increases, then $\delta(s_{i0})$ increases proportionate to the increase in $D_i$. For example, as $D_i$ increased from 1 to 10, $\delta(s_{i0})$ increased by a factor of 10. If $\lambda_{ij} = .25, \lambda_{i0} = 2.5$ and $D_i = 20$, then $\delta(50) = 1.126$ days and $\delta(55) = .412$ days. Thus as $D_i$ increases, while $\lambda_{i0} D_i$ remains constant, we see that $\delta(s_{i0})$ increases and can become a more significant portion of the average resupply time. Hence safety stock at the depot will become more important as $D_i$ becomes larger while $\lambda_{i0} D_i$ remains constant.

In any case, the range of values that need to be evaluated explicitly in an optimization procedure is limited since $\delta(s_{i0})$ approaches small values quite quickly as $s_{i0} > \lfloor \lambda_i D_i \rfloor$. The range can usually be limited to two standard deviations of depot demand over the depot's resupply time. That is, the optimal value that $s_{i0}$ assumes almost always is in the interval $\left[ \lfloor \lambda_i D_i \rfloor, \lceil 2 \cdot (\lambda_i D_i)^{1/2} \rceil + \lfloor \lambda_i D_i \rfloor \right]$, assuming $\lfloor \lambda_i D_i \rfloor$ is the minimum depot stock level that is considered. Furthermore, when solving practical problems, the search for the optimal value of $s_{i0}$ is limited to a subset of these values. Consider the values in Table 5.5. Observe that successive values do not differ by substantial amounts. Hence searches are limited to perhaps every second value in the interval given earlier. Thus when $\lambda_i D_i = 100$, the search for the optimal value of $s_{i0}$ might be restricted to $100, 102, \ldots, 120$. As a consequence of these observations, a maximum of 10 to 15 possible values for $s_{i0}$ are often explicitly considered in practice when employing the optimization methodology we will be discussing in subsequent sections.

### 5.1.2.1 Approximating the Stationary Probability Distribution for the Number of LRUs in Resupply

To calculate the expected number of base $j$ backorders for LRU $i$, we must determine the stationary probability distribution for the number of units in resupply for LRU $i$ at base $j$.

Let $X_{ij}$ = random variable for the number of units in resupply for LRU $i$ at base $j$ in steady state. Clearly,

$$E[X_{ij}] = \lambda_{ij} T_{ij}$$

$$= r_{ij} \lambda_{ij} B_{ij} + (1 - r_{ij}) \lambda_{ij} A_{ij} + (1 - r_{ij}) \frac{\lambda_{ij}}{\lambda_{i0}} \mathcal{B}_D(s_{i0}).$$

To compute the variance of $X_{ij}$ requires some additional analysis. We drop the LRU subscript in this analysis for ease of exposition.

Suppose there are $N_D$ backorders at the depot for some LRU. Let $N_j$ be the number of base $j$ units backordered at the depot. The probability that $N_j = n_j$ when $N_D = n_D$, where $n_j \leq n_D$, is given by

$$P\{N_j = n_j | N_D = n_D\} = \binom{n_D}{n_j} \left( \frac{\hat{\lambda}_j}{\lambda_0} \right)^{n_j} \left( 1 - \frac{\hat{\lambda}_j}{\lambda_0} \right)^{n_D - n_j},$$

where $\hat{\lambda}_j = (1 - r_j)\lambda_j$. Furthermore,

$$E[N_j | s_0] = E[N_j] = E_{N_D} \left[ E_{N_j}[N_j | N_D] \right] = E_{N_D} \left[ \frac{\hat{\lambda}_j}{\lambda_0} N_D \right] = \frac{\hat{\lambda}_j}{\lambda_0} \mathcal{B}_D(s_0).$$

Thus the expected number of LRUs in backorder status corresponding to demands at base $j$ is the fraction of total depot demand due to base $j$, on average, times the expected number of depot backorders given $s_0$.

The variance of $N_j$ also depends on the depot stock level. We know that

$$\text{Var}(N_j | s_0) = E[N_j^2 | s_0] - E[N_j | s_0]^2.$$

We also know that $E[N_j | s_0] = \frac{\hat{\lambda}_j}{\lambda_0} \mathcal{B}_D(s_0)$. Furthermore,

$$
\begin{aligned}
E[N_j^2 | s_0] &= E_{N_D} \left[ E_{N_j}[N_j^2 | N_D] \right] \\
&= E_{N_D} \left[ \text{Var}(N_j | N_D) + E[N_j | N_D]^2 \right] \\
&= E_{N_D} \left[ N_D \cdot \frac{\hat{\lambda}_j}{\lambda_0} \cdot \left( 1 - \frac{\hat{\lambda}_j}{\lambda_0} \right) + \left( \frac{\hat{\lambda}_j}{\lambda_0} N_D \right)^2 \right] \\
&= \frac{\hat{\lambda}_j}{\lambda_0} \left( 1 - \frac{\hat{\lambda}_j}{\lambda_0} \right) \mathcal{B}_D(s_0) + \left( \frac{\hat{\lambda}_j}{\lambda_0} \right)^2 \cdot E_{N_D} \left( N_D^2 | s_0 \right) \\
&= \frac{\hat{\lambda}_j}{\lambda_0} \left( 1 - \frac{\hat{\lambda}_j}{\lambda_0} \right) \mathcal{B}_D(s_0) + \left( \frac{\hat{\lambda}_j}{\lambda_0} \right)^2 \left[ \text{Var}(N_D | s_0) + \left( E_{N_D}(N_D | s_0) \right)^2 \right] \\
&= \frac{\hat{\lambda}_j}{\lambda_0} \left( 1 - \frac{\hat{\lambda}_j}{\lambda_0} \right) \mathcal{B}_D(s_0) + \left( \frac{\hat{\lambda}_j}{\lambda_0} \right)^2 \left[ \text{Var}(N_D | s_0) + (\mathcal{B}_D(s_0))^2 \right].
\end{aligned}
$$

Thus, by combining the above observations, we see that

$$\text{Var}(N_j | s_0) = \frac{\hat{\lambda}_j}{\lambda_0} \left( 1 - \frac{\hat{\lambda}_j}{\lambda_0} \right) \mathcal{B}_D(s_0) + \left( \frac{\hat{\lambda}_j}{\lambda_0} \right)^2 \text{Var}(N_D | s_0).$$

The $\text{Var}(N_D|s_0)$ can be computed as follows. Since

$$\text{Var}(N_D|s_0) = E[N_D^2|s_0] - (\mathcal{B}_D(s_0))^2,$$

we need to determine a method for calculating $E[N_D^2|s_0]$. Recall that $N_D$ measures the number of depot backorders at a random point in time. Hence if $N_D = n_D$, then the number of units of the LRU in depot resupply must exceed the depot stock level by $n_D$. Thus if $x$ measures the number of units of the LRU in depot resupply, there are $n_D$ depot backorders for the LRU if and only if $x - s_0 = n_D$. Hence

$$E[N_D^2|s_0] = \sum_{x \geq s_0} (x - s_0)^2 p(x|\lambda_0 D),$$

where $p(x|\lambda_0 D)$ is the probability that $x$ units are in depot resupply. From Palm's theorem, $p(x|\lambda_0 D) = e^{-\lambda_0 D} \frac{(\lambda_0 D)^x}{x!}$. Then

$$
\begin{aligned}
E[N_D^2|s_0] &= \sum_{x \geq s_0} \left( (x - (s_0 - 1)) - 1 \right)^2 p(x|\lambda_0 D) \\
&= \sum_{x \geq s_0} (x - (s_0 - 1))^2 \, p(x|\lambda_0 D) \\
&\quad - 2 \sum_{x \geq s_0} (x - s_0) p(x|\lambda_0 D) \\
&\quad - \sum_{x \geq s_0} p(x|\lambda_0 D) \\
&= \sum_{x \geq s_0} (x - (s_0 - 1))^2 \, p(x|\lambda_0 D) \\
&\quad - \mathcal{B}_D(s_0) - \sum_{x \geq s_0} (x - (s_0 - 1)) \, p(x|\lambda_0 D) \\
&= E[N_D^2|s_0 - 1] - \mathcal{B}_D(s_0) - \mathcal{B}_D(s_0 - 1).
\end{aligned}
$$

We thus can determine $E[N_D^2|s_0]$ recursively, and therefore can determine $\text{Var}(N_D|s_0)$ recursively. Observe that

$$
\begin{aligned}
E[N_D^2|0] &= \sum_{x \geq 0} x^2 p(x|\lambda_0 D) \\
&= \text{Var [Number of units in Depot resupply]} \\
&\quad + [\text{Expected number of units in depot resupply}]^2 \\
&= \lambda_0 D + (\lambda_0 D)^2
\end{aligned}
$$

when base demand is described by a Poisson process.

Now that we have computed the mean and variance of $N_D$ we are able to compute the variance of $Y_j$, where $Y_j$ is the random variable that represents the demand over the order and ship time $(A_j)$ plus the backordered demand due to the depot stock level $s_0$ at a random point in time at base $j$. Since the demands that cause depot backorders and those occurring during the transportation lead time to base $j$ take place in nonoverlapping intervals, and are therefore independent,

$$\text{Var } Y_j = \text{Var (Demand over } A_j)$$
$$+ \text{ Var (Backordered demand due to depot stock } s_0),$$

$$\text{Var } Y_j = \hat{\lambda}_j A_j + \frac{\hat{\lambda}_j}{\lambda_0}\left(1 - \frac{\hat{\lambda}_j}{\lambda_0}\right)\mathcal{B}_D(s_0) + \left(\frac{\hat{\lambda}_j}{\lambda_0}\right)^2 \text{Var }(N_D|s_0),$$

where as before $\hat{\lambda}_j = (1 - r_j)\lambda_j$.

But to compute the expected number of backorders at base $j$ for LRU $i$ we must know the probability distribution of the number of units in resupply, $X_{ij}$. Rather than computing this distribution exactly, as we did in Chapter 4, we approximate it, as Sherbrooke [227] did, with a negative binomial distribution with parameters

$$\mu_{ij} = r_{ij}\lambda_{ij}B_{ij} + (1 - r_{ij})\lambda_{ij}A_{ij} + (1 - r_{ij})\frac{\lambda_{ij}}{\lambda_{i0}}\mathcal{B}_D(s_{i0})$$

$$\sigma_{ij}^2 = r_{ij}\lambda_{ij}B_{ij} + (1 - r_{ij})\lambda_{ij}A_{ij}$$
$$+ \frac{(1 - r_{ij})\lambda_{ij}}{\lambda_{i0}}\left(1 - \frac{(1 - r_{ij})\lambda_{ij}}{\lambda_{i0}}\right)\mathcal{B}_D(s_{i0})$$
$$+ \frac{\left((1 - r_{ij})\lambda_{ij}\right)^2}{\lambda_{i0}^2}\text{Var }(N_D|s_{i0}).$$

The appropriateness of the negative binomial approximation is discussed in Chapter 6. The negative binomial distribution has two parameters, call them $p$ and $r$. These parameters must satisfy the two equations

$$\mu_{ij} = \frac{r(1 - p)}{p} \text{ and } \sigma_{ij}^2 = \frac{r(1 - p)}{p^2}.$$

Thus

$$1 > \frac{\mu_{ij}}{\sigma_{ij}^2} = p \text{ and } r = \frac{p\mu_{ij}}{1 - p}.$$

The probabilities can be computed recursively with

$$P\{X_{ij} = 0\} = p^r \text{ and } P\{X_{ij} = x\} = P\{X_{ij} = x - 1\}\frac{(r + x - 1)}{x}q,$$

where $q = 1 - p$.

The choice of the negative binomial distribution as an approximation to the probability distribution of $X_{ij}$ was made for two key reasons. First, it is easily computed, which is essential in large scale applications. Second, it is an accurate approximation. A substantial amount of testing has been conducted to verify the validity of the approximation. In Chapter 6, we describe one such validation study.

Additionally, in the Air Force and other implementations of the model we have just described, as well as the one presented in Section 5.3.1, a negative binomial model was used for reasons other than those that have been discussed. A negative binomial model was used to represent the demand process as well in these approximations. That is, the negative binomial probability model replaces the Poisson model in all the demand probability model expressions found throughout this chapter. The negative binomial model was chosen in these applications to accommodate the high variance-to-mean ratios of the demand processes found in practice. A complete discussion of the variance-to-mean ratio of the demand process can be found in Crawford [65].

### 5.1.2.2   Finding Depot and Base LRU Stock Levels

We now turn to finding the best depot and base stock levels for each LRU, which are the stock levels that minimize the expected number of "holes" in aircraft, that is, that minimize the expected number of base LRU backorders at a random point in time subject to a constraint on investment in inventory.

$$\min \sum_i \sum_{j=1}^{m} \mathcal{B}_{ij}(s_{ij}|s_{i0})$$

$$s.t \quad \sum_i c_i(s_{i0} + \sum_{j=1}^{m} s_{ij}) \le b, \qquad s_{ij} = 0, 1, \ldots,$$

(5.1)

where $c_i$ is the cost of one unit of LRU type $i$, $b$ is the available budget to invest in the LRUs, and $m$ is the number of bases.

Unfortunately Problem (5.1) is neither separable nor convex. It is not separable because the variables $s_{ij}$ and $s_{i0}$ interact when computing $\mathcal{B}_{ij}(s_{ij}|s_{i0}) = \sum_{x \ge s_{ij}}(x - s_{ij})p(x|\lambda_{ij}T_{ij}(s_{i0}))$, where $p(x|\lambda_{ij}T_{ij}(s_{i0}))$ is approximated by a negative binomial distribution with mean $\mu_{ij}$ and variance $\sigma_{ij}^2$, as defined previously. Both $\mu_{ij}$ and $\sigma_{ij}^2$ are functions of $s_{i0}$. We will discuss the convexity issue shortly.

There are many different ways to solve (5.1). One way is to use a marginal analysis approach and another is to employ a Lagrange multiplier method.

### 5.1.2.2.1   A Marginal Analysis Algorithm

We will first discuss the marginal analysis approach. Since the objective function is separable by LRU type, we first focus on a single LRU type. As before, we temporarily drop the LRU subscript to reduce the notation.

To determine a solution to (5.1) we analyze the relationships between total system LRU inventory and the expected number of total base backorders. We will construct this function first and will observe that this function need not be convex. Hence, we will construct its convex minorant and use this convex minorant as a basis for making budget allocations among different types of LRUs.

Next, define

$$\alpha(s, s_0) = \text{minimum total expected base backorders for}$$
$$\text{an LRU given that there are } s \text{ units of the}$$
$$\text{LRU in the system and the depot stock is } s_0.$$

Let $S$ represent the set of values of $s_0$ that will be explicitly considered in the optimization process.

Let $s_0 \epsilon S$. Then, given $s_0$, we can compute $\mu_j$ and $\sigma_j^2$ as well as $P\{X_j = x|s_0\} \equiv p_j(x|s_0)$. Recall that

$$B_j(s_j|s_0) = \sum_{x>s_j} (x - s_j) p_j(x|s_0).$$

Also recall that $B_j(s_j|s_0)$ is a convex function, given $s_0$, and satisfies

$$B_j(s_j|s_0) = B_j(s_j - 1|s_0) - (1 - \sum_{x<s_j} p_j(x|s_0)).$$

Thus the reduction in backorders by adding one unit to the inventory at base $j$ is $\left(1 - \sum_{x<s_j} p_j(x|s_0)\right)$. Note also that $B_j(0|s_0) = \mu_j$.

Now suppose that $s = s_0 + 1$. The question we must answer is to which base should this $(s_0 + 1)st$ unit of stock be allocated. To begin the process, set $s_j = 0$ for all $j$. We want to place the unit at the base that has the largest reduction in expected base backorders resulting from the allocation; that is, assign the unit to the base having the largest value of $(1 - p_j(0|s_0))$. Since the functions $B_j(s_j|s_0)$ are convex given $s_0$ and are strictly decreasing as $s_j$ increases, we can use a marginal analysis method to compute the best allocation of a stock of $s - s_0 = a$ units knowing the best allocation of $s - s_0 = a - 1$ units.

If $\tilde{s}_j(a - 1)$ represents the optimal stock level for base $j$ given $a - 1$ units are available for allocation to the bases (given the value of $s_0$), then the base that is allocated the next unit of stock is the one that has the maximum value of

$$\left(1 - \sum_{x<\tilde{s}_j(a-1)+1} p_j(x|s_0)\right).$$

Observe that to calculate the marginal reduction in backorders for base $j$ by adding an additional unit of stock to that base we need to determine

$$\sum_{x<\tilde{s}_j(a-1)+1} p_j(x|s_0) = \sum_{x<\tilde{s}_j(a-1)} p_j(x|s_0) + p_j(\tilde{s}_j(a - 1) + 1).$$

But $\sum_{x<\tilde{s}_j(a-1)} p_j(x|s_0)$ is known so that only $p_j(\tilde{s}_j(a - 1) + 1|s_0)$ must be computed. As observed earlier,

$$p_j(\tilde{s}_j(a-1)+1|s_0) = p(\tilde{s}_j(a-1)|s_0)\frac{\left(r+(\tilde{s}_j(a-1)+1)-1\right)}{\tilde{s}_j(a-1)+1}q.$$

Hence the function $\alpha(s, s_0)$ can be computed very quickly for a large range of $s$ values. For example, we could choose to limit $s$ by setting a minimum value for $\alpha(s, s_0)$. That is, when $\alpha(s, s_0) < \epsilon$, then terminate the calculations.

The functions $\alpha(s, s_0)$ must be calculated for all $s_0 \epsilon S$.

Next, define

$$\hat{\alpha}(s) = \min_{s_0} \alpha(s, s_0).$$

$\hat{\alpha}(s)$ is a strictly decreasing function of $s$; however, it need not be convex. As a consequence, we construct $\hat{\alpha}^c(s)$, where $\hat{\alpha}^c(s)$ is a piecewise linear convex minorant to the function $\hat{\alpha}(s)$. Let $\hat{S}_c$ be the values of $s$ at which the slope of $\hat{\alpha}^c(s)$ changes.

For example, suppose $m = 10$ bases, each of which has a daily expected demand rate of .195 units. Further, assume the depot resupply time is 10 days and the depot to base transit time is 1 day for all bases. Figure 5.2 contains the graph of $\hat{\alpha}(s)$. Table 5.7 shows the values of $\hat{\alpha}(s)$ and $s_0^*$, the latter being the optimal depot stock levels corresponding to the various values of total system stock $s$. Note that the optimal depot stock level is not monotone nondecreasing as the total system stock increases. This illustrates why this type of optimization problem is not easily solved. Note also that $\hat{\alpha}(s)$ is not convex. Thus we construct its convex minorant, which is shown in Figure 5.3. In this case, $\hat{S}_c = \{35, 36, 41, 42, 43, 44, 45, 46, 48, 54, 55\}$.

Total Expected Base Backorders

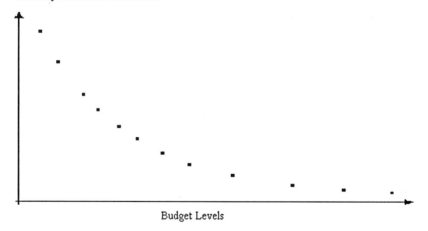

Budget Levels

**Fig. 5.2.** Graph of $\hat{\alpha}(s)$

Suppose we construct functions $\hat{\alpha}_i^c(s_i)$ for all LRU types $i$, where $s_i$ is the total system stock for item type $i$. Rather than solving (5.1) for a specific target

| $s$ | $\hat{\alpha}(s)$ | $s_0^*$ | $s$ | $\hat{\alpha}(s)$ | $s_0^*$ |
|---|---|---|---|---|---|
| 35 | .25798 | 25 | 46 | .01867 | 26 |
| 36 | .22889 | 26 | 47 | .01565 | 27 |
| 37 | .20787 | 26 | 48 | .01234 | 28 |
| 38 | .18685 | 26 | 49 | .01117 | 28 |
| 39 | .16421 | 25 | 50 | .01000 | 28 |
| 40 | .12383 | 20 | 51 | .00882 | 28 |
| 41 | .08727 | 21 | 52 | .00765 | 28 |
| 42 | .06120 | 22 | 53 | .00465 | 23 |
| 43 | .04324 | 23 | 54 | .00285 | 24 |
| 44 | .03128 | 24 | 55 | .00183 | 25 |
| 45 | .02355 | 25 | | | |

**Table 5.7.** $\hat{\alpha}(s)$ and $s_0^*$

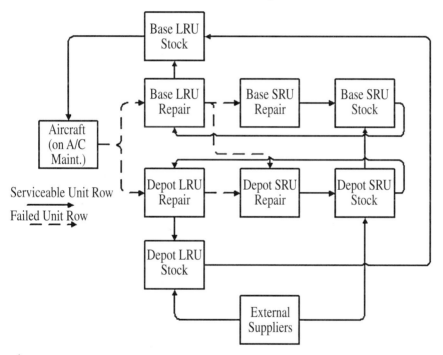

**Fig. 5.3.** Graph of $\hat{\alpha}^c(s)$

budget value $b$, we will construct a tradeoff curve of total base backorders over all LRU types as a function of the total investment in all LRUs. This function is constructed using another marginal analysis algorithm.

Let $s_i^1, \ldots, s_i^{K_i}$ be the elements of $\hat{S}_c^i$, the set of total stock levels that will be considered for LRU type $i$ based on the construction of the convex piecewise linear function $\hat{\alpha}_i^c(s_i)$. That is, the values $s_i^k$ are the stock levels at which the slope of the total base backorder function $\hat{\alpha}_i^c(s_i)$ changes.

To begin the construction process, compute $\hat{\alpha}_i^c(s_i^1)$ for all items $i$. Let $\beta(1) = \sum_i \hat{\alpha}_i^c(s_i^1)$, and let $C(1) = \sum_i s_i^1 c_i$. Next, set $k_i = 1$ for all $i$ and $\ell = 1$.
At the beginning of iteration $\ell \geq 1$, we have $C(\ell)$ and $\beta(\ell)$. For all $i$, compute

$$\Delta_i(s_i^{k_i}) = \frac{\hat{\alpha}_i^c(s_i^{k_i}) - \hat{\alpha}_i^c(s_i^{k_i+1})}{c_i(s_i^{k_i+1} - s_i^{k_i})},$$

the reduction in total expected base backorders by incrementing the stock level from $s_i^{k_i}$ to $s_i^{k_i+1}$ per dollar invested in LRU $i$.
Select $i^*$ such that

$$\Delta_{i^*}(s_{i^*}^{k_{i^*}}) = \max_i \Delta_i(s_i^{k_i}).$$

Set

$$C(\ell+1) = C(\ell) + c_{i^*}(s_{i^*}^{k_{i^*}+1} - s_{i^*}^{k_{i^*}}) \text{ and}$$
$$\beta(\ell+1) = \beta(\ell) - (\hat{\alpha}_{i^*}^c(s_{i^*}^{k_{i^*}}) - \hat{\alpha}_{i^*}^c(s_{i^*}^{k_{i^*}+1})).$$

Increment $k_{i^*} = k_{i^*} + 1$ and $\ell = \ell + 1$, and recompute the value of $\Delta_{i^*}(s_{i^*}^{k_{i^*}})$. The $\Delta_i(\cdot)$ values remain the same for all other values of $i$.
Repeat this process until there are no remaining values of $k_i$ to consider, that is, $k_i = K_i$ for all $i$.
The process we have outlined can be used to compute the total expected base backorder function for a finite set of the investment levels in LRU system inventory, as depicted in Figure 5.4.

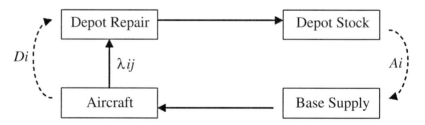

**Fig. 5.4.** Total Expected Base Backorders for Various Budget Levels Generated by the Algorithm

### 5.1.2.2.2 A Lagrangian Method for Computing Depot and Base Stock Levels

Let us now describe an alternative method for computing the depot and base stock levels, that is, another method for solving Problem (5.1). Roughly the algorithm works as follows. First construct a Lagrangian relaxation to Problem (5.1) and then decompose it into a collection of single item problems, which are solved one at a time. The constraint that is relaxed is the investment constraint. Thus there is only one Lagrangian multiplier in the relaxed formulation of our problem. For each item, given the multiplier value, determine the best depot and base stock levels using an enumeration method. This process is repeated over various multiplier values to determine the solution to Problem (5.1).

This Lagrangian relaxation algorithm is based on an observation made in Chapter 3. That is, the optimal solution to a relaxed problem yields an optimal solution to Problem (5.1) for a certain value of $b$. By investigating an appropriate range of values for the Lagrangian multiplier we can provide good, if not necessarily optimal, solutions for Problem (5.1).

Let us begin by constructing the Lagrangian relaxation:

$$\min_{s_{ij}} \sum_{i=1}^{n} \sum_{j=1}^{m} \mathcal{B}_{ij}(s_{ij}, s_{i0}) + \theta \sum_{i=1}^{n} c_i \sum_{j=0}^{m} s_{ij} - \theta b \tag{5.2}$$

subject to $s_{ij} = 0, 1, \ldots$, where we continue to use the same notation as employed in the previous section. $\theta$ represents the multiplier corresponding to the constraint

$$\sum_{i=1}^{n} c_i \sum_{j=0}^{m} s_{ij} \leq b.$$

For this formulation, $\theta > 0$.

The Lagrangian relaxation may also be expressed as

$$\sum_{i=1}^{n} \left\{ \min \sum_{j=1}^{m} \mathcal{B}_{ij}(s_{ij}, s_{i0}) + \theta c_i \sum_{j=0}^{m} s_{ij} \right\} - \theta b$$

$$\text{subject to } s_{ij} = 0, 1, \ldots .$$

Hence the Lagrangian relaxation is separable by item, that is, we can solve

$$\min \sum_{j=1}^{m} \mathcal{B}_{ij}(s_{ij}, s_{i0}) + \theta c_i \sum_{j=0}^{m} s_{ij} \tag{5.3}$$

$$s_{ij} = 0, 1, \ldots ,$$

for each item $i$ independently of all other items.

In Chapter 3 we showed the expected backorder functions $\mathcal{B}_{ij}(s_{ij}, s_{i0})$ are convex functions of $s_{ij}$ for a given value of $s_{i0}$. However, the function is not necessarily jointly convex in $s_{ij}$ and $s_{i0}$, as we observed earlier. Thus to find an optimal solution to (5.1), given $\theta$, we use an exhaustive search on the candidate depot stock levels.

Suppose $\theta$ assumes some value, say $\theta_\ell$, and $s_{i0}$ is equal to $\rho$. Then we want to find the base stock levels that solve

$$W_{\rho\ell}^i = \min_{s_{ij}=0,1,\ldots} \left[ \sum_{j=1}^{m} \left( \mathcal{B}_{ij}(s_{ij}, \rho) + \theta_\ell c_i s_{ij} \right) \right] + \theta_\ell c_i \rho. \tag{5.4}$$

Notice that

$$\min_{s_{ij}=0,1,\dots} \left[ \sum_{j=1}^{m} \left( \mathcal{B}_{ij}(s_{ij}, \rho) + \theta_\ell c_i s_{ij} \right) \right]$$
$$= \sum_{j=1}^{m} \min \left( \mathcal{B}_{ij}(s_{ij}, \rho) + \theta_\ell c_i s_{ij} \right), \quad (5.5)$$

that is, the relaxed optimization problem is also separable by base for each item, given $\theta_\ell$ and $\rho$.

The solution to

$$\min \mathcal{B}_{ij}(s_{ij}, \rho) + \theta_\ell c_i s_{ij} \quad (5.6)$$
$$s_{ij} = 0, 1, \dots$$

can be found quite easily. The objective function to Problem (5.6) is discretely convex in $s_{ij}$. Hence we can easily show, using a first difference argument, that the optimal value for $s_{ij}$ is the smallest nonnegative integer for which

$$\sum_{x \geq s_{ij}+1} p(x|\rho) \leq \theta_\ell c_i \text{ or } \sum_{x \leq s_{ij}} p(x|\rho) \geq 1 - \theta_\ell c_i,$$

where $p(x|\rho)$ is the probability that there are $x$ units in resupply at base $j$ for item $i$ given $s_{i0} = \rho$. These probabilities are computed as described in Section 5.1.2.1.

Observe that whenever $\theta_\ell \geq \frac{1}{c_i}$, the optimal value for $s_{ij}$ is 0. Thus a necessary condition for $s_{ij} > 0$, for all $i$, is $\theta_\ell < \min_i \frac{1}{c_i} = \frac{1}{\max_i c_i}$.

We have shown how to find the optimal values of the base stock levels given $\theta_\ell$ and $\rho$. To find an optimal set of depot and base stock levels for a given value of $\theta_\ell$, find the value of $s_{i0} = \rho$ for which

$$s_{i0} = \arg\min_\rho W^i_{\rho\ell} \quad (5.7)$$

and the corresponding base stock level values, $s_{ij}$. As we have discussed earlier, the range of depot stock levels that need to be examined explicitly is limited because $\frac{\mathcal{B}_{i0}(s_{i0})}{\lambda_{i0}} \to 0$ rapidly as $s_{i0}$ surpasses $\lambda_{i0} D$, the expected depot demand over the depot resupply time. In practical applications $\lfloor \lambda_{i0} D \rfloor$ is often a floor on $s_{i0}$.

Knowing the stock levels for the depot and bases, we also have the corresponding investment level for item $i$,

$$c_i(\theta_\ell) = c_i \sum_{j=0}^{m} s_{ij}. \quad (5.8)$$

The total investment level over all items, given $\theta = \theta_\ell$, is

$$\sum_{i=1}^{n} c_i(\theta_\ell). \quad (5.9)$$

Hence for every choice of $\theta$ there corresponds a required investment level in system stock.

For practical problems there are bounds on the value of $\theta$, too. Recall that for $s_{ij}$ to be positive, $\theta$ must be less than $\frac{1}{c_i}$. Hence a search for the optimal value is normally confined to the range $0 < \theta < \frac{1}{\max c_i}$.

There are two very different ways that a Lagrangian based algorithm can be employed in practice. In one approach a bisection method is used to find the optimal value of $\theta$. Since a range of budget versus expected base level backorder values is often desired, a second approach is sometimes implemented. In this second approach, a set of $M$ values for $\theta$ are prespecified, $0 < \theta_1 < \cdots < \theta_M < \frac{1}{\max c_i}$. These values are set based on experience in solving past problems. We now describe both approaches.

## Algorithm 1

Step 1. Determine $\theta_{\min}$ and $\theta_{\max}$, the minimum and maximum values considered for $\theta$. Set $\theta_1 = \frac{\theta_{\min} + \theta_{\max}}{2}$ and $\ell = 1$

Step 2. Set $\theta = \theta_\ell$. For each item $i$, find the depot and base stock levels that yield $\min_\rho \{W_{\rho\ell}^i\}$.

Step 3. Calculate $C(\theta) = \sum_{i=1}^{n} c_i \sum_{j=0}^{m} s_{ij}(\theta)$. If $|C(\theta) - b| < \epsilon$, stop; otherwise, if $C(\theta) > b$, set $\theta_{\min} = \theta_\ell$; otherwise set $\theta_{\max} = \theta_\ell$. Set $\theta_\ell = \frac{\theta_{\max} + \theta_{\min}}{2}$, $\ell = \ell + 1$, and return to Step 2.

The number of iterations required to find a good solution clearly depends on the initial choices for $\theta_{\min}$ and $\theta_{\max}$ and $\epsilon$. In real problems, when there is experience in choosing the initial range of $\theta$ values and $\epsilon$ is set to be $1/2\%$ of the budget, the number of iterations required normally does not exceed 10. Nonetheless, a very large number of calculations must be made repeatedly. The second algorithm requires fewer calculations, as we will see. The second algorithm is as follows.

## Algorithm 2

Step 1. Select a set of $M$ multiplier values

$$0 < \theta_1 < \theta_2 < \cdots < \theta_M < \frac{1}{\max c_i}$$

Step 2. For each item $i$, for each $\theta_\ell$, solve

$$\min_\rho W_{\rho\ell}^i$$

and obtain stock levels $s_{ij}(\theta_\ell)$ and $\sum_{j=1}^{m} \mathcal{B}_{ij}(s_{ij}(\theta_\ell), s_{i0}(\theta_\ell))$.

Step 3. Compute $C(\theta_\ell)$. Select the solution that has a budget closest to $b$.

As mentioned, it may appear that the two approaches require the same computational effort since the same calculations are performed in Step 2 in both cases. They are not as similar as one might initially believe because of the manner in which the second algorithm is implemented.

For the second algorithm, let us examine how we compute $W_{\rho\ell}^i$. Recall that for a given value of depot stock for item $i$, $s_{i0} = \rho$, we find the optimal base stock levels by determining the smallest values of $s_{ij}$ for which

$$\sum_{x \leq s_{ij}} p(x|\rho) \geq 1 - \theta_\ell c.$$

Observe that when $\ell_1 > \ell_2$,

$$s_{ij}(\theta_{\ell_1}) \leq s_{ij}(\theta_{\ell_2}).$$

Thus, when we implement Algorithm 2, for a given $\rho = s_{i0}$, we make a single pass through the base calculations and determine the optimal base stock levels for each value of $\theta$. Observe that we do not need to repeat the calculations made when finding $s_{ij}(\theta_M)$ when we are finding $s_{ij}(\theta_{M-1})$, since $s_{ij}(\theta_M) \leq s_{ij}(\theta_{M-1})$, given that $\rho = s_{i0}$. In practice, fewer calculations are required to implement the second algorithm. The second algorithm is particularly useful when the goal is to understand the tradeoffs between investment in system stock versus expected base level backorders.

## 5.2  Waiting Time Analysis

The optimization problem we have formulated has as its objective the minimization of the total average number of outstanding LRU backorders at base level at a random point in time. While the average number may be low for a particular LRU at a base and the average waiting time may be low as well, there is a need to know how long a delay could be experienced in responding to a request for resupply. In other words, what is the probability distribution for the LRU resupply waiting time at either a base or the depot. At base level, this time corresponds to the time until an aircraft is again operational; at the depot, this time measures the time a resupply request is delayed prior to shipping a unit to the requesting base.

We will now derive the distribution of waiting times assuming demands for resupply occur according to a simple Poisson process and assuming resupply times are independent and identically distributed random variables with density $\psi(\cdot)$ with mean $T$. Our goal is to compute the probability distribution that a failed LRU will wait longer than a time $u$ given that a demand for resupply occurred at time $t$. We also assume a first-come, first-serve queue discipline is followed.

Let us begin by defining some notation. Let

$W(t)$ be the waiting time random variable for satisfying an LRU demand occurring at time $t$,

$I(t)$ be the net inventory random variable of the LRU just before time $t$,

$X(t)$ be the random variable for the number of LRU units in the resupply process just before time $t$, which does not include the demand occurring at time $t$,

$V_1(t, u)$ be a random variable measuring the number of LRUs completing repair in the interval $(t, t + u]$ that were in the resupply system at time $t$ (i.e. failed during $(0, t)$ but were not repaired by time $t$),

$V_2(t, u)$ be a random variable measuring the number of LRUs completing repair in $(t, t + u]$ that entered the repair cycle during $(t, t + u]$,

$V(t, u) = V_1(t, u) + V_2(t, u)$, and

$F(x) = \int_0^x \psi(y)dy$, which is the probability that a LRU repair is completed in $x$ or fewer time units.

The LRU resupply request at time $t$ will remain unfilled for a period of time greater than $u$ if and only if the number of units in resupply at the point in time $t$ at which the resupply request is made is $s$ units or greater ($s$ is the stock level) and the number of units completing repair during $(t, t + u]$ is insufficient to satisfy that request by time $t + u$. Thus

$$W(t) > u \text{ if and only if } I(t) + V(t, u) + r(u) \leq 0,$$

where $r(u) = 1$ if the repair cycle time of the unit that failed at time $t$ is less than or equal to $u$ and is 0, otherwise.

Since $s = I(t) + X(t)$ for all $t$,

$$I(t) + V(t, u) + r(u) \leq 0 \text{ implies } s - X(t) + V(t, u) + r(u) \leq 0,$$

or $s \leq X(t) - V(t, u) - r(u) = X_1(t + u) - V_2(t, u) - r(u)$, where $X_1(t + u) = X(t) - V_1(t, u)$, that is, the number of LRUs in the resupply system at time $t + u$ that failed during $(0, t)$. Therefore

$$P\{W(t) > u\} = P\{X_1(t + u) - V_2(t, u) - r(u) \geq s\}.$$

Observe that the random variables $X_1(t + u)$, $V_2(t, u)$, and $r(u)$ are independent since LRUs requiring repair that arrive prior to $t$ in no way influence arrivals requiring repair subsequent to time $t$ in terms of timing, quantity or repair times. The repair time of the failed LRU occurring at time $t$ is also unaffected by those arrival times and the repair times of LRUs failing both prior to and subsequent to time $t$.

Let $p_1(t)$ be the probability that an arrival in $(0, t)$ remains in the resupply system at time $t + u$, that is,

$$p_1(t) = \frac{1}{t} \int_0^t (1 - F(t + u - v))dv$$

$$= \frac{1}{t} \int_u^{t+u} (1 - F(v))dv,$$

and $p_0$ be the probability that a LRU that failed during the interval $(t, t + u]$ completes its repair by time $t + u$, that is,

$$p_0 = \frac{1}{u} \int_0^u F(u - v) dv.$$

Then, following the reasoning presented in Chapter 3,

$$P\{X_1(t + u) = k\} = e^{-\lambda t p_1(t)} \frac{(\lambda t p_1(t))^k}{k!}$$

$$= e^{-\lambda \int_u^{t+u}(1 - F(v)) dv} \frac{(\lambda \int_u^{t+u}(1 - F(v)) dv)^k}{k!}$$

and

$$P\{V_2(t, u) = k\} = e^{-\lambda u p_0} \frac{(\lambda u p_0)^k}{k!}$$

$$= e^{-\lambda \int_0^u F(u-v) dv} \frac{(\lambda \int_0^u F(u - v) dv)^k}{k!}.$$

To determine the probability that $W(t) > u$ we consider two cases. In the first case, $X_1(t+u) = s+k, k \geq 1$, and $V_2(t, u) < k$. That is, the number of LRU units remaining in the resupply system at time $t + u$ that correspond to LRU failures that occurred prior to time $t$ is $s + k$, and additionally, less than $k$ LRUs complete repair in $(t, t + u]$ of the units that failed in $(t, t + u]$. Since we assume demands are satisfied on a first-come, first-served basis, the LRU demand that occurred at time $t$ can not be satisfied by time $t + u$.

In the second case, $s + k, k \geq 0$, LRUs are in resupply at time $t + u$ that correspond to resupply requests that were placed prior to $t$. Furthermore, exactly $k$ units completed repair during $(t, t + u]$ that corresponded to LRUs entering the repair cycle during $(t, t + u]$. Thus, in this second case, the resupply request made at time $t$ will remain unfilled if and only if the repair cycle time of the requesting unit exceeds $u$.

Combining these observations we see that

$$P\{W(t) > u\} = \sum_{k=1}^{\infty} \left[ P\{X_1(t + u) = s + k\} \sum_{y=0}^{k-1} P\{V_2(t, u) = y\} \right]$$

$$+ \sum_{k=0}^{\infty} P\{X_1(t + u) = s + k\}\{1 - F(u)\} P\{V_2(t, u) = k\}.$$

Suppose $F(0) = 0$. Then the probability that the waiting time is 0 is given by

$$P\{W(t) = 0\} = \sum_{k=0}^{s-1} P\{X(t) = k\} = P\{I(t) > 0\}.$$

Observe that $P\{X_1(t + u) = k\}$ depends on $t$ and $u$ but $P\{V_2(t, u) = k\}$ is a function of $u$ alone. Let

$$P\{X_1(u) = k\} = \lim_{t \to \infty} P\{X_1(t + u) = k\}$$

$$= e^{-\lambda \int_u^\infty (1 - F(v))dv} \frac{\left[\lambda \int_u^\infty (1 - F(v))dv\right]^k}{k!},$$

which represents the steady state probability that $k$ LRUs that were in the resupply system at the time a LRU fails remain in the resupply system $u$ time units later. Let $W$ be a random variable that measures the steady state waiting time to satisfy a request for resupply. Then

$$P\{W > u\} = \sum_{k=1}^{\infty} \left[ P\{X_1(u) = s + k\} \sum_{y=0}^{k-1} P\{V_2(u) = y\} \right]$$

$$+ \sum_{k=0}^{\infty} P\{X_1(u) = s + k\} \{1 - F(u)\} \cdot P\{V_2(u) = k\},$$

where $V_2(u)$ is the random variable measuring the number of LRUs that both arrive for repair and complete repair in a period of length $u$ following the failure of a LRU.

This waiting time distribution could be used to establish minimum depot LRU stock levels. If there is a desire to ensure that replenishment of base inventories is not delayed by more than $u$ days due to the lack of depot stock with probability $\alpha$, then the minimum depot stock level can be set accordingly. Note that $P\{W(t) > u\}$ requires an explicit statement of the repair cycle time distribution $F(\cdot)$. Thus $F(\cdot)$ will have to be estimated to make the required calculations.

## 5.3  A Multi-Indenture System

To reduce the cost of their maintenance, systems are increasingly being designed in a more modular manner. The expensive systems, often valued at many millions of dollars, contain many repairable assemblies, which we have called LRUs. The objective of the maintenance concept that underlies the product design is to remove defective LRUs from the system quickly so that the system is rapidly returned to a serviceable status. The LRUs are also often very expensive, some valued at as much as several million dollars. Hence keeping them in the repair system for an extended period of time is also costly, since more LRUs are required to keep the systems operational. The system design concept is to avoid lengthy assembly or LRU repair cycles as well as to minimize the length of the system's repair time. This is accomplished by making it possible to detect a faulty subassembly quite quickly and to remove and replace that faulty subassembly rapidly as well. The subassemblies are often expensive, too, and are often subject to repair when they fail. We call these repairable subassemblies shop replaceable units (SRUs). They are called SRUs to reflect the fact that they are removed from their parent assembly in a repair shop and not directly from the system. Thus keeping systems in a serviceable status depends critically on having both LRUs and SRUs available.

Determining the best mix of LRUs and SRUs for given investment in these items is the subject of this section.

### 5.3.1   A Two-Echelon, Two-Indenture Resupply System

An environment that contains LRUs that are assemblies that are removed from and replaced into systems when system failures occur and SRUs that are removed from and replaced into LRUs when they fail are called two-indenture level resupply systems. We will examine in detail a specific two-echelon, two-indenture resupply environment, as depicted in Figure 5.5, consisting of a depot and a set of bases supported by the depot.

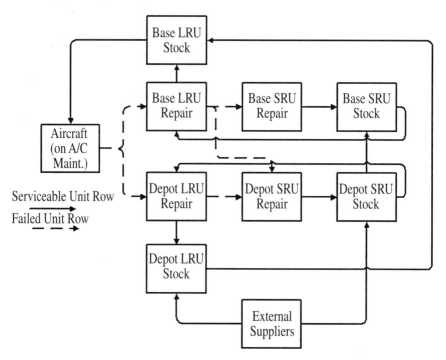

**Fig. 5.5.** A Two-Indenture, Two-Echelon Resupply System

We use the terminology of the U.S. Air Force system for which the model was originally created. Thus the systems are aircraft. The LRUs could be engines or so-called black boxes that make up the avionics subsystem of an aircraft. SRUs could be engine modules or could be boards that go into computers that are the LRUs in an avionics system. The issue then is to determine the depot and base LRU and SRU stock levels that minimize the expected number of base level backorders for LRUs. By doing so, we determine stock levels that will approximately maximize the number of operational aircraft for a range of total investments in LRUs and SRUs.

The optimization methodology is similar to the one we outlined for the two-echelon single indenture level problem discussed in Section 5.1.2.2.1 of this chapter. We first concentrate on constructing the key probability distributions that are the essential building blocks for the optimization models. Thus we will show how to compute the number of units in resupply for an LRU and each of its subordinate SRUs. We then outline a procedure for computing base and depot stock levels for both LRUs and SRUs.

### 5.3.2 Calculating the Stationary Probability Distribution of the Number of Units in Resupply

We begin by outlining the approach for constructing the probability distributions for the number of units in resupply for each LRU and SRU at each base.

A key assumption underlying our analysis is that each LRU failure (or removal from an aircraft) is the result of no more than one type of SRU failure. Furthermore, we assume that each SRU type is found in only one LRU type, which is a reasonable assumption in practice.

We will concentrate on developing the stationary probability distribution for a single LRU type, since the analysis is the same for all LRU types.

We now introduce some notation.

Let $i$ denote the item type. $i = 0$ is the LRU and $i = 1, \ldots, I$ denotes the SRU types. As before, $j$ denotes the location with $j = 0$ representing the depot and $j = 1, \ldots, m$ representing the bases.

Let

$\lambda_{ij}$ = daily demand rate for item type $i$ at location $j$,

$B_{ij}$ = average repair cycle time for item type $i$ at base $j$,

$D_{i0}$ = average depot repair cycle time for an SRU of type $i$,

$D_{00}$ = average depot LRU repair cycle time,

$r_{ij}$ = probability that a removal of SRU $i$ from an LRU at base $j$ results in that SRU being repaired at base $j$,

$r_{0j}$ = probability that a failure of a LRU at base $j$ will be repaired there,

$A_{ij}$ = order and ship time from the depot to base $j$ for item type $i$,

$\alpha_{ij}$ = probability that an LRU being repaired at base $j$ requires SRU $i$ to complete its repair, and

$\alpha_{i0}$ = probability that an LRU being repaired at the depot requires SRU $i$ to complete its repair.

Then the daily demand rate for SRU $i$ at base $j$ is $\lambda_{ij} = r_{0j}\lambda_{0j}\alpha_{ij}$ and the daily demand rate for the LRU at the depot is $\lambda_{00} = \sum_j (1 - r_{0j})\lambda_{0j}$. The depot demand rate for SRU type $i$ results from base and depot requirements and is expressed as

$$\lambda_{i0} = \sum_j \lambda_{ij}(1 - r_{ij}) + \lambda_{00}\alpha_{i0}, \, i > 0.$$

Let us focus on the LRU and compute the mean and variance of the number of LRU units in the depot resupply process.

Let

$$\beta_{i0} = \frac{\lambda_{00}\alpha_{i0}}{\lambda_{i0}} = \text{probability that demand on}$$

depot SRU $i$ stock is due to

depot LRU repairs.

Then, if $X_{00}$ represents the random variable for the number of LRUs in the depot resupply process,

$$E[X_{00}] = \lambda_{00} D_{00} + \sum_i \beta_{i0} \mathcal{B}_{i0}(s_{i0}|\lambda_{i0}D_{i0}) \text{ and}$$

$$\text{Var}(X_{00}) = \lambda_{00} D_{00} + \sum_i \beta_{i0}(1 - \beta_{i0})\mathcal{B}_{i0}(s_{i0}|\lambda_{i0}D_{i0})$$

$$+ \sum_i \beta_{i0}^2 \text{Var of SRU i depot backorders } (s_{i0}|\lambda_{i0}D_{i0}),$$

where the $\text{Var}(s_{i0}|\lambda_{i0}D_{i0})$ is computed in the manner described in the discussion of the METRIC model, that is, in Section 5.1.2.1 of this chapter.

The mean and variance of the number of SRUs in resupply at a base can be computed in a similar manner. Let

$$\beta_{ij} = \text{probability that a unit of demand for SRU } i \text{ at the depot}$$

is attributable to a requirement at base $j$

$$= \frac{\lambda_{ij}(1 - r_{ij})}{\lambda_{i0}}.$$

If $X_{ij}$ is the random variable for the number of units in resupply at base $j$ for SRU type $i$, then

$$E[X_{ij}] = \lambda_{ij}(1 - r_{ij})A_{ij} + r_{ij}\lambda_{ij}B_{ij}$$
$$+ \beta_{ij}\mathcal{B}_{i0}(s_{i0}|\lambda_{i0}D_{i0}),$$

and

$$\text{Var}(X_{ij}) = r_{ij}\lambda_{ij}B_{ij} + (1 - r_{ij})\lambda_{ij}A_{ij}$$
$$+ \beta_{ij}(1 - \beta_{ij})\mathcal{B}_{i0}(s_{i0}|\lambda_{i0}D_{i0})$$
$$+ \beta_{ij}^2 \text{Var of depot SRU i backorders } (s_{i0}|\lambda_{i0}D_{i0}).$$

Finally, let us compute the mean and variance of the number of LRUs that are in the base's resupply process. Following our notation,

$$\beta_{0j} = \text{probability that an LRU being repaired}$$

at the depot corresponds to a demand at base $j$.

Then we have

$$\beta_{0j} = \frac{\lambda_{0j}(1 - r_{0j})}{\lambda_{00}}.$$

Letting $X_{0j}$ be the random variable describing the number of LRUs in resupply at base $j$, we see that

$$E[X_{0j}] = \lambda_{0j}(1 - r_{0j})A_{0j} + r_{0j}\lambda_{0j}B_{0j}$$
$$+ \beta_{0j}\mathcal{B}(s_{00}|E(X_{00}), \text{Var}X_{00})$$
$$+ \sum_i \mathcal{B}(s_{ij}|E[X_{ij}], \text{Var}X_{ij})$$

and
$$\text{Var}X_{0j} = r_{0j}\lambda_{0j}B_{0j} + (1 - r_{0j})\lambda_{0j}A_0$$
$$+ \beta_{0j}(1 - \beta_{0j})\mathcal{B}(s_{00}|E(X_{00}), \text{Var}X_{00})$$
$$+ \sum_i \text{Var of base } j \text{ SRU } i \text{ backorder } (s_{ij}|E(X_{ij}), \text{Var}X_{ij})$$
$$+ \beta_{0j}^2 \text{Var of depot LRU backorder } (s_{00}|E[X_{00}], \text{Var}X_{00}).$$

These means and variances are used to calculate the probability distributions required to determine the respective backorder functions. In each case, we assume the probability distribution is accurately approximated by a negative binomial distribution with the given means and variances.

### 5.3.3  Computing Depot and Base Stock Levels for Each LRU and SRU

Now that we have shown how to construct the approximating probability distributions for the number of units in resupply for each LRU and SRU at each location, we are in a position to develop an optimization model and an algorithm for computing these LRU and SRU stock levels.

Some additional nomenclature is required. Let $\ell$ denote a LRU type, $s_{i0}^\ell$ and $s_{ij}^\ell$ denote the stock levels for the depot and base $j$, respectively, for SRU $i$ found in LRU $\ell$ and $s_{00}^\ell$ and $s_{0j}^\ell$ the depot and base $j$ stock levels for LRU $\ell$, respectively.

Our goal is to select the stock levels $s_{i0}^\ell$ and $s_{ij}^\ell$ for each LRU family $\ell$ - an LRU and its subordinate SRUs - so as to minimize total base level LRU backorders. Assume the unit cost of LRU $\ell$ is $c_0^\ell$ and the unit cost of SRU $i$ in LRU $\ell$ is $c_i^\ell$. Further assume that $b$ represents an investment target. In practical applications we would desire to construct an exchange curve in which minimum total base level backorders is established as a function of various investment budgets, $b$. The algorithm that we will discuss will be used to construct such a trade-off curve.

The optimization problem can be stated as follows for computing the SRU and LRU stock levels

$$\min \sum_{\ell} \sum_{j \neq 0} \mathcal{B}_{0j}^{\ell}(s_{0j}^{\ell} | s_{00}^{\ell}, s_{i0}^{\ell}, s_{ij}^{\ell})$$

$$\sum_{\ell} \left\{ c_0^{\ell} \left( \sum_j s_{0j}^{\ell} + s_{00}^{\ell} \right) + \sum_{i \neq 0} c_i^{\ell} \left\{ \sum_j s_{ij}^{\ell} + s_{i0}^{\ell} \right\} \right\} \leq b$$

(SRUs)

$s_{0j}^{\ell}, s_{ij}^{\ell}$ are integer and nonnegative.

As we discussed earlier, this problem is neither convex nor separable and therefore is not trivial to solve. The solution approach that we describe is similar to the one we presented earlier in Section 5.1.2.2. This method is not guaranteed to find an optimal solution to the original problem since it is a marginal analysis based method. However, it does generate optimal solutions for specific budget levels, as we will see. When we say optimal levels, we mean optimal given the approximations we have made for the stationary probability distributions for the number of units in resupply, as discussed earlier.

We now outline an algorithm for finding the LRU and SRU stock levels. Since the notation required to state the algorithm is cumbersome, we will provide only a nonmathematical description. It will be obvious how the method works once the ideas presented in Section 5.1.2.2. are fully understood. The algorithm constructs a sequence of convex trade-off curves, one for each LRU family, and then combines these curves to generate an exchange curve for the entire system. The outline of the algorithm for each LRU family is as follows.

For Each LRU Family

Step 1. Construct the trade-off curve for subordinate SRUs

(A) For each SRU in the family
(i) For each SRU depot stock level, construct a trade-off curve for investment in base stock for the SRU being examined. (This is accomplished using a marginal analysis method.) The goal is to select base SRU stock levels that minimize the expected LRU waiting time for repairs at the bases.
(ii) Using the results of Step (i), construct a trade-off curve of total expected LRU wait time at the bases and at the depot for LRUs repaired at the depot versus the total system stock. This resulting trade-off curve may not be convex.
(iii) Delete nonconvex points so that the remaining trade-off curve is discretely convex.
(B) Merge trade-off curves for all SRUs in the LRU family
(i) Using the data obtained from constructing the convex minorant of the individual SRU performance versus investment curves in Step 1.A.iii, construct the trade-off curve that relates total expected LRU base and depot waiting times, or delays in average LRU repair cycle times, to specific levels of total investment in the SRUs, within the LRU family. This construction is performed using marginal analysis

and results in a convex trade-off curve. For each of the budget levels generated in the construction of the trade-off curve there is a corresponding minimum total expected LRU waiting time for the bases and depot.

Step 2. For each total SRU budget level generated in Step 1.B.i., (which have corresponding base and depot SRU stock levels)

(A) For all depot LRU stock levels

   (i) Compute the mean and variance of the number of LRUs in resupply for each location using the relationships derived in Section 5.3.2. Using these values, approximate the distribution of the number of LRU units in resupply for each location.

   (ii) Construct the trade-off curve of total base level LRU expected backorders versus total investment. Select the base to which the next unit of LRU stock should be added using the marginal analysis concept. Given the values of the SRU and depot LRU stock levels, the resulting total base level LRU expected backorder curve is a convex function and will be found using this marginal analysis approach.

   (iii) The construction in Step (ii) yields a set of total budget versus total base level LRU expected backorders given the depot LRU stock and the total SRU budget level.

(B) For the given SRU budget level, construct a convex trade-off curve of total investment versus expected LRU base level backorders

   (i) Using the budget/performance pairs generated in Step 2.A.iii., construct a combined trade-off curve considering all depot stock levels. The resulting total expected base LRU backorders versus the budget relationship need not be convex.

   (ii) Delete pairs so that the resulting trade-off curve is discretely convex. This yields a set of total budget versus total base level LRU expected backorders, given the total SRU budget level.

Step 3. Construct the trade-off curve of total investment versus expected LRU base level backorders.

   (i) Using the budget/performance pairs generated in Step 2.B (ii), construct a combined trade-off curve considering all SRU budget levels. The resulting total expected base LRU backorders versus total budget relationship need not be convex.

   (ii) Delete pairs so that the resulting trade-off curve is discretely convex.

The preceding greedy algorithm produces a trade-off curve for a single LRU family. That is, the resulting discretely convex curve indicates the total expected base level backorders that correspond to particular total investment levels in the LRU and its subordinate SRUs. Once these convex functions are constructed for each LRU family, we then combine them to generate a solution to the original problem. This final trade-off curve can be found using the individual trade-off curves using a simple greedy algorithm since each LRU family trade-off curve is convex. Thus, for each budget level that results from applying the marginal analysis based greedy algorithm we obtain the minimum achievable expected total base level backorders. We also obtain the base and depot LRU and SRU stock levels that correspond to each budget.

## 5.4 Problem Set, Chapter 5

**5.1.** Suppose we have a two-echelon base-depot system as discussed in Section 5.1.1. The demand process at each of 10 bases is a Poisson process, where each base has a demand rate of .5 units per day. Suppose the average depot to base order and ship time is either 1, 5 or 10 days. Further assume the average depot resupply time is either 15 or 30 days. All failed parts at the bases are repaired at the depot.

(a) Plot $\delta(s_{i0})$ for each base.
(b) Compute the average base resupply time as a function of $s_{i0}$ for each depot to base order and ship time and each average depot resupply time combination. How important a factor is the depot stock level in each case?
(c) Suppose the demand at each base is described by a Negative Binomial distribution with an average of .5 units per day at each base. Suppose that the variance to mean ratio is identical at all bases, and is equal to either 1.01, 2, 5 or 10. Assuming demands are independent among the bases, what is the distribution of depot demand? Repeat the tasks indicated in (a) and (b) for these cases. How do the values compare with those computed initially?

**5.2.** Suppose we are managing the two-echelon system discussed in Section 5.1.1. Demands at each of 10 bases for an item occur according to a Poisson process with a rate of .5 units per day. The average depot to base resupply time, $A_{ij}$, for the item is 5 days for all 10 bases, and the average depot resupply time, $D$, is 30 days. Suppose the fraction of the failures that arise at each of the bases that are repaired at the bases is either .25, .5 or .75. Further assume the average base repair time for the item is 2 days.

Compute the mean and variance of the number of units in resupply for a base for each case for a range of values of the depot stock level. How does the variance to mean ratio change as the fraction of units repaired at the depot increases when the depot stock is equal to $\lfloor \lambda_0 D \rfloor$, where $\lambda_0$ is the average depot daily demand rate?

**5.3.** In Section 5.1.2.1 we derived expressions for the mean and variance for the number of units in resupply for a LRU at a base. The analysis was based on the assumption that the demand process is a Poisson process. Suppose that the demand process is Negative Binomially distributed rather than Poisson distributed. Construct new equations for the mean and variance of the number of units in resupply for LRU at a base. Carefully state your assumptions as you develop these expressions.

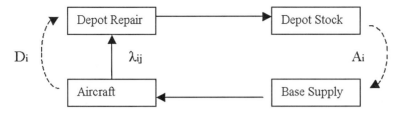

**Fig. 5.6.** Depot/Base Repair Part Flow

**5.4.** Assume there are three items whose stock levels need to be determined in the two-echelon system depicted in the figure below.

Removals of item $i$ at base $j$ occur according to a Poisson process at rate $\lambda_{ij}$ units per day. When a removal occurs at a base, three events take place simultaneously and instantaneously. First, a unit of stock is withdrawn from base stock (if a unit is available); second, the failed unit is sent to the depot where it will be repaired; and third, the depot resupplies the base (if a unit is on-hand). When units of stock are not available, backorders occur. If there is a backorder at a base, then an aircraft is grounded. If a backorder occurs at the depot, then there is a delay in resupplying the requesting base.

The depot repair cycle time for all bases for part type $i$ is denoted by $D_i$, measured in days. The order and shipping time from the depot to each base is $A_i$ days for item $i$. There are 10 bases in the system, numbered 1 through 10. The flying activity at bases 1 through 5 is the same and hence $\lambda_{i1} = \ldots = \lambda_{i5}$ for all $i$. Bases 6 through 8 have identical removal rates for all items, that is, $\lambda_{i6} = \lambda_{i7} = \lambda_{i8}$. Finally, bases 9 and 10 have identical removal rates for each of the three items, $\lambda_{i9} = \lambda_{i\,10}$. The data in the following table indicate the removal rates, depot repair cycle times and unit costs for the three items. The value of $A_i$ is two days for all three items for all bases.

| Item | Base Removal Rate | | | Unit Costs | Depot Repair |
|---|---|---|---|---|---|
| | Base 1–5 | Base 6–8 | Base 9–10 | (1000s of $) | Cycle Times |
| 1 | 1 | 0.5 | 0.25 | 3 | 10 |
| 2 | 1.5 | 0.75 | 0.375 | 4 | 8 |
| 3 | 0.5 | 0.25 | 0.125 | 5 | 10 |

Construct the convex function representing the relationship between minimum total expected base backorders and total system stock for the three items, that is, construct the functions $\hat{\alpha}_i^c(s_i)$.

Once these three functions have been created, construct the convex function that represents the relationships between minimum total base backorders (across items) and investment in stock for these three items.

What would the impact be if the values of $D_i$ and $A_i$ were reduced or increased by a factor of 50%?

**5.5.** Resolve Problem 4 using a Lagrangian relaxation algorithm.

**5.6.** In Section 5.3 we discussed a multi-indenture system consisting of LRUs and subordinate SRUs. In the analysis we defined $X_{0j}$ to be the random variable describing the number of LRUs in resupply at base $j$. Verify the expressions given in Section 5.3.2 for the $E[X_{0j}]$ and the $\text{Var}[X_{0j}]$.

**5.7.** We outlined an algorithm for computing both depot and base stock levels for an LRU and SRU in Section 5.3.3. State this algorithm precisely by using mathematical notation.

**5.8.** The algorithm described in Section 5.3.3 that can be employed to find depot and base stock levels for each LRU and SRU is a marginal analysis method. For the same environment, develop a Lagrangian based algorithm for computing these stock levels.

# 6

# A Continuous Time, Multi-Echelon, Multi-Item System with Time-Based Service Level Constraints

We now examine a different environment that exists in practice for a wide variety of products. In these situations, suppliers and customers often establish service agreements that apply to a product (or group of products) that the customer has purchased from the supplier. These agreements extend over a period of months or years and normally specify the type of service that will be provided, as well as the timing with which the service will take place. The details of these service agreements vary in nature, often involving specific time-based guarantees, and often covering multiple pieces of equipment across multiple customer locations. The increasing complexity of these agreements has led to a new set of challenges with which managers must contend. Specifically, setting system stock levels and positioning individual item types to satisfy these service guarantees at minimum investment is exceptionally difficult.

Suppliers must recognize that *the customer's concern is the maintenance of the product*, not the maintenance of the individual component items. By understanding the customer's service level requirements in terms of the *product*, as well as the *timing* with which the customer is willing to receive service, suppliers of service parts can achieve considerable savings in inventory investment and operational overhead.

In this chapter we consider a multi-item, multi-echelon distribution system in which general service level requirements have been established between the supplier and its customers. We assume that locations at the lowest level, or echelon, of the distribution network experience demand for parts on a continual basis. As in the other situations we have studied, the topology of the system is such that each location on a particular level is replenished from a unique location at the next-higher level over a constant transport lead time. The location at the top level is replenished via a process that has a known and constant lead time. Demands that cannot be fulfilled immediately are backordered. The objective is to construct a tactical planning model to establish target inventory levels for each item type at each location so that all service level requirements stipulated by the agreements are satisfied while minimizing the total system inventory investment.

The model we present captures a rich and realistic class of service level constraints that allow target service levels to be specified across multiple item types *and* multiple locations, in any combination. The models we have discussed and most of the models presented in the literature have equated "service level" with "item fill rate." However, single item fill rates or backorder rates may not reflect the service levels with which the customer is concerned. The model presented in this chapter views service agreements from the customer's perspective, and not the supplier's.

The time-based service level constraints reflect the contractual agreements made by the supplier for various items and locations. For instance, for a critical item at a particular customer location, we can specify the desired fill rate to be 90% instantaneously, 95% within 8 hours, and 98% within 2 days. In keeping with practice, our model represents time-based service levels in which the required service times coincide with the transportation times from replenishment sites within the distribution network.

To model these time-based fill rates, we provide an exact characterization of what we call *channel fill rates*. In our distribution network, each demand is replenished via a unique path from the top level location in the network. For each intermediate location along the replenishment path from the top location to the base, we define the associated *channel fill rate* to be the probability that an arriving order for the item at the base can be fulfilled within the transportation time along the replenishment path from that location to the base. By allowing for these time-based service level constraints in our framework, we capture response time requirements that are an integral part of many real customer service agreements.

We develop an optimization procedure to minimize overall system inventory investment while meeting all service level requirements. Note that the general nature of the service level constraints makes this system-wide optimization problem considerably more difficult than the ones we have discussed in earlier chapters, as this problem may not be separable by item or by location.

The remainder of this chapter is organized as follows. We begin by describing our modeling framework in detail and formulate the optimization problem. We then derive exact expressions for the channel fill rates that are key to analyzing overall service level fulfillment. Finally, we describe an iterative approximation scheme for solving a common class of problems.

The contents of this chapter follow Caggiano, Muckstadt, Jackson and Rappold [37].

## 6.1  The Model

Here, we state the assumptions upon which our model is based and illustrate the types of service level requirements that can be represented within the modeling framework. We also define the notation and present a mathematical programming formulation of the problem.

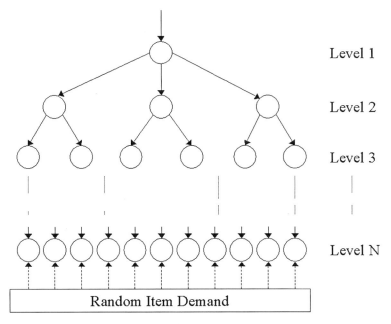

**Fig. 6.1.** Example item distribution network

### 6.1.1 Modeling Assumptions

For our purposes, the multi-item, multi-echelon distribution system has the following properties:

1. The distribution system is the composition of its item distribution networks. Each item distribution network has a tree-like structure, where each location in the network is replenished from a unique parent location at the next-higher level. The sole location at the top level of an item network is replenished via a process that has a known and constant lead time. See Figure 6.1.

2. Demand for a particular item occurs only at the lowest echelon of its item network. We refer to locations in the lowest echelon as *bases*. We assume this without loss of generality, since dummy locations and arcs with negligible lead times can be added to achieve this structure. In the same manner, we assume that all bases are on the same level in the item distribution network.

3. The demand processes for all items at all bases are mutually independent Poisson processes with known demand rates. Thus, demands arise for one unit of an item at a time.

4. All items are replenished on a one-for-one basis at all locations.

5. Transportation times for each item between adjacent network locations are known and constant.

6. Orders that cannot be fulfilled immediately are backordered.

7. Orders are filled at all locations on a first-come, first-serve basis.

For notational convenience only, we assume that all items share a common distribution network, which will alleviate the need to define a separate network structure for each item.

### 6.1.2   Service Level Requirements

We will illustrate the types of service level requirements that may be represented in our modelling framework with an example.

Suppose a regional supplier of office equipment has as its main business the leasing of photocopiers. Included in each lease is a service agreement that stipulates the timing with which equipment breakdowns will be addressed by the supplier. As part of the lease agreements, the supplier owns and is responsible for providing any service parts that are needed to repair malfunctioning equipment.

As it happens, most photocopier breakdowns are caused by worn or overused parts. Many of these parts, such as toner cartridges, document feed rollers, xerographic modules, and staples, can be swapped-out quickly and easily, without the aid of a trained technician. When a breakdown occurs and the needed parts are stocked and available at the customer location, then repair can commence immediately. If the needed parts are not available at the customer location, they must be obtained from a regional warehouse. Parts can be transported from the warehouse to any customer location within 24 hours. Hence, as long as the needed parts are available either at the customer location or at the warehouse (or are en-route) at the time a breakdown occurs, the repair can be completed within a 24-hour time window. Accordingly, the standard service agreement offered by the supplier is based upon a 24-hour window. Specifically, the agreement stipulates that all copier breakdowns will be investigated by a service technician within 24 hours, and that 95% of all copier breakdowns will be fixed within the same period.

Many customers find that the standard service agreement is sufficient to meet their needs. Some customers, however, depend heavily on the photocopiers and cannot afford to have their operations disrupted for up to 24 hours on a regular basis. For this type of customer, the supplier typically agrees to stock some parts at the customer site so that a portion of the customer's breakdowns can be remedied immediately. Recall that the supplier, not the customer, owns and is responsible for providing the service parts. Each time a customer uses a part from their on-site supply to fix a breakdown, a replacement order is placed immediately with the warehouse. Once the order is filled at the warehouse, the replacement part will be delivered to the customer site within 24 hours.

There are clearly tradeoffs for the supplier in agreeing to accommodate the second type of customer. On one hand, stocking parts on-site for a customer will keep the customer satisfied and will result in fewer service calls that require a technician to be dispatched to that customer site. Also, if the majority of breakdowns require only inexpensive items for repair, notable improvements in customer service may be achieved with relatively little investment. On the other hand, parts that are stocked at the customer site are not available to service other customer demands. Depending on the demand patterns and costs of items and the extent to which customers require instantaneous service, this could mean a huge investment in service parts inventory in order to honor all service commitments.

Suppose we have two offices, $a$ and $b$, that lease photocopiers from the supplier. These offices receive service parts from the supplier's regional warehouse, denoted by $r$. In office $a$, the leased copier is lightly used, and breakdowns are infrequent. Furthermore, when the copier does break down, alternative means of photocopying are readily available on a temporary basis. Hence, while office $a$ certainly has no objection to having parts stocked on-site, the 24-hour service agreement stipulated in the lease is sufficient to meet its needs. When stock is not on-hand at $a$, then inventory stocked at $r$ is used to achieve the desired service level stipulated in the contract.

In office $b$, however, the leased copier is heavily used, and breakdowns are more frequent. While a potential 24-hour delay is tolerable once in a great while, frequent delays of this length would be too disruptive to the operation of the office. Thus, in addition to the 24-hour service agreement stipulated in the lease, the supplier has agreed to place enough stock at office $b$ so that 90% of office $b$'s photocopier breakdowns can be repaired immediately. Note that this is very different from agreeing to stock the office so that *each* photocopier part is immediately available for 90% of all breakdowns in which the item is required.

For purposes of describing the service level constraints associated with the two offices, we will use the following notation: Let

$I$ denote the set of photocopier component items, indexed by $i$.

$\lambda_a$ denote the rate at which office $a$ experiences copier breakdowns, and let $\lambda_{ia}$ denote the rate at which office $a$ experiences copier breakdowns that require item $i$ for repair. The ratio $\frac{\lambda_{ia}}{\lambda_a}$ then represents the fraction of breakdowns at office $a$ that require item $i$ for repair. Define $\lambda_b$ and $\lambda_{ib}$ similarly.

$s_{ia}$ and $s_{ib}$ denote the stock levels for item $i$ at locations $a$ and $b$, respectively. Let $s_{ir}$ denote the stock level for item $i$ at the regional warehouse $r$.

$f_{ia}^2$ denote the probability that a breakdown at location $a$ requiring item $i$ can be fixed immediately. That is, $f_{ia}^2$ is the probability that item $i$ is available on-site at location $a$ when it is needed. The superscript "2" refers to the level of the (two-level) network with which the fill rate is associated. Define $f_{ib}^2$ similarly.

$f_{ia}^1$ denote the probability that a breakdown at location $a$ requiring item $i$ can be filled within 24 hours. That is, $f_{ia}^1$ is the probability that item $i$ is either available on-site at location $a$, or it is available at the regional warehouse, or it is en route from the warehouse to location $a$ when it is needed. Define $f_{ib}^1$ similarly.

The probabilities $f_{ia}^2$ and $f_{ia}^1$ are called *channel fill rates* for item $i$ at location $a$, and are used as building blocks in constructing service level constraints. Both of these fill rates are functions of the stock levels $s_{ia}$ and $s_{ir}$, although the impact of $s_{ir}$ on the instantaneous fill rate $f_{ia}^2$ is very different from its impact on the 24-hour fill rate $f_{ia}^1$. We will explain this difference shortly.

To demonstrate the different types of service level constraints that may arise under different operating conditions, we present three scenarios.

### 6.1.2.1  Scenario 1

In Scenario 1, offices $a$ and $b$ each have their own lease and service agreement with the supplier, and the stock placed on-site at either of the office locations cannot be shared by the other. Thus, from a distribution viewpoint, offices $a$ and $b$ are distinct stocking locations. The service level requirements for offices $a$ and $b$ under Scenario 1 are depicted in Figure 6.2, and the corresponding constraints are given in 6.1- 6.3.

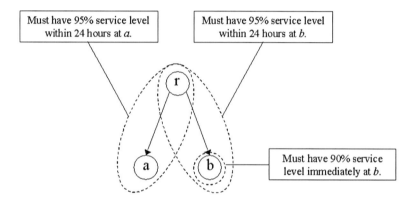

**Fig. 6.2.** Service Level Requirements for Scenario 1

Constraints 6.1 and 6.2 represent the 24-hour service level guarantees stipulated in the service agreements for offices $a$ and $b$, respectively. Constraint 6.3 represents the instantaneous service level requirement of office $b$.

$$\sum_{i \in I} \frac{\lambda_{ia}}{\lambda_a} f_{ia}^1(s_{ia}, s_{ir}) \geq .95 \tag{6.1}$$

$$\sum_{i \in I} \frac{\lambda_{ib}}{\lambda_b} f_{ib}^1(s_{ib}, s_{ir}) \geq .95 \tag{6.2}$$

$$\sum_{i \in I} \frac{\lambda_{ib}}{\lambda_b} f_{ib}^2(s_{ib}, s_{ir}) \geq .90 \tag{6.3}$$

Note that increasing the stock level $s_{ia}$ contributes only to the satisfaction of constraint 6.1, and that increasing $s_{ib}$ contributes to the satisfaction of 6.2 and 6.3, but not 6.1. This agrees with our intuition, since any stock placed at one of the office locations cannot be used to service the other, and hence raising the stock level at one office site should not have any impact on the other office's service.

By contrast, an increase in $s_{ir}$, the replenishment stock level at the warehouse, contributes to the satisfaction of all three constraints since the fill rates $f_{ia}^1$, $f_{ib}^1$, and $f_{ib}^2$ all depend upon $s_{ir}$. The dependency, however, is different for $f_{ib}^2$ than it is for $f_{ia}^1$ and $f_{ib}^1$. $f_{ib}^2$ depends in part on the *timeliness with which replenishment orders placed by b (to the regional warehouse) are filled*, and this timeliness is fundamentally a function of $s_{ir}$. Also, $s_{ir}$ *only* affects $f_{ib}^2$ through its impact on the replenishment lead time. Hence, while it is possible (if $s_{ib} > 0$) to increase the instantaneous fill rate $f_{ib}^2$ by raising the warehouse stock level $s_{ir}$, there is a limit to the increase that can be achieved by this method. Beyond this limit, the *only* way to increase $f_{ib}^2$ is to increase the local stock level $s_{ib}$. For the 24-hour fill rates $f_{ia}^1$ and $f_{ib}^1$, there is no such limitation. That is, for any $\epsilon > 0$, it is always possible to achieve $f_{ia}^1 \geq 1 - \epsilon$ (and/or $f_{ib}^1 \geq 1 - \epsilon$) by raising the stock level $s_{ir}$ high enough.

### 6.1.2.2 Scenario 2

In Scenario 2, offices $a$ and $b$ each have their own lease and service agreement with the supplier, but stock placed on-site at either office location can be shared. That is, from a distribution viewpoint, there is a *single* stocking location from which offices $a$ and $b$ draw needed parts. The service level requirements for offices $a$ and $b$ under Scenario 2 are depicted in Figure 6.3. In the corresponding constraints 6.4–6.6, $\overline{ab}$ is used to denote the common stocking location for offices $a$ and $b$.

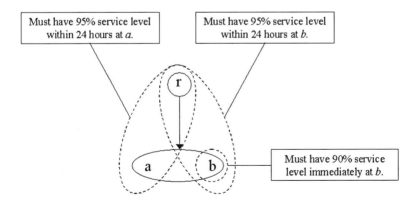

**Fig. 6.3.** Service Level Requirements for Scenario 2

$$\sum_{i \in I} \frac{\lambda_{ia}}{\lambda_a} f_{iab}^1(s_{i\overline{ab}}, s_{ir}) \geq .95 \tag{6.4}$$

$$\sum_{i \in I} \frac{\lambda_{ib}}{\lambda_b} f_{iab}^1(s_{i\overline{ab}}, s_{ir}) \geq .95 \tag{6.5}$$

$$\sum_{i \in I} \frac{\lambda_{ib}}{\lambda_b} f_{iab}^2(s_{i\overline{ab}}, s_{ir}) \geq .90 \tag{6.6}$$

Note that the fill rates and stock levels are indexed by item and *stocking* location, not item and customer location. In this case, increasing the stock level $s_{i\overline{ab}}$ contributes to the satisfaction of all three constraints, as we would expect. At first glance, one might think that the common stocking location makes the constraints in this scenario a relaxed version of the constraints in Scenario 1. That is, one might suppose that any stock levels that satisfy 6.1-6.3 would also satisfy 6.4-6.6 if we make the substitution $s_{i\overline{ab}} = s_{ia} + s_{ib}$. In fact, this is not the case for any of the constraints. This is most easily seen for constraint 6.6.

In Scenario 2, office $a$ will draw stock from location $\overline{ab}$ to fix its breakdowns (provided the stock is available), even though it has no instantaneous service level requirement. The presence of the common stocking location makes the instantaneous fill rate $f_{i\overline{ab}}^2$ a function of *both* $\lambda_{ia}$ and $\lambda_{ib}$. As a consequence, the satisfaction of constraint 6.6 depends upon the item demand rates at office $a$, even though the instantaneous service level requirement exists at office $b$ only. To satisfy 6.6, enough stock must be held at location $\overline{ab}$ to make the fill rates $f_{i\overline{ab}}^2$, $i \in I$, sufficiently high. A high demand rate $\lambda_{ia}$ (relative to $\lambda_{ib}$) means that $s_{i\overline{ab}}$ may have to be significantly higher than Scenario 1's $s_{ib}$ in order for the fill rate $f_{i\overline{ab}}^2$ to be as high as Scenario 1's $f_{ib}^2$. We are assuming, of course, that a demand is always satisfied if there is stock on-hand when it arises. In practice, it is possible to withhold stock from $a$ when the stock on-hand at location $\overline{ab}$ is too low; however, we assume that this type of rationing does not occur.

This scenario highlights the fact that strategic decisions, such as the placement of stocking locations and rationing rules, can greatly affect the types of service agreements that can be satisfied by a supplier in a cost-effective manner. We have just seen that promising a high level of service to a low-demand customer that draws stock from a high-demand stocking location can be costly. Since suppliers cannot always avoid such situations, it is important to establish operating policies that are designed to help achieve the promised customer service levels. For instance, careful prioritization of customer orders and replenishment orders, as opposed to a simple first-come-first-served scheme, can improve system performance. As stated, we do not address these issues here.

### 6.1.2.3   Scenario 3

In Scenario 3, offices $a$ and $b$ share a *common* lease and service agreement with the supplier, so the 95% service level applies to the two offices *together*, not separately. However, stock placed on-site at either of the office locations cannot be shared. (One may imagine two offices that are not physically close to one another, but are owned and managed jointly.) The service level requirements for offices $a$ and $b$ under Scenario 3 are depicted in Figure 6.4, and the corresponding constraints are given by 6.7 and  6.8.

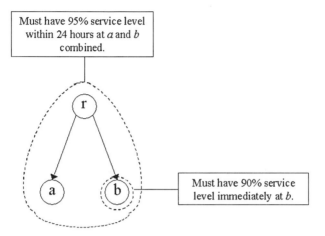

**Fig. 6.4.** Service Level Requirements for Scenario 3

$$\sum_{i \in I} \left( \frac{\lambda_{ia}}{(\lambda_a + \lambda_b)} f_{ia}^1(s_{ia}, s_{ir}) + \frac{\lambda_{ib}}{(\lambda_a + \lambda_b)} f_{ib}^1(s_{ib}, s_{ir}) \right) \geq .95 \qquad (6.7)$$

$$\sum_{i \in I} \frac{\lambda_{ib}}{\lambda_b} f_{ib}^2(s_{ib}, s_{ir}) \geq .90 \qquad (6.8)$$

Unlike Scenario 2, the constraints of Scenario 3 truly are a relaxed version of the constraints in Scenario 1. Upon inspection, it is easy to see that any stock levels that satisfy 6.1-6.3 will also satisfy 6.7 and 6.8. The common service agreement provides the supplier more flexibility than Scenario 1 in fulfilling the service level requirements.

The preceding scenarios illustrate the types of time-based constraints that may be considered within the modeling framework presented subsequently in this chapter.

### 6.1.3  Notation and Problem Statement

For the remainder of this chapter, we will use the following notation:

### Distribution Network Parameters

$I$ the set of items, indexed by $i$.

$J$ the set of locations, indexed by $j$.

$J^v$ the set of locations at level $v$, $v = 1, 2, \ldots, N$. $\bigcup_{v=1}^{N} J^v = J$, and $J^{v_1} \cap J^{v_2} = \emptyset$, $v_1 \neq v_2$.

$P_j$ the set of locations in the unique path from base $j \in J^N$ to the top level location in the distribution network, inclusive. This is called the *channel* associated with location $j$.

$P_j(v)$ the unique location in the channel $P_j$ at level $v$.

$p(j)$ the parent location of location $j$ in the distribution network, $j \notin J^1$.

$A_{ij}$ the transportation time for item $i$ from location $p(j)$ to location $j$.

$T_{ij}$ the expected replenishment lead time for item $i$ from location $p(j)$ to location $j$. $T_{ij} \geq A_{ij}$.

$c_i$ the unit investment cost of item $i$.

### Service Level Requirement and Demand Parameters

$K$ the set of service level constraints, indexed by $k$.

$F_k$ the established service level of service level constraint $k$. For all $k \in K$, $F_k \leq 1$.

$\lambda_{ij}$ the rate at which orders for item $i$ arrive at location $j$.

$\lambda_{ijk}$ the rate at which orders for item $i$ that are associated with service level constraint $k$ arrive at location $j$.

$\lambda_k$ the total rate at which orders for service parts associated with service level constraint $k$ are placed. That is, $\lambda_k = \sum_{i \in I, j \in J^N} \lambda_{ijk}$.

$w_{ijk}$ the fraction of orders for service parts associated with service level constraint $k$ that are for item $i$ at location $j$. $w_{ijk} = \lambda_{ijk}/\lambda_k$.

$v_{ijk}$ the level of the distribution network with which service level constraint $k$ is concerned for item $i$ at location $j \in J^N$. $v_{ijk} \in \{1, 2, \ldots, N\}$.

$w_{ijk}^v$ the relative weight of channel fill rate $f_{ij}^v$ in service level constraint $k$. That is, $w_{ijk}^v = w_{ijk}$ for $v = v_{ijk}$, and $w_{ijk}^v = 0$ otherwise.

### Stock Levels and Fill Rates

$s_{ij}$ the stock level of item $i$ at location $j$.

$\mathbf{s^v}$ the vector of stock levels of all items $i \in I$ at all network locations $j \in J^v$.

$\mathbf{s_i}$ the vector of stock levels of item $i$ at all network locations.

$\mathbf{s_i^v}$ the vector of stock levels of item $i$ at all network locations $j \in J^v$.

$\mathbf{s_{iP_j}}$ the vector of stock levels of item $i$ at the locations in the channel $P_j$.

$f_{ij}^v(\mathbf{s_{iP_j}})$ the probability that an incoming order for item $i$ at location $j \in J^N$ can be filled within the transportation time from location $P_j(v)$.

Given the defined notation, we state the *Service Level Satisfaction* problem, or (**SLS**) as:

$$(\textbf{SLS}) \qquad \text{minimize} \qquad \sum_{i \in I} \sum_{j \in J} c_i s_{ij} \qquad\qquad (6.9)$$

subject to

$$\sum_{v=1}^{N} \sum_{i \in I} \sum_{j \in J^N} w_{ijk}^{v} f_{ij}^{v}(\mathbf{s_{iP_j}}) \geq F_k \quad \forall k \in K, \qquad (6.10)$$

$$s_{ij} \geq 0 \text{ and integer} \quad \forall i \in I, j \in J. \qquad (6.11)$$

There are two sources of complexity in these service level constraints. The first is that each fill rate function $f_{ij}^{v}$ may appear in multiple service level constraints in combination with other fill rate functions, so the constraint set may not be separable by item or by location. Most practical problem instances, however, will have constraint sets that are *separable by network level*. That is, most instances will be such that each constraint $k \in K$ is concerned with channel fill rates $f_{ij}^{v}$ at one and only one network level $v$ (i.e., $v_{ijk}$ is the same for all items $i$ and all bases $j$ with which constraint $k$ is concerned). For such instances, we can rewrite constraints (6.10) as:

$$\sum_{i \in I} \sum_{j \in J^N} w_{ijk} f_{ij}^{v}(\mathbf{s_{iP_j}}) \geq F_k \quad \forall k \in K^v, v = 1, \ldots, N, \qquad (6.12)$$

where the sets $K^v$, $v = 1, \ldots, N$, partition the constraint set $K$ by network level.

The second source of complexity is the fill rate functions themselves. For a given item $i$ and a given base $j \in J^N$, each channel fill rate $f_{ij}^{v}$, $v = 1, \ldots, N$, depends in a highly nonlinear way on the $N$ stock levels in the channel $P_j$, as was the case in earlier chapters.

## 6.2   Channel Fill Rate Functions

For ease of exposition, we focus on deriving channel fill rates in a three-level system, although the analysis extends easily to systems with more than three levels.

Consider a particular item $i$ in the channel composed of locations 1, 2, and 3 in the distribution network, as shown in Figure 6.5. Location 3 is the demand location, or base, for which we explicitly derive the probability expressions for the channel fill rates. Let location $a$ represent all locations that are replenished by location 1 *except* for location 2, and let location $b$ represent all locations replenished by location 2 *except* for location 3.

For notational clarity, we will suppress the item subscript $i$ on all variables and parameters. The following variable definitions will be helpful in our discussion:

$Y_j$ the number of units on order at location $j$, $j = 1, 2, 3, a, b$.
$N_j$ the number of units backordered at location $j$, that is, $N_j = [Y_j - s_j]^+$, $j = 1, 2, 3, a, b$.
$E_j$ the number of units en route from location $p(j)$ to location $j$, $j = 2, 3, a, b$.

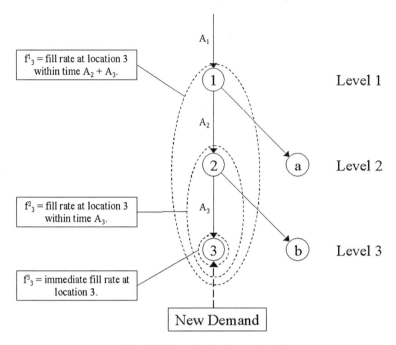

**Fig. 6.5.** Item distribution network

$Z_j$ the number of units on order at location $j$ that are still backordered at location $p(j)$, that is, $Z_j = (Y_j - E_j)$, $j = 2, 3, a, b$. This also represents the number of units currently on order at location $j$ that will not arrive at location $j$ within $A_j$ units of time.

Our primary goal in this section is to derive exact expressions for the channel fill rates at location 3 in terms of the probability distributions of $Y_1$, $Y_2$, and $Y_3$. Although the distributions of $Y_2$ and $Y_3$ are difficult to characterize exactly, for given stock levels $(s_1, s_2, s_3)$ and transportation times $(A_1, A_2, A_3)$, the means and variances of these two distributions can be easily approximated using ideas from Chapter 5. In Section 6.2.5, we review our method for approximating these distributions, and we validate our approximations by comparing our analytically-computed fill rates based on these distributions with those obtained from simulation experiments. Together, the channel fill rate expressions and the distribution approximation method yield a mechanism for evaluating the service level constraints (6.12) presented in the previous section.

### 6.2.1  The Channel Fill Rate $f_3^3(s_3, s_2, s_1)$

We begin with $f_3^3(s_3, s_2, s_1)$, since this is the simplest case. In the context of our network, $f_3^3(s_3, s_2, s_1)$ is the probability that an incoming order (for item $i$) at location 3 can be filled immediately. An instantaneous fill can occur if and only if there is stock on-hand at location 3 when the order arrives. Since a one-for-one replenishment policy is followed in the network, this is equivalent to having strictly less than $s_3$ units on order at location 3 at the time the new order arrives. Hence,

$$f_3^3(s_3, s_2, s_1) = \Pr[Y_3 < s_3]. \tag{6.13}$$

When $s_3 = 0$, the instantaneous fill rate is also 0, as we would expect.

Although we have not made any explicit statements yet about the distribution of $Y_3$, we can easily derive an upper bound for $f_3^3$. Note that the distribution of $Y_3$ depends only on the demand process at location 3 and the order replenishment lead time at location 3. That is, $Y_3$ is a function of $s_2$, and $s_1$, but not $s_3$. For finite values of $s_2$, it is clear that the distribution function of $Y_3$ is monotonically increasing in $s_2$. When $s_2 = \infty$, the replenishment lead time for location 3 is exactly the transportation time $A_3$. In this case, $Y_3$ is a Poisson distributed random variable with mean $\lambda_3 A_3$. Hence, for any values of $s_2$ and $s_1$, we have that

$$\Pr[Y_3 < s_3] \leq \sum_{x=0}^{s_3-1} \frac{(\lambda_3 A_3)^x e^{-\lambda_3 A_3}}{x!}. \tag{6.14}$$

Hence, there is a limit to the impact that increasing $s_2$ can have on $f_3^3$. Indeed, increasing $s_2$ will tend to drive the distribution of $Y_3$ towards a Poisson distribution with mean $\lambda_3 A_3$, but this is the extent of its impact on $f_3^3$. In general, $Y_3$ will have a distribution with mean $\lambda_3 T_3 = \lambda_3 A_3 + \frac{\lambda_3}{\lambda_2} E[N_2]$, where $T_3 = A_3 + \frac{E[N_2]}{\lambda_2}$ denotes the expected replenishment lead time, as we discussed in Chapter 5.

## 6.2.2   The Channel Fill Rate $f_3^2(s_3, s_2, s_1)$

Next, let us determine the probability that an incoming order at location 3 can be filled within time $A_3$, the transportation time from location 2 to location 3. We will consider two cases: $s_3 = 0$ and $s_3 > 0$.

When $s_3 = 0$, all orders arriving at location 3 effectively are filled from stock at location 2. That is, each order that arrives at location 3 waits *at least* $A_3$ units of time until it is filled, since there is never any stock on-hand at location 3, and any units en-route from location 2 to location 3 at the time an order arrives are already claimed by existing backorders at location 3. Hence, a new order arriving at location 3 will be filled within $A_3$ units of time if and only if there is stock on-hand at location 2 when the order arrives. That is,

$$f_3^2(s_3, s_2, s_1) = \Pr[Y_2 < s_2], \text{ if } s_3 = 0. \tag{6.15}$$

Observe that this fill rate will be 0 when $s_2 = s_3 = 0$.

Now consider the case where $s_3 > 0$. Recall that $Z_3$ represents the number of units currently on order at location 3 that will not arrive at location 3 within $A_3$ units of time. Hence, a newly arriving order to location 3 will be filled within $A_3$ units of time if and only if $Z_3 < s_3$. That is,

$$f_3^2(s_3, s_2, s_1) = \Pr[Z_3 < s_3] \text{ , if } s_3 > 0. \tag{6.16}$$

The above expression is not in a usable form, however, since $Z_3$ is a function of $Y_3$ and $E_3$. In order to complete the analysis, we will consider $N_2$, the number of units backordered at location 2. Each of the $N_2$ backordered units is owed to either location 3 or location $b$. Since location 2 is the unique supplier to locations 3 and $b$, and since no other locations place orders with location 2, we have that

$$N_2 = (Y_3 - E_3) + (Y_b - E_b) = Z_3 + Z_b. \tag{6.17}$$

Rewriting equation (6.16) and conditioning on $N_2$, we have that when $s_3 > 0$,

$$\begin{aligned} f_3^2(s_3, s_2, s_1) &= \sum_{x=0}^{s_3-1} \Pr[Z_3 = x] \\ &= \sum_{x=0}^{s_3-1} \sum_{y=x}^{\infty} \Pr[Z_3 = x | N_2 = y] \Pr[N_2 = y]. \end{aligned} \tag{6.18}$$

The lower limit on $y$ in the second summation follows from the fact that $N_2$ and $Z_3$ are both nonnegative random variables, and $N_2 \geq Z_3$. Indeed, $Z_3$ is the portion of $N_2$ that is owed to location 3.

As we discussed in earlier chapters, since orders arriving at location 2 are filled on a first-come-first-serve basis, and since the arrival process to location 2 is a Poisson process with arrival rate $\lambda_2 = \lambda_3 + \lambda_b$, the conditional probability $\Pr[Z_3 = x | N_2]$ follows a binomial distribution with parameters $n = N_2$ and $p = \frac{\lambda_3}{\lambda_2}$. That is,

$$\Pr[Z_3 = x | N_2 = y] = \binom{y}{x} \left(\frac{\lambda_3}{\lambda_2}\right)^x \left(1 - \frac{\lambda_3}{\lambda_2}\right)^{y-x}. \tag{6.19}$$

Also, note that

$$\Pr[N_2 = y] = \begin{cases} \Pr[Y_2 \leq s_2], & \text{if } y = 0. \\ \Pr[Y_2 = s_2 + y], & \text{if } y > 0. \end{cases} \tag{6.20}$$

Combining and simplifying, we have that

$$f_3^2(s_3, s_2, s_1) = \begin{cases} \Pr[Y_2 < s_2], & \text{if } s_3 = 0. \\ \Pr[Y_2 < s_2 + s_3] + \displaystyle\sum_{x=0}^{s_3-1} h_2(s_3, x), & \text{if } s_3 > 0, \end{cases} \tag{6.21}$$

where

$$h_2(u, x) = \sum_{z=u}^{\infty} \binom{z}{x} \left(\frac{\lambda_3}{\lambda_2}\right)^x \left(1 - \frac{\lambda_3}{\lambda_2}\right)^{z-x} \Pr[Y_2 = s_2 + z] \tag{6.22}$$

denotes the probability that *there are at least u backorders at location 2 and exactly x of these are owed to location 3.*

### 6.2.3 The Channel Fill Rate $f_3^1(s_3, s_2, s_1)$

Finally, we derive the probability that an incoming order at location 3 can be filled within time $A_2 + A_3$, the transportation time from location 1 to location 3. We will consider three cases: $s_3 = s_2 = 0$; $s_3 = 0$ and $s_2 > 0$; and $s_3 > 0$.

When $s_3 = s_2 = 0$, all orders arriving at location 3 effectively are filled from stock at location 1. Since each order that arrives at location 3 waits *at least* $A_2 + A_3$ units of time until it is filled, a new order arriving at location 3 will be filled within $A_2 + A_3$ units of time if and only if there is stock on-hand at location 1 when the order arrives. Hence,

$$f_3^1(s_3, s_2, s_1) = \Pr[Y_1 < s_1], \quad \text{if } s_3 = s_2 = 0. \tag{6.23}$$

Recall that a new order arriving at location 3 instantly triggers corresponding orders to be placed to locations 2 and 1. If $s_3 = 0$ and $s_2 > 0$, then a new order arriving at location 3 will be filled within $A_2 + A_3$ units of time if and only if the corresponding order that location 3 places on location 2 is filled by location 2 (i.e., sent out to location 3) within $A_2$ units of time after the order is placed. Hence, we need to derive the *probability that a newly arriving order to location 2 can be filled at location 2 within $A_2$ units of time after the order is placed.* Consider the previous sentence. If we simply replace the "2"s with "3"s, this is precisely the probability we derived for the fill rate $f_3^2(s_3, s_2, s_1)$ (for the case $s_3 > 0$). Thus, by a completely parallel argument, we have that $f_3^1(s_3, s_2, s_1) = \Pr[Z_2 < s_2]$ when $s_3 = 0$ and $s_2 > 0$, or:

$$f_3^1(s_3, s_2, s_1) = \Pr[Y_1 < s_1 + s_2] + \sum_{x=0}^{s_2-1} h_1(s_2, x), \quad \text{if } s_3 = 0, s_2 > 0, \tag{6.24}$$

where

$$h_1(u, x) = \sum_{z=u}^{\infty} \binom{z}{x} \left(\frac{\lambda_2}{\lambda_1}\right)^x \left(1 - \frac{\lambda_2}{\lambda_1}\right)^{z-x} \Pr[Y_1 = s_1 + z] \tag{6.25}$$

denotes the probability that *there are at least $u$ backorders at location 1 and exactly $x$ of these are owed to location 2.*

For the last case, $s_3 > 0$, we define two more variables:

$N_{12} - [Z_2 - s_2]^+$, the number of units backordered at location 2 that are still backordered at location 1. This also represents the number of units currently backordered at location 2 that will not arrive at location 2 within $A_2$ units of time.

$W_j$ – the number of units on order at location $j$ that are still backordered at location 2 and at location 1, $j = 3, b$ (i.e., the portion of $N_{12}$ that is owed to location $j$). This also represents the number of units currently on order at location $j$ that will not arrive at location $j$ within $A_2 + A_j$ units of time.

Given these definitions, it is clear that $N_{12} = W_3 + W_b$. Also, a new order arriving at location 3 will be filled within $A_2 + A_3$ units of time if and only if $W_3 < s_3$. Hence,

$$f_3^1(s_3, s_2, s_1) = \Pr[W_3 < s_3]$$
$$= 1 - \Pr[W_3 \geq s_3], \text{ if } s_3 > 0. \qquad (6.26)$$

We will analyze this expression by expanding $\Pr[W_3 \geq s_3]$, which is slightly easier to characterize when $s_3 > 0$. Note that $W_3 \geq s_3 > 0$ implies that $N_{12} > 0$, which implies that $N_{12} = Z_2 - s_2 > 0$. Rewriting equation (6.26) and conditioning on $N_{12}$, we have that when $s_3 > 0$,

$$f_3^1(s_3, s_2, s_1) = 1 - \sum_{x=s_3}^{\infty} \Pr[W_3 = x]$$

$$= 1 - \sum_{x=s_3}^{\infty} \sum_{y=x}^{\infty} \Pr[W_3 = x | N_{12} = y] \Pr[N_{12} = y]$$

$$= 1 - \sum_{x=s_3}^{\infty} \sum_{y=x}^{\infty} \Pr[W_3 = x | N_{12} = y] \Pr[Z_2 = y + s_2]. \quad (6.27)$$

Following the same line of reasoning that we did for $f_3^2(s_3, s_2, s_1)$, the conditional probability $\Pr[W_3 = x | N_{12}]$ follows a binomial distribution with parameters $n = N_{12}$ and $p = \frac{\lambda_3}{\lambda_2}$. Also, since $N_1 = Z_2 + Z_a$, we can expand the term $\Pr[Z_2 = y + s_2]$ by conditioning on $N_1$. As before, the conditional probability $\Pr[Z_2 = y + s_2 | N_1]$ follows a binomial distribution with parameters $n = N_1$ and $p = \frac{\lambda_2}{\lambda_1}$. The conditioning will also result in expressions of the form $\Pr[N_1 = z]$ for values of $z \geq y + s_2$. However, for $z > 0$, $\Pr[N_1 = z] = \Pr[Y_1 = s_1 + z]$. We are left with:

$$f_3^1(s_3, s_2, s_1)$$
$$= 1 - \left[ \sum_{x=s_3}^{\infty} \sum_{y=x}^{\infty} \binom{y}{x} \left( \frac{\lambda_3}{\lambda_2} \right)^x \left( 1 - \frac{\lambda_3}{\lambda_2} \right)^{y-x} h_1(s_2 + y, s_2 + y) \right], \quad (6.28)$$

if $s_3 > 0$.

Summarizing the three cases, the exact fill rate expressions are:

$$f_3^1(s_3, s_2, s_1) =$$
$$\begin{cases} \Pr[Y_1 < s_1], & \text{if } s_3 = s_2 = 0, \\[2mm] \Pr[Y_1 < s_1 + s_2] + \sum_{x=0}^{s_2-1} h_1(s_2, x), & \text{if } s_3 = 0, s_2 > 0, \\[2mm] 1 - \left[ \sum_{x=s_3}^{\infty} \sum_{y=x}^{\infty} \binom{y}{x} \left( \frac{\lambda_3}{\lambda_2} \right)^x \left( 1 - \frac{\lambda_3}{\lambda_2} \right)^{y-x} h_1(s_2 + y, s_2 + y) \right], & \\ & \text{if } s_3 > 0, \end{cases} \qquad (6.29)$$

where

$$h_1(u, x) = \sum_{z=u}^{\infty} \binom{z}{x} \left(\frac{\lambda_2}{\lambda_1}\right)^x \left(1 - \frac{\lambda_2}{\lambda_1}\right)^{z-x} \Pr[Y_1 = s_1 + z].$$

### 6.2.4   The Channel Fill Rate $f_n^v(s_n, s_{n-1}, \ldots, s_1)$

Following the same type of argument employed in the previous section, we can show that in an $n$-level channel where the locations in the channel are labelled from 1 to $n$, with location $n$ representing the demand location, the channel fill rates at all levels $v = 1, \ldots, n$ are given by:

$$f_n^v(s_n, \ldots, s_1) =$$

$$1 - \left[ \sum_{x_v = \mathcal{S}_v}^{\infty} \sum_{x_{v+1} = \mathcal{S}_{v+1}}^{x_v - s_v} \cdots \sum_{x_n = \mathcal{S}_n}^{x_{n-1} - s_{n-1}} B(x_v - s_v) B(x_{v+1} - s_{v+1}) \right.$$

$$\left. \cdots B(x_{n-1} - s_{n-1}) \Pr[Y_v = x_v] \right] \qquad (6.30)$$

where

$$\mathcal{S}_l = \sum_{j=l}^{n} s_j$$

is the total installation stock at and below network level $l$ in the channel $P_n$, and

$$B(x_l - s_l) = \binom{x_l - s_l}{x_{l+1}} \left(\frac{\lambda_{l+1}}{\lambda_l}\right)^{x_{l+1}} \left(1 - \frac{\lambda_{l+1}}{\lambda_l}\right)^{(x_l - s_l) - x_{l+1}}$$

is the binomial probability that exactly $x_{l+1}$ of the $(x_l - s_l)$ backorders at location $l$ are owed to location $l + 1$.

### 6.2.5   The Distributions of $Y_1$, $Y_2$, $Y_3$, and Fill Rate Accuracy

Under the assumptions of our problem, it is clear that $Y_1$ has a Poisson distribution with mean $\lambda_1 A_1$. Hence, the fill rate $f_3^1$ can be computed exactly using (6.29) or (6.30). The distributions of $Y_2$ and $Y_3$, however, are difficult to characterize in general and must be approximated by some means. (The approximation method, in turn, will impact the accuracy of the fill rate calculations for $f_3^2$ and $f_3^3$. We will come back to this point shortly.) As we saw in Chapter 5, it is possible to calculate the mean and variance of $Y_2$ for a given stock level, $s_1$, and fixed transportation times, $A_1$ and $A_2$. Specifically:

$$E[Y_2] = \lambda_2 A_2 + \frac{\lambda_2}{\lambda_1} E[N_1], \text{ and} \qquad (6.31)$$

$$\text{Var}[Y_2] = \lambda_2 A_2 + \frac{\lambda_2}{\lambda_1}\left(1 - \frac{\lambda_2}{\lambda_1}\right) E[N_1] + \left(\frac{\lambda_2}{\lambda_1}\right)^2 \text{Var}[N_1]. \qquad (6.32)$$

Using these moments, we approximate the distribution of $Y_2$ with a negative binomial (since this distribution also has a variance-to-mean ratio that is greater than 1). Under this assumption, we compute $E[N_2]$ and $\text{Var}[N_2]$, and then, using the same relationships given in (6.31) and (6.32), we recursively compute moments for $Y_3$. While the moment calculations for $Y_2$ are exact, those for $Y_3$ are only approximate because of the negative binomial assumption for $Y_2$. For implementation purposes, we also assume that the distribution of $Y_3$ is negative binomial.

The validation of the use of negative binomial distributions for $Y_2$ and $Y_3$ in making fill rate calculations is required. A continuous-time, discrete-event simulator of the three-echelon system shown in Figure 6.6 was used to estimate the channel fill rates achieved at location 3. We will compare the simulated channel fill rates at location 3 with those that were analytically computed using (6.30) with negative binomial distributions assumed for $Y_2$ and $Y_3$.

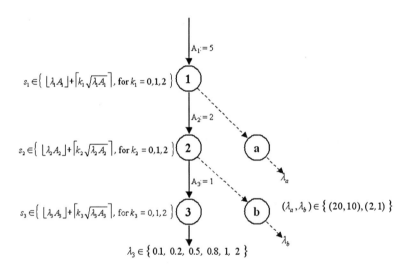

**Fig. 6.6.** Overview of the three-echelon simulated supply chain.

For our purposes, a system scenario is defined by a specific set of location demand rates, $\lambda_a$, $\lambda_b$, $\lambda_3$, and a set of location stock levels, $s_1$, $s_2$, and $s_3$. Our test scenarios capture different absolute and relative mean demand rates at location 3 over the transit lead time, as well as different safety stock placement strategies within the channel. The values of $k_1$, $k_2$, and $k_3$ correspond to the amounts of safety stock carried at each of the echelons (in this case, the number of standard deviations of lead time demand). The transit lead times between the echelons, $A_1$, $A_2$, and $A_3$, are fixed to values of 5, 2, and 1, respectively. A full factorial experiment was run using the parameter values shown in Figure 6.6, for a total of 324 different system scenarios. The computed immediate fill rates (other than 0%) ranged from 22.49% to 99.53%; the computed fill rates within $A_3$ ranged from 25.40% to 100.00%; and the computed fill rates within $A_2 + A_3$ ranged from 82.62% to 100.00%. For each scenario, fifty independent replications, plus their antithetic streams, were simulated for time durations equivalent to 20,000 demand events at location 3. This was done in order to equate the number of demand observations across the different scenarios for the fill rate estimates.

**Table 6.1.** The overall average and standard deviation of channel fill rate errors.

| Channel Fill Rate | Average Error ($\pm$ Standard Deviation) |
|---|---|
| Fill Rate within $A_2 + A_3$ | $-0.004\%$ ($\pm$ 0.043%) |
| Fill Rate within $A_3$ | $-0.026\%$ ($\pm$ 0.227%) |
| Immediate Fill Rate | $-0.044\%$ ($\pm$ 0.274%) |

The average absolute error for each channel fill rate across all scenarios is listed in Table 6.1, along with its standard deviation. Since the computed channel fill rate at the highest echelon is exact (i.e., the fill rate within $A_2 + A_3$), we expect to see extremely low deviations here. The magnitude of the average error increases at the second echelon and is highest at the third echelon (i.e., for the immediate fill rate), supporting the conjecture that the estimation error resulting from using negative binomial approximations compounds as the number of echelons increases; however, the approximations are still very accurate.

We note that there were several test scenarios in which no stock was held at location 3 (i.e, $s_3 = 0$), resulting in computed and simulated immediate fill rates of 0% for these scenarios, with 0% error. These scenarios are included in the average error statistic for the immediate fill rate shown in Table 6.1. When we exclude these scenarios and consider only those in which $s_3 \geq 1$, the immediate fill rate error averaged $-0.057\%$ and ranged between $-1.073\%$ and 0.558%. There were no test scenarios in which $s_2 = 0$ or $s_1 = 0$.

For each scenario, the average absolute errors for the immediate fill rate and the fill rate within $A_3$ are shown in Table 6.2. The largest errors were observed for scenarios in which there was little or no safety stock at locations 1 and 2. This makes sense, since decreasing the amount of safety stock at location 1 raises the variance of $Y_2$, and decreasing the amount of safety stock at location 2 raises the variance of $Y_3$.

From a practical standpoint, the errors we observed in this simulation study were extremely small. We conclude that the negative binomial approximation method described is appropriate for our model, as it is not likely to materially affect the quality or the feasibility of a solution resulting from an optimization routine using it.

We close this section by making two important observations. First, from (6.13), (6.21), and (6.29), it is clear that in a 3-level system, all of the channel fill rates ($f_3^3$, $f_3^2$, and $f_3^1$) can be made arbitrarily close to 100% by raising the base stock level $s_3$, regardless of the stock levels $s_2$ and $s_1$. The implication for problem **SLS** is that a feasible solution can always be found by adjusting the stock

| | | | | $\lambda_a=2,\ \lambda_b=1$ | | | | | | $\lambda_a=20,\ \lambda_b=10$ | | | | | | |
| | | | | $\lambda_3$ | | | | | | $\lambda_3$ | | | | | | |
| $k_1$ | $k_2$ | $k_3$ | | 0.1 | 0.2 | 0.5 | 0.8 | 1 | 2 | 0.1 | 0.2 | 0.5 | 0.8 | 1 | 2 | Average |
|---|---|---|---|---|---|---|---|---|---|---|---|---|---|---|---|---|
| 0 | 0 | 0 | Average Error of Immediate Fill Rate | 0.000% | 0.000% | 0.000% | 0.000% | 0.008% | -0.169% | 0.000% | 0.000% | 0.000% | 0.000% | -0.036% | -0.087% | -0.024% |
| | | | Average Error of Fill Rate Within A3 | 0.297% | 0.252% | -0.009% | 0.277% | -0.432% | -0.793% | -0.183% | -0.228% | -0.459% | -0.204% | -0.151% | -0.087% | -0.143% |
| | | 1 | Average Error of Immediate Fill Rate | 0.009% | -0.161% | -0.238% | -0.348% | -0.492% | -1.073% | -0.228% | -0.095% | -0.077% | -0.069% | -0.089% | -0.299% | -0.263% |
| | | | Average Error of Fill Rate Within A3 | 0.090% | -0.072% | -0.291% | -0.386% | -0.281% | -0.060% | -0.044% | -0.068% | -0.140% | -0.143% | 0.043% | 0.104% | -0.104% |
| | | 2 | Average Error of Immediate Fill Rate | 0.039% | 0.043% | -0.227% | -0.417% | -0.338% | -0.073% | 0.024% | -0.017% | -0.142% | -0.085% | -0.040% | 0.109% | -0.094% |
| | | | Average Error of Fill Rate Within A3 | -0.006% | 0.012% | -0.123% | -0.315% | -0.031% | 0.229% | 0.015% | 0.042% | 0.018% | 0.022% | 0.065% | 0.030% | -0.003% |
| | 1 | 0 | Average Error of Immediate Fill Rate | 0.000% | 0.000% | 0.000% | 0.000% | 0.558% | 0.321% | 0.000% | 0.000% | 0.000% | 0.000% | 0.023% | 0.016% | 0.076% |
| | | | Average Error of Fill Rate Within A3 | -0.359% | -0.348% | -0.524% | -0.588% | -0.250% | 0.053% | -0.592% | -0.664% | -0.613% | -0.665% | 0.036% | 0.160% | -0.363% |
| | | 1 | Average Error of Immediate Fill Rate | 0.013% | -0.055% | 0.145% | 0.271% | -0.454% | -1.058% | -0.238% | -0.085% | -0.075% | 0.006% | -0.008% | -0.037% | -0.131% |
| | | | Average Error of Fill Rate Within A3 | -0.021% | -0.085% | -0.165% | -0.336% | -0.008% | 0.259% | 0.006% | 0.056% | 0.083% | 0.019% | 0.070% | 0.077% | -0.004% |
| | | 2 | Average Error of Immediate Fill Rate | -0.002% | 0.004% | -0.372% | -0.601% | -0.482% | 0.080% | 0.048% | 0.065% | 0.083% | -0.023% | 0.029% | 0.105% | -0.103% |
| | | | Average Error of Fill Rate Within A3 | -0.002% | 0.003% | -0.004% | -0.046% | 0.076% | 0.135% | 0.000% | 0.021% | 0.028% | 0.058% | 0.045% | 0.008% | 0.027% |
| | 2 | 0 | Average Error of Immediate Fill Rate | 0.000% | 0.000% | 0.000% | 0.000% | 0.390% | 0.269% | 0.000% | 0.000% | 0.000% | 0.000% | -0.101% | 0.056% | 0.051% |
| | | | Average Error of Fill Rate Within A3 | -0.265% | -0.047% | -0.132% | -0.330% | 0.059% | 0.290% | -0.004% | 0.064% | 0.135% | 0.096% | 0.172% | 0.149% | 0.016% |
| | | 1 | Average Error of Immediate Fill Rate | 0.001% | -0.023% | 0.045% | 0.258% | -0.181% | -0.456% | -0.207% | -0.022% | -0.120% | -0.086% | 0.084% | 0.110% | -0.050% |
| | | | Average Error of Fill Rate Within A3 | -0.031% | 0.022% | 0.031% | -0.011% | 0.120% | 0.186% | 0.026% | 0.043% | 0.101% | 0.123% | 0.068% | 0.022% | 0.058% |
| | | 2 | Average Error of Immediate Fill Rate | 0.002% | 0.024% | -0.148% | -0.301% | -0.200% | 0.139% | 0.038% | 0.031% | -0.043% | 0.050% | 0.138% | 0.049% | -0.018% |
| | | | Average Error of Fill Rate Within A3 | -0.008% | 0.001% | 0.030% | 0.070% | 0.067% | 0.050% | 0.000% | 0.005% | 0.020% | 0.053% | 0.016% | 0.002% | 0.025% |
| 1 | 0 | 0 | Average Error of Immediate Fill Rate | 0.000% | 0.000% | 0.000% | 0.000% | 0.118% | -0.020% | 0.000% | 0.000% | 0.000% | 0.000% | 0.013% | -0.004% | 0.009% |
| | | | Average Error of Fill Rate Within A3 | 0.262% | 0.142% | -0.105% | 0.086% | -0.215% | -0.396% | -0.179% | -0.111% | -0.062% | -0.037% | | 0.029% | -0.068% |
| | | 1 | Average Error of Immediate Fill Rate | 0.020% | -0.094% | -0.109% | -0.216% | -0.253% | -0.690% | -0.172% | -0.025% | -0.093% | 0.001% | 0.011% | -0.157% | -0.148% |
| | | | Average Error of Fill Rate Within A3 | 0.058% | -0.016% | -0.209% | -0.236% | -0.129% | -0.004% | -0.036% | -0.021% | 0.021% | -0.050% | 0.004% | 0.044% | -0.048% |
| | | 2 | Average Error of Immediate Fill Rate | 0.023% | 0.036% | -0.181% | -0.282% | -0.232% | -0.027% | 0.023% | -0.012% | -0.133% | 0.013% | 0.010% | 0.057% | -0.059% |
| | | | Average Error of Fill Rate Within A3 | 0.000% | -0.001% | -0.072% | -0.162% | -0.018% | 0.084% | 0.012% | 0.035% | 0.006% | -0.033% | 0.018% | 0.010% | -0.010% |
| | 1 | 0 | Average Error of Immediate Fill Rate | 0.000% | 0.000% | 0.000% | 0.000% | 0.179% | 0.064% | 0.000% | 0.000% | 0.000% | 0.000% | 0.008% | -0.017% | 0.016% |
| | | | Average Error of Fill Rate Within A3 | -0.352% | -0.198% | -0.297% | -0.306% | -0.103% | 0.068% | -0.105% | -0.064% | -0.127% | -0.146% | 0.051% | 0.073% | -0.125% |
| | | 1 | Average Error of Immediate Fill Rate | -0.002% | -0.084% | -0.036% | 0.073% | -0.228% | -0.383% | -0.232% | -0.096% | -0.184% | 0.020% | 0.020% | 0.002% | -0.094% |
| | | | Average Error of Fill Rate Within A3 | -0.050% | -0.072% | -0.075% | -0.137% | 0.001% | 0.087% | 0.011% | 0.060% | 0.039% | 0.013% | 0.034% | 0.014% | -0.006% |
| | | 2 | Average Error of Immediate Fill Rate | 0.005% | 0.015% | -0.144% | -0.315% | -0.188% | 0.067% | 0.005% | 0.027% | -0.041% | -0.018% | 0.027% | 0.033% | -0.040% |
| | | | Average Error of Fill Rate Within A3 | -0.002% | -0.010% | 0.000% | -0.027% | 0.014% | 0.041% | 0.001% | 0.004% | 0.015% | 0.012% | 0.010% | 0.002% | 0.005% |
| | 2 | 0 | Average Error of Immediate Fill Rate | 0.000% | 0.000% | 0.000% | 0.000% | 0.047% | -0.002% | 0.000% | 0.000% | 0.000% | 0.000% | -0.062% | -0.075% | -0.008% |
| | | | Average Error of Fill Rate Within A3 | -0.259% | 0.005% | -0.045% | -0.114% | 0.032% | 0.123% | -0.003% | 0.025% | 0.058% | 0.063% | 0.050% | 0.038% | -0.002% |
| | | 1 | Average Error of Immediate Fill Rate | -0.010% | -0.052% | -0.146% | -0.017% | -0.049% | -0.112% | -0.193% | -0.049% | -0.128% | -0.050% | 0.010% | 0.087% | -0.059% |
| | | | Average Error of Fill Rate Within A3 | -0.031% | -0.003% | 0.023% | -0.005% | 0.041% | 0.054% | 0.007% | 0.005% | 0.024% | 0.036% | 0.016% | 0.005% | 0.014% |
| | | 2 | Average Error of Immediate Fill Rate | 0.031% | 0.055% | 0.004% | -0.087% | -0.032% | 0.017% | 0.039% | 0.008% | -0.083% | 0.000% | 0.069% | 0.024% | 0.004% |
| | | | Average Error of Fill Rate Within A3 | -0.001% | 0.001% | 0.007% | 0.024% | 0.023% | 0.015% | -0.001% | 0.000% | 0.001% | 0.014% | 0.001% | 0.001% | 0.007% |
| 2 | 0 | 0 | Average Error of Immediate Fill Rate | 0.000% | 0.000% | 0.000% | 0.000% | 0.141% | 0.045% | 0.000% | 0.000% | 0.000% | 0.000% | 0.038% | 0.038% | 0.022% |
| | | | Average Error of Fill Rate Within A3 | 0.100% | -0.083% | -0.197% | -0.133% | -0.015% | -0.115% | -0.147% | -0.132% | -0.167% | -0.067% | 0.036% | 0.070% | -0.071% |
| | | 1 | Average Error of Immediate Fill Rate | 0.050% | -0.025% | -0.035% | -0.163% | -0.127% | -0.458% | -0.147% | -0.017% | -0.060% | 0.035% | 0.058% | -0.021% | -0.084% |
| | | | Average Error of Fill Rate Within A3 | 0.128% | 0.069% | -0.072% | -0.120% | 0.024% | -0.046% | -0.007% | 0.008% | 0.065% | -0.001% | 0.022% | 0.027% | 0.008% |
| | | 2 | Average Error of Immediate Fill Rate | 0.029% | 0.058% | -0.090% | -0.157% | -0.099% | -0.021% | 0.018% | 0.003% | -0.113% | 0.045% | 0.031% | 0.059% | -0.020% |
| | | | Average Error of Fill Rate Within A3 | 0.012% | 0.018% | -0.011% | -0.031% | 0.016% | -0.013% | 0.012% | 0.031% | 0.004% | -0.028% | 0.008% | 0.004% | 0.002% |
| | 1 | 0 | Average Error of Immediate Fill Rate | 0.000% | 0.000% | 0.000% | 0.000% | 0.148% | -0.018% | 0.000% | 0.000% | 0.000% | 0.000% | 0.006% | -0.024% | 0.009% |
| | | | Average Error of Fill Rate Within A3 | -0.083% | 0.041% | -0.023% | -0.083% | 0.035% | -0.006% | 0.172% | 0.160% | 0.066% | 0.072% | 0.056% | 0.044% | 0.038% |
| | | 1 | Average Error of Immediate Fill Rate | -0.014% | -0.038% | -0.020% | 0.083% | -0.113% | -0.312% | -0.225% | -0.072% | -0.166% | 0.021% | 0.014% | 0.008% | -0.070% |
| | | | Average Error of Fill Rate Within A3 | -0.014% | -0.025% | -0.011% | 0.006% | 0.018% | -0.024% | 0.013% | 0.054% | 0.031% | 0.035% | 0.008% | 0.004% | 0.009% |
| | | 2 | Average Error of Immediate Fill Rate | 0.040% | 0.028% | -0.107% | -0.189% | -0.110% | -0.028% | 0.053% | 0.027% | -0.052% | 0.000% | 0.040% | 0.025% | -0.023% |
| | | | Average Error of Fill Rate Within A3 | 0.000% | -0.008% | -0.001% | 0.010% | 0.001% | -0.001% | 0.001% | 0.007% | 0.010% | 0.003% | 0.000% | 0.000% | 0.002% |
| | 2 | 0 | Average Error of Immediate Fill Rate | 0.000% | 0.000% | 0.000% | 0.000% | -0.154% | -0.067% | 0.000% | 0.000% | 0.000% | 0.000% | -0.070% | -0.090% | -0.032% |
| | | | Average Error of Fill Rate Within A3 | -0.068% | 0.057% | 0.017% | 0.041% | 0.027% | -0.008% | 0.083% | 0.052% | 0.051% | 0.067% | 0.017% | 0.007% | 0.029% |
| | | 1 | Average Error of Immediate Fill Rate | 0.003% | -0.054% | -0.122% | -0.040% | -0.045% | -0.166% | -0.200% | -0.062% | -0.136% | -0.055% | -0.004% | 0.059% | -0.069% |
| | | | Average Error of Fill Rate Within A3 | -0.020% | -0.007% | 0.008% | 0.010% | 0.010% | -0.001% | 0.008% | -0.002% | -0.001% | 0.007% | 0.005% | 0.001% | 0.001% |
| | | 2 | Average Error of Immediate Fill Rate | 0.026% | 0.044% | -0.020% | -0.068% | 0.048% | -0.047% | 0.038% | 0.008% | -0.093% | -0.015% | 0.056% | 0.012% | -0.001% |
| | | | Average Error of Fill Rate Within A3 | 0.001% | -0.001% | 0.001% | 0.003% | 0.004% | 0.001% | -0.001% | -0.002% | -0.002% | 0.002% | 0.000% | 0.000% | 0.001% |
| | | | Overall Error of Immediate Fill Rate | 0.010% | -0.010% | -0.067% | -0.093% | -0.079% | -0.156% | -0.056% | -0.014% | -0.067% | -0.008% | 0.010% | -0.002% | -0.044% |
| | | | Overall Error of Fill Rate Within A3 | -0.023% | -0.013% | -0.083% | -0.105% | -0.034% | 0.008% | -0.035% | -0.023% | -0.036% | -0.026% | 0.025% | 0.031% | -0.026% |

**Table 6.2.** Average absolute error between the computed and simulated fill rates

level vector $\mathbf{s}^3$, regardless of the vectors $\mathbf{s}^2$ and $\mathbf{s}^1$. Generalizing this concept to an $n$-level network where the constraint set $K$ is separable by level, the channel fill rates $f_n^v, f_n^{v-1}, \ldots, f_n^1$, can be made arbitrarily close to 100% by raising the stock level $s_v$. Thus, we can always find a solution to **SLS** that satisfies the service level constraints $K^1 \cup K^2 \cup \cdots \cup K^v$ by adjusting the stock level vector $\mathbf{s}^v$, regardless of the vectors $\mathbf{s}^{v-1}, \ldots, \mathbf{s}^1$. (When $v = n$, this means that all constraints will be satisfied.)

Second, recall that when the stock levels $s_1$ and $s_2$ are fixed to values that are at least $\lfloor \lambda_1 A_1 \rfloor$ and $\lfloor \lambda_2 T_2 \rfloor$, respectively, the channel fill rates given by (6.13), (6.21), and (6.29) are concave functions in $s_3$ for $s_3 \geq \lfloor \lambda_3 T_3 \rfloor$. (Note that $T_2$ is a function of $s_1$, and $T_3$ is a function of $s_1$ and $s_2$, so the stock level combinations must be chosen carefully in order for this concavity property to hold.) Our solution approach, which we describe next, makes use of both of these facts.

## 6.3 Solution Approach

Now that we have defined the channel fill rate functions, let us turn our attention to solving the optimization problem **SLS**. It is clear that **SLS** cannot be solved to optimality easily for realistically-sized problem instances due to the fact that the fill rate functions are not jointly concave in their arguments. Even if we had a very restricted set of candidate stock levels for each item at each location in the network, exhaustive enumeration would be virtually impossible, even in a modest-sized network. Hence, we focus our attention on developing a practical heuristic approach. The algorithm we describe here is a column generation procedure that finds stock level vectors for each item type that collectively give a near-optimal solution to **SLS**.

We give an overview of the column generation procedure in Section 6.3.1 and define additional notation that will be useful in our explication. In Section 6.3.2, we define the master problem that will be augmented and re-solved in each iteration of the procedure. In Section 6.3.3, we describe three potential column generation techniques, any one of which can be used to generate a new set of columns within an iteration of the procedure. Next, we discuss methods for generating an initial set of columns to seed the procedure.

### 6.3.1 Procedure Overview

The idea behind column generation is to repeatedly solve a *master problem*, a restricted version of the overall problem that is relatively easy to solve. The master problem is restricted in the sense that it only considers a subset of the feasible solution space (i.e, the subset spanned by its columns). Adding columns to the master problem is equivalent to expanding the subset of the solution space that is considered. If the optimal solution to the overall problem lies within the space of solutions covered by the master problem, then solving the master problem solves the overall problem. A typical column generation procedure begins with an initial set of columns, solves the master problem associated with this set of columns, and then uses this solution to generate one or more new columns. The process is repeated until the master problem produces a satisfactory solution, or until the solution to the master problem does not change (i.e., the most-recently added columns are of no benefit). The latter may indicate that an optimal solution has been found, depending on the formulation of the master problem and the column generation technique employed. Specifically, the algorithm will terminate with an optimal solution if the column generation technique, by its construction, is guaranteed to find an improving column if one exists.

The fundamental elements of our column generation procedure are a collection of stock level vectors $\Gamma_i$ for each item $i \in I$, where each vector $\gamma_i \in \Gamma_i$

contains a candidate stock level entry $s_{ij}$ for each location $j \in J$. Associated with each stock level vector $\gamma_i$ is a cost element, $c_{\gamma_i}$, and a contribution column, $\beta_{\gamma_i}$. The term $c_{\gamma_i}$ represents the total cost associated with the vector $\gamma_i$; that is, the unit cost $c_i$ multiplied by the sum of the stock level entries in the vector $\gamma_i$. The column $\beta_{\gamma_i}$ has entries representing the contribution that $\gamma_i$ makes to the fulfillment of each service level constraint $k \in K$. Specifically, the entry of $\beta_{\gamma_i}$ that corresponds to service constraint $k$, denoted by $[\beta_{\gamma_i}]_k$ is given by:

$$[\beta_{\gamma_i}]_k = \sum_{v=1}^{N} \sum_{j \in J^N} w_{ijk}^v f_{ij}^v(\mathbf{s_{iP_j}}),$$

where the channel stock level vectors $\mathbf{s_{iP_j}}$ are determined by $\gamma_i$. $B_i$ will be used to denote the set of all contribution columns $\beta_{\gamma_i}$ for item $i$.

In our procedure, the master problem is a linear programming approximation to **SLS**. The formulation is based on the existing stock level vector sets $\Gamma_i, i \in I$, the corresponding costs $c_{\gamma_i}, \gamma_i \in \Gamma_i, i \in I$, and the corresponding contribution column sets $B_i, i \in I$. We use the solution to the master problem to generate $|I|$ new stock level vectors $\gamma_i$, one for each item $i \in I$ (each having an associated cost $c_{\gamma_i}$ and a contribution column $\beta_{\gamma_i}$). The procedure terminates when the master problem's solution does not change from the previous iteration. However, because our master problem is a linearized approximation to **SLS**, and because the column generation techniques we employ are not guaranteed to find improving columns if they exist, we cannot claim that the resulting solution is optimal. Empirical evidence suggests, however, that these techniques work very well in practice.

The new notation is summarized below, along with other notation that will be used in describing the various column generation techniques we employ:

$\Gamma_i$ the set of candidate stock level vectors for item $i$, indexed by $\gamma_i$.

$B_i$ the set of contribution columns associated with candidate stock level vectors for item $i$, indexed by $\beta_{\gamma_i}$.

$c_{\gamma_i}$ the total cost associated with the stock level vector $\gamma_i \in \Gamma_i$.

$\alpha^*$ a (possibly fractional) solution vector for the master problem, whose elements are indexed by $\gamma_i$.

$\theta_k^*$ with respect to the master problem solution vector $\alpha^*$, the dual variable corresponding to service level constraint $k \in K$.

### 6.3.2  Master Problem for SLS

Given stock level vector sets $\Gamma_i, i \in I$, the corresponding costs $c_{\gamma_i}, \gamma_i \in \Gamma_i, i \in I$ and the corresponding contribution column sets $B_i, i \in I$, we define the *Service Level Satisfaction Master Problem*, or **SLSMP** as:

$$\textbf{(SLSMP)} \qquad \text{minimize} \qquad \sum_{i \in I} \sum_{\gamma_i \in \Gamma_i} c_{\gamma_i} \alpha_{\gamma_i} \tag{6.33}$$

subject to

$$\sum_{i \in I} \sum_{\gamma_i \in \Gamma_i} [\beta_{\gamma_i}]_k \alpha_{\gamma_i} \geq F_k \quad \forall k \in K, \tag{6.34}$$

$$\sum_{\gamma_i \in \Gamma_i} \alpha_{\gamma_i} = 1 \quad \forall i \in I, \tag{6.35}$$

$$0 \leq \alpha_{\gamma_i} \leq 1 \quad \forall \gamma_i \in \Gamma_i, i \in I. \tag{6.36}$$

Observe that the columns of the constraints (6.34) of the above linear program correspond to the elements of the vector sets $B_i$, $i \in I$. Hence, the optimal solution $\alpha^*$ to **SLSMP** corresponds to the following (possibly fractional) solution to **SLS**:

$$\mathbf{s_i} = \sum_{\gamma_i \in \Gamma_i} \alpha^*_{\gamma_i} \gamma_i, \quad i \in I. \tag{6.37}$$

Note that, even if we round up the fractional elements of the vectors (6.37) to their integer ceilings, the resulting solution will not necessarily be feasible for **SLS** since the channel fill rate functions are not jointly concave in their arguments.

Most of the column generation techniques we employ require the master problem solution to be integral (though not feasible and integral) in order to use it to generate new columns. For these purposes, simply rounding the fractional solution given by (6.37) suffices. When the column generation procedure terminates, however, we do need to convert the fractional solution into a feasible, integral solution. To accomplish this, take the integer ceiling of the fractional elements of the vectors given by (6.37), and then use a greedy heuristic to increment stock levels until all constraints are satisfied. This greedy heuristic works as follows:

### Construct-Feasible-Solution

Input:     An instance of problem **SLS**;
           **s**, an integral solution to **SLS** that is not necessarily feasible.
Output:    $\tilde{\mathbf{s}}$, a feasible integral solution to **SLS**.

1. $\tilde{s}_{ij} \leftarrow s_{ij}$ for all $i \in I, j \in J$.
2. Update $\bar{K} \subseteq K$, the set of all *unsatisfied* service level constraints with respect to the current stock level vector $\tilde{\mathbf{s}}$.
3. For all $i \in I, j \in J^N$ (i.e., all bases), compute:

$$\Delta_{ij} = \sum_{k \in \bar{K}} \left( \min \{F_k, \sum_{v=1}^{N} w^v_{ijk} f^v_{ij}(\tilde{s}_{ij} + 1, \tilde{\mathbf{s}}_{i\mathbf{P_j}} \backslash \tilde{s}_{ij})\} - \sum_{v=1}^{N} w^v_{ijk} f^v_{ij}(\tilde{\mathbf{s}}_{i\mathbf{P_j}}) \right).$$

4. For all $i \in I, j \in J^v, v < N$ (i.e., all nonbase locations), compute:

$$\Delta_{ij} = \sum_{k \in \bar{K}} \sum_{j' \in J^N : j \in P_{j'}} \left( \min \left\{ F_k, \sum_{v=1}^{N} w_{ij'k}^v f_{ij'}^v (\tilde{s}_{ij} + 1, \tilde{s}_{i\mathbf{P}_j'} \backslash \tilde{s}_{ij}) \right\} \right.$$

$$\left. - \sum_{v=1}^{N} w_{ij'k}^v f_{ij'}^v (\tilde{s}_{i\mathbf{P}_j'}) \right). \quad (6.38)$$

5. Find the pair $(i, j)^*$ such that:

$$(i, j)^* = \arg \max_{(i,j)} \frac{\Delta_{ij}}{c_i}.$$

6. $\tilde{s}_{ij^*} \leftarrow \tilde{s}_{ij^*} + 1$.
7. If all service level constraints $k \in K$ are satisfied, then STOP and return $\tilde{s}$. Otherwise, go to step 2.

Note that the solution returned by **Construct-Feasible-Solution** may overstate some of the stock levels needed to satisfy the service level constraints. This overstatement may be due to the starting solution used, the greedy order in which the stock levels were incremented, or a combination of the two. Although we do not give the details here, a second-pass greedy heuristic may be used to improve the current solution by reducing stock levels while maintaining feasibility.

### 6.3.3  Column Generation

Given a solution $\alpha^*$ to the master problem, there are a multitude of ways in which new columns can be generated for **SLSMP**. We discuss three possibilities here.

#### 6.3.3.1  Technique 1 (Simple Rounding)

The most obvious approach is to parse the master problem solution given by $\alpha^*$ into a new set of stock level vectors by item type, employing some rounding technique to ensure that each stock level vector is integral. That is, for each item type $i$, define a new stock level vector:

$$\tilde{\gamma}_i = R(\sum_{\gamma_i \in \Gamma_i} \alpha_{\gamma_i}^* \gamma_i), \quad (6.39)$$

where the rounding function $R(\cdot)$ is user-defined (e.g., take the floor, the ceiling, or the nearest integer value of each fractional element). Different definitions of $R(\cdot)$ can be used to give rise to multiple new stock level vectors $\tilde{\gamma}_i$ for each item type $i$. For each new vector $\tilde{\gamma}_i$ that is generated, we can now compute its associated cost $c_{\tilde{\gamma}_i}$ and its contribution column $\beta_{\tilde{\gamma}_i}$, which we can add to the master problem.

It is important to note that in generating new columns this way, we are *not* adding redundant columns to the master problem, even if no rounding is required.

This is due to the fact that, even though each new stock level vector $\tilde{\gamma}_i$ is a (possibly rounded) convex combination of existing stock level vectors, its associated contribution column $\beta_{\tilde{\gamma}_i}$ is *not* a convex combination of the corresponding contribution columns. In fact, the contribution elements of the column $\beta_{\tilde{\gamma}_i}$ will in general be higher, due to the concavity of the fill rate functions.

Although the above technique is an extremely useful way to generate new columns quickly, it is limited by the fact that the individual stock level elements of an existing $\gamma_i$ vector will always be given the same relative weight $\alpha^*_{\gamma_i}$ when it comes to generating new columns. For instance, suppose our initial vector set for some item contains one stock level vector where every element has the value 5, and a second stock level vector where every element has the value 1. Taking convex combinations of these vectors, it is possible to generate a new integral vector having all 2's, another having all 3's, and yet another having all 4's. However, we cannot generate new vectors where the individual elements differ from one another. Hence, for this technique to be effective, we need to make sure that the initial sets of vectors that we generate, $\Gamma_i$, $i \in I$, contain combinations of stock levels that are very different from one another. Alternatively, we can employ column generation techniques, such as the ones described next, that provide mechanisms for making relative increases and decreases among the elements of a stock level vector.

### 6.3.3.2  Technique 2 (Fix Stock Values Level-by-Level)

Recall that $\mathbf{s}^v$ is defined to be a vector of stock levels for all items $i \in I$ at the network locations at level $v$. That is, $\mathbf{s}^v = (s_{ij} : i \in I, j \in J^v)$.

This technique assumes that the constraint set $K$ of the problem instance in question is separable by network level (i.e., $v_{ijk}$ is the same for all items $i$ and all bases $j$ with which constraint $k$ is concerned). Thus, we can partition the constraint set $K$ into $N$ subsets, $K^v$, $v = 1, \ldots, N$.

Consider an integral vector of stock levels $\mathbf{s}^*$ that has been derived (via a chosen rounding scheme) from the solution $\alpha^*$ to the master problem. That is, for each item $i \in I$:

$$\mathbf{s_i^*} = R(\sum_{\gamma_i \in \Gamma_i} \alpha^*_{\gamma_i} \gamma_i). \tag{6.40}$$

Beginning with stock levels $s_{ij} = s^*_{ij}$ for every item $i \in I$ and every location $j \in J$, the idea behind this technique is to devise a new solution, $\tilde{\mathbf{s}}$, to the problem by solving a sequence of $N$ subproblems, one for each network level $v$, beginning with level $v = 1$. Each subproblem works as follows: Keep all stock levels $s_{ij}$ fixed at their current values *except* for those stock levels at level-$v$ network locations. For all items $i \in I$ at level-$v$ network locations, set $s_{ij} = \lfloor \lambda_{ij} T_{ij} \rfloor$, where $T_{ij}$ is a function of the fixed stock levels for item $i$ at locations in the channel $P_j$ that are above location $j$ in the network. (This will ensure that the channel fill rate functions are concave in $\mathbf{s}^v$.) Next, use a greedy heuristic to increment the level-$v$

stock levels until the service level constraints $k \in (K^1 \cup K^2 \cup \cdots \cup K^v)$ are satisfied. That is, in the $v$th subproblem, the only constraints that are considered are those associated with levels 1 through $v$.

The greedy heuristic is essentially just a version of **Construct-Feasible-Solution** that is restricted to level-$v$ locations and service level constraints $k \in K^1 \cup \cdots \cup K^v$. For each item at each level-$v$ location, compute the total incremental contribution made to all *unsatisfied* service level constraints $k \in K^1 \cup \cdots \cup K^v$ and divide this incremental contribution by the item's unit cost. Select the highest ratio, and increment the corresponding stock level. Once all constraints $k \in K^1 \cup \cdots \cup K^v$ are satisfied, a second phase may be performed to adjust level-$v$ stock levels downward and reduce investment while maintaining constraint satisfaction.

The result is a new feasible solution vector $\tilde{s}$ that may be parsed into $|I|$ stock level vectors, $\tilde{\gamma}_i$, $i \in I$, which in turn may be added to the master problem.

### 6.3.3.3  Technique 3 (Decomposition by Item)

Suppose we have a set of dual variables, $\{\theta_k^*, k \in K\}$, that correspond to a particular primal solution, $\alpha^*$, to the master problem. In this method, we use these multiplier values to dualize the service level constraints and construct a Lagrangian relaxation of **SLS**. As we will show, the relaxed problem, denoted **SLS-LR**, may be decomposed by item type, so that we are left with a set of $|I|$ independent item subproblems to solve, one for each $i \in I$. The solution to each item subproblem yields a new column $\tilde{\gamma}_i$ that may be added to the master problem.

We will discuss a method for solving the item subproblems shortly. First, we detail the decomposition of **SLS** by constructing the following Lagrangian relaxation:

**(SLS-LR)**

$$\min_{s_{ij} \geq 0, \text{integer}} \left( \sum_{i \in I} \sum_{j \in J} c_i s_{ij} + \sum_{k \in K} \theta_k^* \left( F_k - \sum_{v=1}^{N} \sum_{i \in I} \sum_{j \in J^N} w_{ijk}^v f_{ij}^v (S_{i P_j}) \right) \right). \quad (6.41)$$

Since the terms $\theta_k^* F_k$ are constant, we may ignore them without affecting the optimal solution to **SLS-LR**. Letting

$$\overline{w_{ij}^v} = \sum_{k \in K} \theta_k^* w_{ijk}^v, \quad (6.42)$$

and leaving off the constant terms, (6.41) becomes

$$\min_{s_{ij} \geq 0, \text{integer}} \left( \sum_{i \in I} \sum_{j \in J} c_i s_{ij} - \sum_{i \in I} \sum_{v=1}^{N} \sum_{j \in J^N} \overline{w_{ij}^v} f_{ij}^v (S_{i P_j}) \right)$$

$$= \min_{s_{ij} \geq 0, \text{integer}} \sum_{i \in I} \left( \sum_{v < N} \sum_{j \in J^v} c_i s_{ij} + \sum_{j \in J^N} \left( c_i s_{ij} - \sum_{v=1}^{N} \overline{w_{ij}^v} f_{ij}^v (S_{i P_j}) \right) \right). \quad (6.43)$$

Since each weight $\overline{w_{ij}^v}$ and each channel fill rate $f_{ij}^v(s_{iP_j})$ corresponds to a single item, the minimization problem is separable by item. Thus, we are left with solving

**(SLS-LR-SI)**

$$\min_{s_{ij} \geq 0, \text{integer}} \left( \sum_{v < N} \sum_{j \in J^v} c_i s_{ij} + \sum_{j \in J^N} \left( c_i s_{ij} - \sum_{v=1}^{N} \overline{w_{ij}^v} f_{ij}^v(s_{iP_j}) \right) \right). \quad (6.44)$$

for each item type $i$ independently.

We are now ready to describe a procedure for solving **SLS-LR-SI** for a particular item type $i$. The algorithm involves repeatedly fixing the stock levels $s_{ij}$ at all locations $j \in J^v$, $v < N$, to predetermined values (i.e., all *nondemand* locations). When we do this, the first term in (6.44) becomes a constant, and the second term becomes a convex function that is separable by location (provided that we restrict $s_{ij} \geq \lfloor \lambda_{ij} \tau_{ij} \rfloor$, for all $j \in J^N$). That is, letting $\hat{\gamma}_i$ denote a specified vector of stock levels at all nondemand locations, the problem becomes:

$$G_i(\hat{\gamma}_i, \theta^*) = \min_{s_{ij} \geq 0, \text{integer}} \left( \sum_{v < N} \sum_{j \in J^v} c_i s_{ij} + \sum_{j \in J^N} \left( c_i s_{ij} - \sum_{v=1}^{N} \overline{w_{ij}^v} f_{ij}^v(s_{ij}, \hat{\gamma}_i) \right) \right)$$

$$= c_{\hat{\gamma}_i} + \min_{s_{ij} \geq 0, \text{integer}} \left( \sum_{j \in J^N} \left( c_i s_{ij} - \sum_{v=1}^{N} \overline{w_{ij}^v} f_{ij}^v(s_{ij}, \hat{\gamma}_i) \right) \right)$$

$$= c_{\hat{\gamma}_i} + \sum_{j \in J^N} \min_{s_{ij} \geq 0, \text{integer}} \left( c_i s_{ij} - \sum_{v=1}^{N} \overline{w_{ij}^v} f_{ij}^v(s_{ij}, \hat{\gamma}_i) \right)$$

$$= c_{\hat{\gamma}_i} + \sum_{j \in J^N} \min_{s_{ij} \geq 0, \text{integer}} g_i(j, \hat{\gamma}_i, \theta^*), \quad (6.45)$$

where $c_{\hat{\gamma}_i} = \sum_{v<N} \sum_{j \in J^v} c_i \hat{\gamma}_{ij}$, and $g_i(j, \hat{\gamma}_i, \theta^*) = \left( c_i s_{ij} - \sum_{v=1}^{N} \overline{w_{ij}^v} f_{ij}^v(s_{ij}, \hat{\gamma}_i) \right)$ is discretely convex in $s_{ij} \geq \lfloor \lambda_{ij} \tau_{ij}(\hat{\gamma}_i) \rfloor$. Hence, by restricting our search to $s_{ij} \geq \lfloor \lambda_{ij} \tau_{ij}(\hat{\gamma}_i) \rfloor$ for $j \in J^N$, the stock levels minimizing $g_i(j, \hat{\gamma}_i, \theta^*)$, $j \in J^N$, can be found quickly and easily using marginal analysis. That is, beginning with $s_{ij} = \lfloor \lambda_{ij} \tau_{ij}(\hat{\gamma}_i) \rfloor$, simply increase $s_{ij}$ until $c_i > \sum_{v=1}^{N} \overline{w_{ij}^v} (f_{ij}^v(s_{ij} + 1, \hat{\gamma}_i) - f_{ij}^v(s_{ij}, \hat{\gamma}_i))$. Thus, given a candidate set of stock level vectors for item $i$ at all nondemand locations, $\hat{\Gamma}_i$, the following rudimentary algorithm can be used to find a solution for **SLS-LR-SI**.

**Construct-Single-Item-Solution**

Input:    An instance of problem **SLS**;
          An item type $i$;
          Lagrange multipliers $\{\theta_k^* : k \in K\}$;

A set of stock level vectors $\hat{\Gamma}_i$, where each $\hat{\gamma}_i \in \hat{\Gamma}_i$ represents a vector of fixed stock levels for item $i$ at all nondemand locations;

Output:    A feasible solution to **SLS-LR-SI** for item $i$, $\tilde{\gamma}_i$.

1. $U \leftarrow \infty$.
2. For each $\hat{\gamma}_i \in \hat{\Gamma}_i$:
   a) For each location $j \in J^N$, find the stock level $s_{ij}$ that minimizes $g_i(j, \hat{\gamma}_i, \theta^*)$ using marginal analysis.
   b) Using the computed stock level values $s_{ij}$ for all $j \in J^N$, compute:

$$G_i(\hat{\gamma}_i, \theta^*) = c_{\hat{\gamma}_i} + \sum_{j \in J^N} \min_{s_{ij} \geq 0, \text{integer}} g_i(j, \hat{\gamma}_i, \theta^*).$$

   c) If $G_i(\hat{\gamma}_i, \theta^*) < U$, then $U \leftarrow G_i(\hat{\gamma}_i, \theta^*)$, and $\tilde{\gamma}_i \leftarrow (\hat{\gamma}_i; s_{ij}, j \in J^N)$.
3. Return $U$ and $\tilde{\gamma}_i$.

The resulting solution $\tilde{\gamma}_i$ may be added to the master problem.

There are many ways to construct the vector set $\hat{\Gamma}_i$ for a particular $i \in I$. The most obvious way is simply to use the existing vector set $\Gamma_i$ as a basis, leaving off the entries that correspond to the demand locations $j \in J^N$.

Another way is to partially enumerate all candidate stock levels at the upper echelons of the distribution network. For a 3-level network, this involves generating combinations of stock levels at all level-1 and level-2 network locations. While this may sound computationally intractable if there are many level-2 locations, it is actually doable for problem instances in which the constraint set is *decomposable by subsystem*, where a subsystem is defined to be the level-1 location, a level-2 location, and all level-3 locations that are children of the specified level-2 location. That is, as long as each service level constraint is confined to locations in a single subsystem, then for a given stock level at the level-1 location, each subsystem may be considered independently. This substantially reduces the computational effort required to solve **SLS-LR-SI** to optimality.

Why would we want to solve **SLS-LR-SI** to optimality? Certainly the vector $\tilde{\gamma}_i$ resulting from **Construct-Single-Item-Solution** does not need to be an optimal solution to **SLS-LR-SI** to continue the column generation procedure. In fact, $\tilde{\gamma}_i$ will not be optimal for **SLS-LR-SI** unless one of the vectors, $\hat{\gamma}_i$, in our candidate set happens to partially describe an optimal solution. The reason for wanting the optimal solution to all **SLS-LR-SI** item subproblems for a given set of multipliers is that the collection of these optimal item vectors, together, yield an objective function value that is guaranteed to be a lower bound for the original **SLS** problem. Thus, for many practical problem instances, this technique can be used to provide an optimality gap for the problem.

### 6.3.4  Constructing the Initial Vector Sets $\Gamma_i$

To get started, the column generation procedure requires an initial set of stock levels vectors, $\Gamma_i$, for each item $i \in I$. The initial number of vectors in each $\Gamma_i$

need not be big; however, it is desirable to have a relatively diverse set of columns so that a larger portion of the solution space is collectively spanned in the master problem.

Here are some ways in which initial vector sets can be generated:

Beginning with level $v = 1$ up to $N$, do the following: For $i \in I$, $j \in J^v$, set $s_{ij} = \lfloor \lambda_{ij} A_{ij} \rfloor$. These vectors represent lower bound stock levels at every location. One could also use $s_{ij} = \lfloor \lambda_{ij} T_{ij} \rfloor$, where $T_{ij}$ is a function of the fixed stock levels for item $i$ at locations in the channel $P_j$ that are above location $j$ in the network.

For level $v = 1$, set $s_{ij} = \infty$ for $i \in I$. (This guarantees that level-2 locations will have no delay in replenishment.) Beginning with level $v = 2$ up to $N$, set $s_{ij} = \lfloor \lambda_{ij} T_{ij} \rfloor$, where $T_{ij}$ is a function of the fixed stock levels for item $i$ at locations in the channel $P_j$ that are above location $j$ in the network.

Using the lower bound stock level vectors as a starting solution, execute Technique 2 to create a new set of vectors.

Using the lower bound stock level vectors as a starting solution, execute **Construct-Feasible-Solution** to create a new set of vectors. (Since this may be computationally intensive, it can be partially executed.)

At this time we have no knowledge of which of these methods, or others that can easily be constructed, will work best for different problem instances. Furthermore, other heuristics may be developed for generating good solutions for the **SLS** problem. Heuristics may be effectively used to generate not only initial vectors of stock levels, that is, initial sets $\Gamma_i$, but also good starting estimates of multiplier values, $\theta_k$. The Lagrangian relaxation approach (Technique 3) requires values for each $\theta_k$. Obviously, **SLS-LR** is computationally demanding to solve and hence should be solved as few times as possible. Thus having good initial estimates of the $\theta_k^*$ will greatly reduce the total computation time needed to solve the **SLS** problem.

## 6.4  Problem Set, Chapter 6

**Problem 1.** In Section 6.3.2, a first-pass greedy heuristic was described for converting a fractional solution found when solving the master problem into a feasible, integral solution to the original problem. This method may overstate the stock levels needed to satisfy some of the constraints, as mentioned in that section. Formally state a second-pass greedy heuristic that improves upon the solution obtained when employing the first-pass greedy heuristic. Ensure that feasibility is maintained.

**Problem 2.** The general Service Level Satisfaction (SLS) problem was stated in Section 6.1.3. Let us now consider a specific instance of the problem. Suppose the network consists of but two echelons. Echelon 1 locations are called bases and the top echelon is called the depot. At each base there is only a single fill rate constraint, which measures the aggregate fill rate across all items. Furthermore, for each item $i$ there is a channel fill rate constraint at the depot specifying that item's fill rate requirement for base $j$ within the transportation lead time, $A_{ij}$. The other assumptions underlying the model developed in Section 6.1.3 remain. Construct a method for solving this special instance of the SLS problem. Construct heuristics that could be used to find good solutions for this specific problem.

**Problem 3.** Verify the channel fill rate expressions given in Section 6.2.4.

# 7

# Lateral Resupply and Pooling in Multi-Echelon Systems

The tactical planning models presented in previous chapters were constructed based on the assumption that each location in the multi-echelon resupply network had a sole source of resupply. Furthermore, the replenishment lead times were not sensitive to the amount of on-hand stock, in-transit stock, or backorders existing at a receiving location. This sole source assumption has a very significant effect on the total amount of inventory required to meet customer service objectives or to contain expected backorder costs. However, in many, if not in most of the existing real world multi-echelon resupply systems, the sole source assumption is violated.

There are numerous examples of sharing inventories among locations in industrial, retail and military settings. For example, automotive dealers share parts on a daily basis so that they can complete the repairs of their customer's vehicles quickly. The parts distribution centers operated by each of the car companies routinely provide stocks from one center to satisfy demands placed on another center when the latter center is out of stock. In another setting, service technicians for computer and photocopier equipment share inventories in emergency situations. Lateral resupply of parts among military bases within a geographic region occurs on a regular basis as well.

The improvements in information technology coupled with the substantial reduction in the cost of processing, storing, and analyzing data have made sharing of inventories more attractive. Furthermore, logistics companies, such as UPS and Federal Express, have made the rapid movement of parts from one place to another possible and more affordable.

While we find firms commonly engaging in lateral resupply activities, the underlying question that must be addressed is: what is the impact of lateral resupply on inventory levels and operations, and is it worth the inherent cost?

A number of authors have addressed this issue. First, there have been simulation studies that have demonstrated the effects of lateral resupply in multi-echelon systems [81, 82, 198, 219]. While the environments that were examined by these authors did differ, their results showed that in a wide variety of circumstances, lateral resupply among locations is a very effective way to improve customer service and to lower inventory investments. In the case of repairable items, Pyke [198], Sherbrooke [219], Scudder and Hausman [215], Scudder [214], and others have also shown that priority scheduling of repairs and priority allocation of stock to bases can improve system performance, as would be expected. For example, in some situations the effect of using rules other than the first-come, first-serve rule for allocating stock to bases proved to be highly beneficial.

Second, authors have presented and tested many analytic models that explicitly consider the possibility of supplying locations through lateral shipments. For example, see Archibald et al. [12]; Lee [156]; Lee and Billington [157]; Alfredsson and Verrijdt [6]; Axsäter [17]; Cohen, Kleindorfer, Lee [57]; Dada [66]; Das [69]; Gross [101]; Tagaras [243, 244]; Tagaras and Cohen [245]; Bowman [29]; Hoadley and Heyman [130]; Sherbrooke [219]; Yanagi and Sasaki [254]. These models differ in many ways. Some are stationary, continuous-time models while others are periodic review, both infinite and finite horizon. Essentially, though, all these analytic models are tactical planning models. They are either economic models that suggest what quantities of material to buy or they are models used to determine the probabilities of various events occurring. We say that these models are tactical models because, for the most part, they do not consider the possibility of using all state-of-the-world information when representing the operational environment. For example, first-come first-serve inventory allocation rules are often assumed as the basis for shipping parts from a central location (depot) to field stocking locations (bases). Real-time execution systems would take more information into account. The Hoadley and Heyman model is a single period planning model that does take many operational details into account. However, lead times are assumed to be zero. Such real-time models are discussed in Chapter 10.

Rather than describing all the approaches for incorporating lateral resupply into models, we will focus on just a few in this chapter. We begin by developing two stationary, continuous-time models. The first is a queuing like model based on the assumption that the underlying system is governed by a continuous-time Markov process. The second model is an extension of the ones developed earlier in Chapter 5. Both are approximations. In the first model we focus on computing probabilities of system performance resulting from given stock levels. In the second model, we also construct probability distributions for key random variables; but, we also construct an economic model that can be used to set stock levels in a two-echelon system. The models pertain to repairable items.

We ultimately develop a periodic review model that can be used to establish stock levels for repairable items in a three echelon environment. But to begin the section on periodic review models, we will first establish why a multi-echelon operating environment may be desirable. To do this, we construct a particular environment that demonstrates analytically why an intermediate stocking location could be cost effective to create. While the environment we will examine is simple in nature, it does suggest why it may be advantageous to create a multi-echelon inventory system.

## 7.1 Continuous Time Models for Lateral Resupply

As stated, we will discuss two continuous-time models. The system analyzed in both cases contains two-echelons, and, as in earlier chapters, we call the upper echelon location the depot and the lower echelon locations the bases. We assume there are $n$ bases in the system and that the items can be managed independently. Thus we focus on a single part type. Demands for this part type arise only at the bases and occur according to a Poisson process with rate $\lambda_j$ at base $j$, $j = 1, \ldots, n$.

Each demand at a base immediately triggers a replenishment order for a part from the depot. Each demand occurs because a part has failed. All failed parts are sent to the depot for repair.

Up to this point, our discussion is the same as given in earlier chapters. Where things differ is the manner in which bases interact. We assume, as shown in Figure 7.1, that the $n$ bases are divided into pools. In practice, a set of bases that are in the same geographic region might form a pool. Suppose there are $P$ such pools. We assume that a base is a member of exactly one such pool. Hence the collection of $P$ pools partitions the bases into mutually exclusive and exhaustive sets, as implied in Figure 7.1.

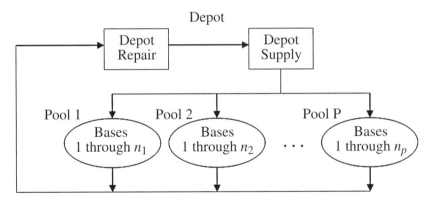

**Fig. 7.1.** A Depot-Base Two-Echelon System with Pools

Furthermore, inventory within each pool can be shared among the bases in the pool as follows. If a demand arises at a base in a pool and that base has no stock on hand, then the demand can be satisfied from stock on-hand at some other base in the pool. Thus when stock is required at a base and none is available there, a lateral resupply event will take place assuming, of course, that some other base in the pool has stock on-hand. Obviously, there are many mechanisms for choosing the base that should laterally resupply the base in need. In practice, it might be the nearest base with stock on-hand or it might be the base that has the maximum number of days of stock on-hand. In the models that we will discuss in this section, we assume that the base providing the lateral resupply is randomly chosen from the set of bases with a positive amount of stock on-hand. In the case where a demand arises at a base and no stock is on-hand at any base in the pool, we assume the depot is used to satisfy the demand.

We are now ready to present the details of each model. The first is based on Axsäter [17] and the second on Lee [156].

### 7.1.1  Model 1

Suppose we are given stock levels for the depot, $s_0$, and for the bases, $s_j$. Our goal in this section is to show how to estimate the fraction of demand arising at a base that is satisfied from the stock on-hand at that base, the fraction met from a lateral resupply event, and finally, the fraction backordered, that is, awaiting arrival of inventory from the depot.

Let us define notation used in this section. We let

$P$ = number of pools
$n_i$ = number of bases in pool $i$,
$\lambda_j$ = demand rate at base $j$,
$\lambda_0 = \sum \lambda_j$, the depot demand rate,
$A$ = average order and ship time from the depot to a base,
$D$ = average depot repair cycle time (the average transportation time to the depot from a base plus the average depot repair time), and
$\mathcal{B}(s_0)$ = average number of outstanding depot backorders given the depot stock level is $s_0$.

Other definitions will be presented as we proceed.

Since we are assuming that demands at the bases are independent Poisson processes, the depot demand process is also a Poisson process. Correspondingly, the units entering the depot's repair process are also governed by the same Poisson process. Let us assume that the repair cycle times are independent and identically distributed. Since we are following an (s–1,s) replenishment policy, the number of units in repair at a random point in time has a Poisson distribution with mean $\lambda_0 D$, from Palm's theorem.

In Chapter 5 we developed the probability distribution for the number of units in depot resupply for a base. This same idea holds for a pool of bases. Thus we can compute the mean resupply time for a base as

$$T = A + \mathcal{B}(s_0)/\lambda_0.$$

Let us now focus on a single pool. Let $\overline{N}$ be a random variable describing the number of units in depot resupply for this pool. Then the $E[\overline{N}] = (\sum_{j \in \overline{P}} \lambda_j) \cdot T$, where $\overline{P}$ is the set of bases in the pool. The variance of $\overline{N}$ can be found using the procedure developed in Section 5.1.2.1. Furthermore, we may approximate the probability distribution for $\overline{N}$ by a negative binomial distribution having these means and variances as shown in Section 5.1.2.1.

Let $\bar{s} = (\sum_{j \in \bar{P}} s_j)$ be the total stock for the pool of interest. Then $P[\bar{N} = \bar{s} - k]$ also measures the probability that the total on-hand stock in that pool is $k$ units at a random point in time, $k > 0$.

In problems of interest, $T$ is usually many days in length whereas the lateral resupply time among bases in a pool may be measured in hours or perhaps a day. Because $T$ is normally at least an order of magnitude greater than the lateral resupply time, we will assume that this lateral resupply time is of length zero. The implication of this assumption is that there can never be a base with a backorder when there is another base in the pool with positive stock. Suppose a demand arises at a base in the pool and there is no stock on-hand at any base in the pool. Then a backorder occurs at that base. Subsequently, a unit arrives from the depot to a base at which there are no backorders. In this case, that newly arriving unit would be laterally resupplied immediately to fill the outstanding backorder. Hence we assume that whenever pool net inventory is nonnegative there are no backorders at any base in the pool.

Next, we want to compute

$\beta_j$, the probability that a demand at base $j$ is satisfied from stock on-hand at that base,

$\alpha_j$, the probability that the demand at base $j$ is satisfied by a lateral resupply action from another base in the same pool, and

$\Theta_j$, the probability that a demand at base $j$ in the pool is backordered.

Since a demand must be satisfied either from its on-hand stock or lateral resupply or must be backordered, $\alpha_j + \beta_j + \Theta_j = 1$. Obviously if $s_j = 0$, then $\beta_j = 0$.

To illustrate our modelling concepts, we assume all the bases in the pool are identical. Hence we assume that all bases have the same demand rate, $\lambda_j$, and stock levels, and that $\Theta_j, \alpha_j$ and $\beta_j$ are the same for all bases, and, consequently, we drop the base subscript. Additionally, we assume that the depot to base resupply time for all bases in the group are independent and identically exponentially distributed random variables with mean $T$.

Recall that we assume that each base has a probability $\beta$ that a demand will be met from on-hand inventory at that base. $\beta$ also measures the fraction of time that the base stock is positive. Hence the fraction of time the base stock is negative or zero is $1 - \beta = \alpha + \Theta$.

Suppose the base stock is positive. During these periods of time, the base satisfies its local demands, which arrive at a rate of $\lambda$. But it also may receive lateral resupply requests from other bases in the pool. Since all bases in the pool are identical and lateral resupply requests are met in a random manner, $\alpha\lambda$ is the expected long term rate at which lateral resupply requests are satisfied by each base in the pool. But lateral resupply requests are satisfied only when there is positive stock on hand at a base. Clearly no lateral resupply occurs from a base when its on-hand stock is zero. For the average lateral resupply rate to be $\alpha\lambda$, the average lateral resupply rate must be $\alpha\lambda/\beta$ when there is stock on hand. When there is no stock on hand at the base, the demand rate is $\lambda$. But a portion of that demand is satisfied from lateral resupply. Since the bases are identical, the long term average rate of material that this base receives via lateral resupply is $\alpha\lambda$. Thus during the period of time when the base is out of stock, the incoming lateral resupply rate must be $\alpha\lambda/(1-\beta)$.

Axsäter makes the assumption that the demand and resupply processes at each base are continuous-time Markov processes. Based on this assumption, we can represent the environment as a queuing system, where the arrival rate is a function of whether or not there is on hand stock at the base. When the inventory on hand is positive, let the demand rate be given by

$$g = \lambda + \alpha\lambda/\beta = \lambda(1+\alpha/\beta); \tag{7.1}$$

when the on-hand inventory is zero, the demand rate is

$$\begin{aligned} h &= \lambda - \alpha\lambda/(1-\beta) = \lambda(1-\alpha/(1-\beta)) \\ &= \lambda(1-\beta-\alpha)/(1-\beta) \\ &= \lambda\Theta/(1-\beta). \end{aligned} \tag{7.2}$$

It is clearly possible to describe the system more accurately by representing it using a state space that captures the inventory levels at each location and the transitions that can occur from state to state. This more exact representation is far more complex. We note that although (7.1) and (7.2) are correct on average, they are approximations of the system's operation at any point in time. We leave the more exact development for a special case as an exercise.

We are now ready to construct the simple queuing model for a base. The following graph displayed in Figure 7.2 describes the transition in the system. Let $\pi_k$ represent the stationary probability that net inventory for this queuing system is equal to $k$.

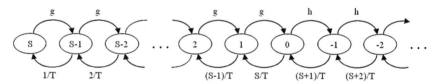

**Fig. 7.2.** State Transition Diagram

Then we can easily see that

$$\pi_s \cdot g = \pi_{s-1} \cdot \frac{1}{T},$$

$$\pi_{s-k} \cdot (g + \frac{k}{T}) = \pi_{s-k+1} \cdot g + \pi_{s-k-1} \cdot \frac{k+1}{T}, \quad \text{for } k = 1, \ldots, s-1,$$

$$\pi_0 \cdot (h + \frac{s}{T}) = \pi_1 \cdot g + \pi_{-1} \cdot \frac{s+1}{T}, \quad \text{and}$$

$$\pi_{s-k}(h + \frac{k}{T}) = \pi_{s-k+1} \cdot h + \pi_{s-k-1} \cdot \frac{(k+1)}{T}, \quad \text{for } k > s.$$

The solution to this set of equations is

$$\pi_{s-k} = \pi_0 \frac{s!}{k!} \frac{1}{(gT)^{s-k}}, \quad k = 0, 1, \ldots, s-1, \tag{7.3}$$

and

$$\pi_{s-k} = \pi_0 \frac{s!}{k!} (hT)^{k-s}, \quad k \geq s. \tag{7.4}$$

Since $\sum \pi_k = 1$,

$$\pi_0 = \frac{1}{\sum_{k=0}^{s-1} \frac{s!}{k!} \cdot \frac{1}{(gT)^{s-k}} + \sum_{k=s}^{\infty} \frac{s!}{k!} \cdot (hT)^{k-s}}, \tag{7.5}$$

we have the means to compute the stationary distribution of the base level probabilities. Note, however, that both $g$ and $h$ depend on $\Theta$ and $\beta$.

Recall that $\Theta$ measures the probability that an arriving demand can not be satisfied from the pool's collective inventory. But this can happen only if all the pool's stock is on order from the depot. That is, the probability that an arriving demand will be backordered is the probability that $\bar{s}$ or more units are on order from the depot by the bases in the pool. Thus

$$\Theta = P[\overline{N} \geq \bar{s}].$$

We also know that

$$\beta = \sum_{k=1}^{s} \pi_k \tag{7.6}$$

represents an approximation to the probability that a base can satisfy its demand from on-hand stock. Unfortunately, to solve for $\beta$ we must have the values of the $\pi_k$, which, in turn, can only be obtained knowing $\beta$. Let us now see how to find $\beta$ approximately.

We know the value of $\Theta$ and know that $g$ increases as $\beta$ decreases (and vice versa) and that $h$ increases as $\beta$ increases (and vice versa). For systems in which we desire to satisfy most of the demand from on-hand base stock, the values of $s$ will be set so that $\beta$ is large compared with $\Theta$. $\beta > .7$ would be likely with $\Theta < .1$. Since $\beta$ increases, the value of $g$ decreases and $h$ increases, and therefore the overall effect would be to increase the value of $\sum_{k=1}^{s} \pi_k$, or $\beta$. Thus, if we start with an initial value of $\beta$ and compute $g$ and $h$, we can determine a set of values for the $\pi_k$. Using these values, we can use (7.6) to determine a new value of $\beta$. We will then use this value to recompute $g$ and $h$. This process results in a bounded monotone sequence of values for $\beta$, which means that the sequence will converge. Axsäter [17] reports that this process converges in relatively few iterations and is therefore easy and effective to implement.

### 7.1.2  Model 2

Lee [156] proposed another way to represent the problem analyzed in the previous section. His method differs substantially from the previous one even though most of the assumptions underlying the development of the two models are the same. This alternative modeling approach is very similar to the one used in earlier chapters.

Let us begin by summarizing the main assumptions. We assume the system's design and operation are the same as the ones discussed in the development of Model 1. We continue to assume that there are $P$ pooling groups and that the bases in the same pool are identical. That is, the stock levels are the same for all bases within a pool and the demand processes are Poisson processes with a common rate $\lambda_i$. Since there are $n_i$ bases in pool $i$, the aggregate demand process for pool $i$ is a Poisson process with rate $n_i\lambda_i$. Furthermore, the aggregate depot demand process for serviceable parts, and the corresponding arrival process into the depot repair center, are Poisson processes with rate $\lambda_0 = \sum_i n_i\lambda_i$ since we are employing (s–1,s) policies at each base. The depot to base order and ship time is $A$. However, we now assume that $A$ is a constant. We again assume first-come, first-serve policies are used to satisfy demands at both the depot and the bases. Furthermore, we assume that the average depot repair cycle time is $D$, and these repair cycle times are independent and identically distributed. Thus we continue to assume that the depot has an infinite number of servers. (This assumption permits us to invoke Palm's theorem.)

Finally, we assume that the lateral resupply time is zero for all bases within a pool. While this is not true in reality, as we observed earlier, these times are normally very short. Furthermore, since demand rates for most parts are quite low, the likelihood of a demand occurring over a short lateral resupply time is also very low. Hence we continue to make this assumption.

The remainder of our discussion pertaining to this model is divided into two parts. First, we will construct probability distributions of key random variables that are used to calculate $\alpha$, $\beta$, and $\Theta$, as defined previously. Second, using these probability distributions, we will construct an optimization model that can be used to determine the stock levels for both the bases and the depot. We also will discuss an approach for finding these values.

#### 7.1.2.1  Determining $\alpha$, $\beta$ and $\Theta$

Our immediate goal is to establish a method for calculating (1) $\beta = \beta_i$, the fraction of demands arriving at a base in pool $i$ that are satisfied from on-hand base stock, (2) $\alpha = \alpha_i$, the proportion of demands arising at a base in pool $i$ that are satisfied by lateral resupply from another base in the pool, and (3) $\Theta = \Theta_i$, the proportion of demands that are backordered in pool $i$.

We begin our analysis by determining the steady state distribution of the number of units on-order from the depot by a base in a pool at a random point in time. Let $N_j$ represent the random variable for this quantity at base $j$. Furthermore, let

$\overline{N}$ and $N_D$ represent the random variables for the number of units in resupply for the pool we are examining and the depot, respectively.

We now employ the logic used in Chapter 5 to find the distribution of $\overline{N}$. Suppose that $N_D = n_D$. Since the arrival process to the depot is the sum of the independent base Poisson demand processes, the probability that any of the $n_D$ units is attributable to pool $i$ is $\frac{n_i \lambda_i}{\lambda_0}$. Thus

$$P[\overline{N} = k | N_D = n_D] = \binom{n_D}{k} \left( \frac{n_i \lambda_i}{\lambda_0} \right)^k \left( 1 - \frac{n_i \lambda_i}{\lambda_0} \right)^{n_D - k}$$

and hence

$$P[\overline{N} = k] = \sum_{n_D = k}^{\infty} \binom{n_D}{k} \left( \frac{n_i \lambda_i}{\lambda_0} \right)^k \left( 1 - \frac{n_i \lambda_i}{\lambda_0} \right)^{n_D - k} \cdot P[N_D = n_D]. \qquad (7.7)$$

By invoking Palm's theorem, we know that

$$P[N_D = n_D] = e^{-\lambda_0 D} \frac{(\lambda_0 D)^{n_D}}{n_D!}.$$

Therefore we have established the distribution for $\overline{N}$.

Next let us consider a specific base within the pool, say base $j$. Applying the same argument, we may approximate the steady state probability that base $j$ has $k$ units in a backorder status at the depot as

$$P[N_j = k] = \sum_{\overline{n} = k}^{\infty} \binom{\overline{n}}{k} \left( \frac{\lambda_i}{n_i \lambda_i} \right)^k \left( 1 - \frac{\lambda_i}{n_i \lambda_i} \right)^{\overline{n} - k} P[\overline{N} = \overline{n}]$$

$$= \sum_{\overline{n} = k}^{\infty} \binom{\overline{n}}{k} \left( \frac{1}{n_i} \right)^k \left( 1 - \frac{1}{n_i} \right)^{\overline{n} - k} P[\overline{N} = \overline{n}]. \qquad (7.8)$$

This expression is based on the assumption that it is equally likely that a backordered demand comes from any base in the pool. This may not be the case since the demands at bases, including lateral resupply shipments are clearly not independent and hence the number of units a base has on-order depends on the demands at it and the other bases in its pool, and the lateral resupply priority rules that are employed. We leave an analysis of the accuracy of this approximation to the reader as an exercise. Observe that we obtain the same expression for $P[N_j = k]$ when there is no pooling of inventory.

We are now ready to estimate the value of $\beta$ for each base in the pool. Given the approximating values for $P[N_j = k]$ and the values of $P[\overline{N} = \overline{n}]$, as computed in (7.7) and (7.8), respectively, we see that $\beta$ is approximated by

$$\beta = \sum_{k=0}^{s_j - 1} P[N_j = k] = \sum_{k=0}^{s_j - 1} \sum_{\overline{n} = k}^{\infty} \binom{\overline{n}}{k} \left( \frac{1}{n_i} \right)^k \left( 1 - \frac{1}{n_i} \right)^{\overline{n} - k} P[\overline{N} = \overline{n}]$$

since $s_j = s_i$ for all bases in pool $i$.

As we did in earlier chapters, we approximate the distributions of the probability distributions of the random variables $\overline{N}$ and $N_j$ with negative binomial distributions for computational reasons. This requires computing both the mean and variance of these random variables and using these values to estimate the parameter values of the corresponding negative binomial distributions.

The probability that an arriving demand to a pool can not be satisfied from pool stock is the probability that the aggregate number of units on order for the pool is greater than or equal to the aggregate pool stock. But, this probability is $P[\overline{N} \geq n_i s_i]$. Therefore $\Theta = P[\overline{N} \geq n_i s_i]$, since we assume that lateral resupply between bases within a pool is instantaneous and occurs whenever needed.

Since we have shown how $\Theta$ and $\beta$ can be estimated, we are also able to estimate $\alpha$ for a particular pool, that is, $\alpha = 1 - (\beta + \Theta)$. Thus the expected number of lateral resupply shipments corresponding to a base per unit time is $\lambda_i \alpha$ and $n_i \lambda_i \alpha$ for the entire pool.

Lee [156] conducted experiments showing that the approximations are accurate when service levels are high. Axsäter [17] developed the alternative model, Model 1, for estimating $\alpha, \beta, \Theta$. The two models fundamentally differ in only a couple of ways. The most important difference is as follows. In Lee's model the base demand rate is implicitly independent of whether or not there is stock on hand at the base. Recalling the definitions of $g$ and $h$ in Section 7.1.1 of this chapter, Lee's model, that is Model 2, is based on the assumption that $\lambda_i = g = h$. The other difference between the models arises because Axsäter [17] represents the entire operating environment as a continuous-time Markov process. In certain situations, Model 1 gives better estimates of $\alpha$ and $\beta$. See Axsäter [17] for a detailed discussion of the numerical tests and comparisons.

Unfortunately, while more accurate, Model 1 is not easily employed to find optimal stock levels. In Lee's model, however, a negative binomial distribution is used to approximate the required probability distributions and thus a computationally tractable method exists for finding the stock levels, as we now show.

### 7.1.2.2   Finding the Optimal Stock Levels

There are many different optimization models that may be stated for finding the values of $s_0$ and the $s_i$. Clearly both backorder and lateral resupply costs must be considered. Holding costs might be included as well. (They would be required to extend the model to consider multiple items.) If they are included in the single item case, then fill rate constraints are often included in the model's statement. If they are not included, then an investment budget constraint is present in the model's formulation.

We now state one possible economic model for finding the values of $s_0$ and the base stock levels for each base in each pool, $s_i$, for pool $i$. Let

$\bar{s}_i = n_i s_i,$

$b$ = backorder cost rate incurred per unit time per unit backordered at a base,

$a_i$ = cost of lateral resupply shipments within pool $i$ (per unit shipped),

$B$ = investment limitation,

$c$ = unit cost,

$s_i^\ell$ = lower bound on the stock level for bases in pool $i$,

$s_0^\ell$ = lower bound on the depot stock level,

$\beta_i$ = probability that a base in pool $i$ satisfies its demands from the base's stock on hand,

$\Theta_i$ = probability that a base in pool $i$ backorders its demands.

$A_i$ = depot to pool $i$ transportation time.

$D$ = depot resupply time.

The objective in this model is to find stock levels that minimize the expected costs per unit time subject to an investment constraint, while recognizing minimum stocking constraints at the bases and the depot. The corresponding model is

$$\min C = b \sum_{i=1}^{P} \sum_{k > \bar{s}_i} (k - \bar{s}_i) P[\overline{N}_i = k]$$

$$+ \sum_{i=1}^{P} a_i n_i \lambda_i (1 - \Theta_i - \beta_i)$$

subject to

$$
\begin{aligned}
s_i &\geq s_i^\ell & (\text{or } \bar{s}_i \geq n_i s_i^\ell), \\
s_0 &\geq s_0^\ell, \\
c s_0 + \textstyle\sum_i c \bar{s}_i &\leq B \ (\text{or } s_0 + \textstyle\sum \bar{s}_i \leq \lfloor B/c \rfloor),
\end{aligned}
$$

and $s_0$, $s_i$ are nonnegative integers, where $\overline{N}_i$ is the number of units in pool $i$ that are on order (in resupply) at a random point in time.

We assume the lower bounds on the stock levels are set so that $\bar{s}_i \geq \lfloor n_i \lambda_i A_i + \frac{n_i \lambda_i}{\lambda_0} B_D(s_0^\ell) \rfloor$ and $s_0 \geq \lfloor \lambda_0 D \rfloor$. Obviously, $\lfloor B/c \rfloor$ must be at least as large as $\sum_i n_i s_i^\ell + s_0^\ell$ for a feasible solution to exist.

This problem can be solved using a marginal analysis algorithm similar to the one described in earlier chapters. We now outline such an algorithm.

For each depot stock level $s_0 \in [s_0^\ell, \lfloor B/c \rfloor - \sum_i s_i^\ell n_i]$ and integer, we can solve

$$C(s_0) = \min \sum_i C_i(s_i | s_0)$$

subject to

$$\sum_{i=1}^{P} \bar{s}_i \leq \lfloor B/c \rfloor - s_0,$$

$$s_i \geq s_i^\ell \geq 0 \text{ and integer,}$$

where

$$C_i(s_i|s_0) = b \sum_{k > \bar{s}_i} (k - \bar{s}_i) P[\overline{N}_i = k|s_0]$$

$$+ a_i n_i \lambda_i (1 - \Theta_i - \beta_i).$$

Recall that we approximated $\beta_i$ as $\sum_{k=0}^{s_i-1} P[N_j = k]$, and that we further approximated $P[N_j = k]$ by assuming this distribution for $N_j$ has a negative binomial distribution. Similarly, we assumed that $P[\overline{N} = \bar{n}]$ could be approximated by a negative binomial distribution. Thus $\Theta_i$ can be approximated, too.

The reason for stipulating the specific values for lower bounds on $s_i$ and $\bar{s}_i$ was to ensure that the functions

$$C_i(s_i|s_0)$$

are convex. We leave the convexity proof as an exercise.

Since the objective function $C(s_0)$ is the sum of convex functions, it is convex in the values of the $s_i$. Thus given $s_0$, we can find the best allocation among the pools using a straight-forward marginal analysis approach. That is, at each step, add the incremental stock to the base and pool that reduces the cost by the greatest amount. The process begins with $s_i = s_i^\ell$ for all bases. Continue until all the stock is allocated at which time we have also computed $C(s_0)$. Repeat this process for all values of $s_0$. The optimal solution is the $s_0^* = \arg\min_{s_0} C(s_0)$ and the corresponding values $s_i(s_0^*)$. Ties can be broken arbitrarily throughout the implementation of this algorithm.

## 7.2  Risk Pooling

As we have stated, the design of logistics resupply systems for service parts has a substantial impact on operating costs, investment costs in facilities and inventories, and, of course on customer service. Conventional wisdom suggests that as the number of echelons increases, operating and investment costs increase, and, in particular, investment in inventory grows. Nonetheless, in practice we see that many service parts resupply systems often have many echelons. There are obvious reasons for having two echelon systems. But real systems often contain more than two echelons. Is there a reason, from an inventory investment perspective, to have more than two echelons? Conventional wisdom says that there will be more inventory in systems with more echelons. Is this conventional wisdom correct?

We have two purposes in this section. First, we will address these questions directly. As we will see, there may be reasons for constructing systems containing several echelons so that system inventory levels are actually lowered. To demonstrate this fact, we will analyze a system consisting of a supplier, depots, and warehouses. In this system material flows from the supplier's warehouse to a set of depots and then to the warehouses that are supplied by the particular depot. Thus this resupply system is an arborescence. Demand occurs only at the warehouse echelon in this system. The environment is one in which these demands occur in each period of an infinite horizon. The analysis we present is based on that given by Eppen and Schrage [78].

Our second purpose is to analyze a three echelon system consisting of a depot, several intermediate stocking locations, and a set of bases that are resupplied by the intermediate stocking locations. This system is again an arborescence. However, the items in this system are repairable. Demands, which occur at the bases, correspond to failures and require repairs. Demands also correspond to the immediate need for serviceable stock. This environment is a three echelon extension of the continuous review environment we analyzed in detail in Chapter 5. A model similar to the one we will present was used by the US Air Force to determine the effect of consolidating repairs while stocking inventories in three echelons (see Muckstadt [178]).

### 7.2.1  A Periodic Review Pooling Environment

The fundamental question we address in this section is whether or not we will increase system inventories by adding echelons into a system. To address this question, we will study a system in which orders are placed on the external supplier in each period of an infinite horizon by every depot. These depot orders are placed in response to demands that have arisen in the previous period at the warehouses that are supplied by that depot. We assume that after $D$ periods, which is the supplier to depot lead time, the entire quantity that was ordered by each depot is delivered to it by the supplier. By making this assumption, every depot subsystem is, in effect, independent of every other depot subsystem. Hence, we will concentrate our analysis on a single depot and warehouse subsystem.

The depot in this subsystem resupplies $m$ warehouses. We assume that items are also managed independently so that we may analyze them one item at a time. We will further assume that the demand for an item at warehouse $j$ in period $t$ is denoted by the random variable $d_{jt}$. We assume the random variables $d_{jt}$ are independent and identically distributed and normally distributed random variables in each period of the infinite planning horizon. Furthermore, demands are assumed to be independent across warehouses. We let the parameters of the per period distributions be $\mu_j = E[d_{jt}]$ and $\sigma_j^2 = \text{Var}\,[d_{jt}]$ at warehouse $j$.

We assume that each warehouse places an order on the depot in every period. The transportation lead time from the depot to each warehouse is denoted by $A$, which is measured in periods.

The design of real world resupply networks of the type we have described would exist when $D$ is much greater than $A$. That is, the procurement lead time is much greater than the distribution lead time, which is certainly the case for most service parts. We assume $D \gg A$, and that $A$ is not trivial in length. $A$ may be several days or a week where $D$ may be many weeks or months.

We assume that the holding cost for a unit is the same at the warehouses and the depot. The holding cost at either location type is denoted by $h$ dollars per unit per period. If, at the end of a period, the net inventory at a warehouse is negative, backorders will exist. We let $b$ represent the backorder cost per unit per period at each warehouse.

Let $I_{jt}$ be the inventory position for warehouse $j$ in time period $t$ just after the depot to warehouse allocations have been made.

### 7.2.1.1 Imbalance Assumption

We make one additional important assumption concerning the system's dynamic operation, which we call the Imbalance Assumption. By this we imply that in each period there is enough inventory at the depot so that following allocation of this inventory to the warehouses, all warehouses will have an equal fractional value of its distribution of demand over the $A$ period lead time. Since we have assumed that the demands are independent and identically distributed normal random variables and that $b$ and $h$ are the backorder and holding costs, respectively, for all locations, the imbalance assumption implies that $\Phi\left(\frac{I_{jt} - A\mu_j}{\sqrt{A}\sigma_j}\right)$ is the same for all $j$ and $t$, where $\Phi(\cdot)$ is the standard normal distribution function. Thus we assume that the probability of a stockout occurring in period $t + A$ is the same at all warehouses.

Based on the imbalance assumption and the equal holding cost assumption at all locations, we can easily prove that all units available at the depot at the beginning of a period should be sent to the warehouses. That is, it is always best to hold no stock at the depot in this system.

Let us now establish how likely it is to achieve a balanced allocation. Remember that in each period the depot orders the total demand incurred in the previous period at the warehouses. This quantity arrives $D$ periods later, at which time we have assumed that the amount on hand at the depot is sufficient to ensure an equal fractile allocation. Suppose we placed an order in period $t$. The order would be of size $\sum_{j=1}^{m} d_{j,t-1}$. Further assume that the system was in balance at the beginning of period $t + D - 1$. Demands then arise in period $t + D - 1$ at each warehouse. To remain in balance, the amount received at the depot at the beginning of period $t + D$, namely $\sum_{j=1}^{m} d_{j,t-1}$, must be large enough to compensate for the demands arising at the warehouses during period $t + D - 1$. Let us now prove the following result.

**Lemma 3.** *Suppose the system is in balance at the beginning of period $t + D - 1$. Then it will be in balance following the depot allocation to the warehouses in period $t + D$ if*

$$\sum_{j=1}^{m} d_{j,t-1} \geq \max_{i=1,\ldots,m} \left\{ \sum_{j \neq i} d_{j,t+D-1} + d_{i,t+D-1}\left(1 - \frac{\sum \sigma_j}{\sigma_i}\right) \right\}.$$

*Proof.* Since the system is in balance at the beginning of period $t + D - 1$, and demand is normally distributed at each warehouse, there exists a value $k$ such that

$$I_{j,t+D-1} = A\mu_j + k\sqrt{A}\sigma_j, \quad j = 1, \ldots, m.$$

Then demands occur at each warehouse in period $t+D-1$. Next, the order placed by the depot in period $t$, for $\sum_{j=1}^{m} d_{j,t-1}$ units, arrives at the depot and is available to be allocated to the warehouses. Since the depot does not keep stock on hand, all $\sum_{j=1}^{m} d_{j,t-1}$ units are allocated to the warehouses. Furthermore, these are the only units that are available for distribution.

Suppose $x_j$ units are allocated to warehouse $j$ in period $t + D$. Then the inventory position for warehouse $j$ is

$$I_{j,t+D} = I_{j,t+D-1} + x_j - d_{j,t+D-1}$$
$$= A\mu_j + k\sqrt{A}\sigma_j + x_j - d_{j,t+D-1}.$$

Hence the system will remain in balance following the allocation only if there exist $x_j \geq 0$ such that

$$\sum_j x_j = \sum_j d_{j,t-1}$$

and there exists $k'$ for which

$$I_{j,t+D} = A\mu_j + k'\sqrt{A}\sigma_j.$$

Since

$$I_{j,t+D} = A\mu_j + k'\sqrt{A}\sigma_j$$
$$= A\mu_j + k\sqrt{A}\sigma_j + x_j - d_{j,t+D-1},$$

we have

$$x_j = (k' - k)\sqrt{A}\sigma_j + d_{j,t+D-1},$$

for $j = 1, \ldots, m$. Furthermore, we know that

$$\sum_j x_j = \sum_j d_{j,t-1} = \sum_j \left\{ (k' - k)\sqrt{A}\sigma_j + d_{j,t+D-1} \right\}$$
$$= (k' - k)\sqrt{A}\sum_j \sigma_j + \sum_j d_{j,t+D-1}.$$

Hence

$$k' - k = \frac{\left( \sum_j d_{j,t-1} - \sum_j d_{j,t+D-1} \right)}{\sqrt{A}\sum_j \sigma_j}$$

and

$$x_j = \frac{\left(\sum_i d_{i,t-1} - \sum_i d_{i,t+D-1}\right)}{\sqrt{A}\sum_i \sigma_i}\sqrt{A}\sigma_j + d_{j,t+D-1}$$

$$= \frac{\left(\sum_i d_{i,t-1} - \sum_i d_{i,t+D-1}\right)}{\sum_i \sigma_i}\sigma_j + d_{j,t+D-1}.$$

For $x_j$ to be nonnegative, we see that the right hand side must be nonnegative as well. Thus

$$\sum_i d_{i,t-1} \geq \left(\sum_i d_{i,t+D-1}\right) - d_{j,t+D-1} \cdot \left\{\frac{\sum_i \sigma_i}{\sigma_j}\right\}.$$

Therefore $x_j \geq 0$ for all $j$ if

$$\sum_i d_{i,t-1} \geq \max_j \left\{\sum_{i\neq j} d_{i,t+D-1} + d_{j,t+D-1}\left\{1 - \frac{\sum_i \sigma_i}{\sigma_j}\right\}\right\}$$

or

$$\sum_i d_{i,t-1} \geq \sum_i d_{i,t+D-1} + \max_j \left[-d_{j,t+D-1}\left\{\frac{\sum_i \sigma_i}{\sigma_j}\right\}\right].$$

This concludes the proof.                                                        □

This lemma provides us with a means for estimating the probability that the imbalance assumption is satisfied. Note that the lemma is based on the hypothesis that the system was not in a state of imbalance when a period began.

Let us now compute a bound on the following probabilities

$$P\left[\left\{\sum_i d_{i,t-1} - \sum_{i\neq j} d_{i,t+D-1} - d_{j,t+D-1}\left[1 - \frac{\sum_i \sigma_i}{\sigma_j}\right]\right\} \geq 0\right]$$

for all warehouses $j$. Note that the random variable

$$X_j \equiv \sum_i d_{i,t-1} - \sum_{i\neq j} d_{i,t+D-1} - d_{j,t+D-1}\left[1 - \frac{\sum_i \sigma_i}{\sigma_j}\right]$$

is the sum of independent, normally distributed random variables and therefore also has a normal distribution with expectation $\mu_j \frac{\sum_i \sigma_i}{\sigma_j}$ and variance $2\sum_{i\neq j}\sigma_i^2 + \sigma_j^2\left(1 + \left(1 - \frac{\sum_i \sigma_i}{\sigma_j}\right)^2\right)$. From Bonferroni's inequality,

$$P\left\{X_j \geq 0(\text{for all } j)\right\} \geq 1 - \sum_{j=1}^m P\left\{X_j < 0\right\}.$$

Suppose we consider the case where the warehouse demand distributions are identical. Thus $\mu_j = \mu$ and $\sigma_j = \sigma$. Then the $E[X_j] = m\mu$ and $\text{Var}[X_j] = m^2\sigma^2$, for all $j$. Thus

$$P\{X_j < 0\} = P\left\{\frac{X_j - \mu}{\sigma} \leq -\frac{\mu}{\sigma}\right\} = \Phi\left(-\frac{\mu}{\sigma}\right),$$

where $\Phi(\cdot)$ is the distribution function of a standard normal distribution. Suppose $\mu = 100$ and $\sigma = 10$, yielding a coefficient of variation of .1. Also suppose that $m = 10$. Note this would correspond to the situation where we are approximating a demand process that is a Poisson process with a mean of 100 units per period for each of the 10 warehouses. In this case,

$$1 - \sum_{j=1}^{m} P\{X_j < 0\} = 1 - 10 \cdot \Phi(-10) \approx 1.$$

As $\frac{\mu}{\sigma}$ decreases the quality of the approximation will decrease as well. Thus as long as $\mu/\sigma$ is large the imbalance assumption is quite reasonable to make.

### 7.2.1.2  System Analysis

In this section we establish methods for computing the desired system stock level, $s$, and determining how stock arriving at the depot should be allocated among the various warehouses. Let us assume the system operates in the following manner.

At the beginning of each period, as we stated, the depot places an order for the total amount ordered by warehouse customers on the previous day. Next, the orders placed previously (a lead time ago) arrive at the depot. The depot then allocates this inventory to the regional warehouses. Then demands occur at each warehouse and are satisfied to the maximum extent possible. Finally, holding and backorder costs are incurred at the warehouses.

Let us define two random variables. First, let $Y_0$ represent the total system demand over a depot lead time, that is

$$Y_0 = \sum_{t=1}^{D} \sum_{j=1}^{m} d_{jt}.$$

Second, let $Y_j$ represent the demand at warehouse $j$ over the replenishment lead time of length $A$ plus one period, that is,

$$Y_j = \sum_{t=D+1}^{D+A+1} d_{jt}.$$

We will now see why the upper limit is $D + A + 1$.

Assume the system's inventory position is $s$ units following the placement of a depot order in period 1. Then the net system inventory in period $D + 1$

prior to allocating depot stock is $s - Y_0$ units. Furthermore, the net inventory at warehouse $j$ at the end of period $D + A + 1$, given an allocation of $x_j$ units in period $D + 1$, has an expected value of $x_j - (A + 1)\mu_j$ units and a standard deviation of $\sqrt{A + 1}\sigma_j$ units. This is the case because when units are shipped from the depot to a warehouse in period $t$, they arrive at that warehouse at the beginning of period $t + A$. Suppose $A$ is three periods. An allocation from the depot in period $t$ arrives at warehouse $j$ at the beginning of period $t + 3$. Demands occurring in periods $t, t + 1, t + 2$, and $t + 3$ reduce the on hand stock level. The warehouse's allocated level, $x_j$, at the beginning of period $t$ results in a net inventory level of $x_j - d_{j,t} - d_{j,t+1} - d_{j,t+2} - d_{j,t+3}$ at the beginning of period $t + 4$ or at the end of period $t + 3$. Therefore, the ending inventory at warehouse $j$ is reduced by the demand in $A + 1$ periods.

We have assumed that an imbalance of inventory among the warehouses will not occur. Hence the $s - Y_0$ units can be allocated in a manner that yields the same probability of running out of stock a lead time in the future at all the warehouses. This assumption implies that there exists a $k$ such that

$$\sum_{j=1}^{m} x_j = \sum_j \left\{ (A + 1)\mu_j + k\sqrt{A + 1}\sigma_j \right\} = s - Y_0$$

with $x_j \geq 0$. Then

$$x_j = (A + 1)\mu_j + (s - Y_0 - (A + 1)\sum_{i=1}^{m} \mu_i)\sigma_j / \sum_{i=1}^{m} \sigma_i.$$

The random variable $z_j = x_j - Y_j$ measures the net inventory at $j$ at the end of a period.

Observe that

$$z_j = (A + 1)\mu_j + (s - (A + 1)\sum_{i=1}^{m} \mu_i)\frac{\sigma_j}{\sum_{i=1}^{m} \sigma_i}$$

$$- \left( Y_j + Y_0 \cdot \frac{\sigma_j}{\sum_i \sigma_i} \right).$$

Hence $z_j$ has a normal distribution with mean

$$E[z_j] = (A + 1)\mu_j + (s - (A + 1)\sum_{i=1}^{m} \mu_i)\frac{\sigma_j}{\sum_i \sigma_i}$$

$$-((A + 1)\mu_j + D\sum_{i=1}^{m} \mu_i \cdot \frac{\sigma_j}{\sum_i \sigma_i})$$

$$= (s - (D + A + 1)\sum_{i=1}^{m} \mu_i) \cdot \left\{ \frac{\sigma_j}{\sum_{i=1}^{m} \sigma_i} \right\}$$

and variance

$$\text{Var}[z_j] = \text{Var}[Y_j] + \left[\frac{\sigma_j}{\sum_i \sigma_i}\right]^2 \text{Var}[Y_0],$$

since $Y_j$ and $Y_0$ are independent random variables. But $\text{Var}[Y_j] = (A+1)\sigma_j^2$ and $\text{Var}[Y_0] = D\sum_{i=1}^m \sigma_i^2$. Consequently

$$\text{Var}[z_j] = (A+1)\sigma_j^2 + \left[\frac{\sigma_j}{\sum_{i=1}^m \sigma_i}\right]^2 \cdot D \cdot \sum_{i=1}^m \sigma_i^2.$$

Let $F_{z_j}(\cdot)$ be the distribution function for the random variable $z_j$. Then the expected per period cost at warehouse $j$ is given by

$$h\int_0^\infty z\, dF_{z_j}(z) + b\int_{-\infty}^0 z\, dF_{z_j}(z).$$

But $z_j$ is a function of $s$ and hence the expected cost per period is a function of $s$. Note that the $\text{Var}[z_j]$ is independent of $s$, however.

We can write

$$z_j = s\cdot\frac{\sigma_j}{\sum_{i=1}^m \sigma_i} + c_j - \bar{z}_j, \quad \text{where}$$

$$c_j = (A+1)\mu_j - (A+1)\sum_{i=1}^m \mu_i \cdot \frac{\sigma_j}{\sum_{i=1}^m \sigma_i}, \quad \text{a constant,}$$

and

$$\bar{z}_j = Y_j + Y_0\cdot\frac{\sigma_j}{\sum_i \sigma_i}, \quad E[\bar{z}_j] = (A+1)\mu_j + \frac{\sigma_j}{\sum_i \sigma_i}\cdot D\cdot\sum_{i=1}^m \mu_i$$

and

$$\text{Var}[\bar{z}_j] = \text{Var}[z_j] = (A+1)\sigma_j^2 + \left[\frac{\sigma_j}{\sum_{i=1}^m \sigma_i}\right]^2 \cdot D\cdot\sum_{i=1}^m \sigma_i^2.$$

Furthermore, let $F_{\bar{z}_j}(\cdot)$ be the distribution function for the random variable $\bar{z}_j$.

The expected cost per period at warehouse $j$ is then expressed as

$$h\int_{-\infty}^{s\cdot\sigma_j/\sum_{i=1}^m \sigma_i + c_j}\left(s\cdot\frac{\sigma_j}{\sum_{i=1}^m \sigma_i} + c_j - z\right)dF_{\bar{z}_j}(z)$$

$$+ b\int_{s\cdot\frac{\sigma_j}{\sum_{i=1}^m \sigma_i} + c_j}^\infty\left(z - \left(s\cdot\frac{\sigma_j}{\sum_{i=1}^m \sigma_i} + c_j\right)\right)dF_{\bar{z}_j}(z).$$

This is a news-vendor function which has its minimum occurring at

$$F_{\bar{z}_j}\left(s\frac{\sigma_j}{\sum_i \sigma_i} + c_j\right) = \frac{b}{b+h}.$$

Since $\bar{z}_j$ has a normal distribution,

$$F_{\bar{z}_j}\left(s\frac{\sigma_j}{\sum_i \sigma_i} + c_j\right) = \Phi(z)$$

where

$$z = \frac{s\frac{\sigma_j}{\sum_i \sigma_i} + c_j - E(\bar{z}_j)}{\sigma_{\bar{z}_j}}$$

$$= \frac{\left(\frac{\sigma_j}{\sum_{i=1}^m \sigma_i}\right)\left[s - (D + A + 1)\sum_{i=1}^m \mu_i\right]}{\sigma_j\left[(A+1) + D\cdot\frac{\sum_{i=1}^m \sigma_i^2}{\left(\sum_{i=1}^m \sigma_i\right)^2}\right]^{1/2}}$$

$$= \frac{s - (D + A + 1)\sum_{i=1}^m \mu_i}{\left[(A+1)\left(\sum_{i=1}^m \sigma_i\right)^2 + D\cdot\sum_{i=1}^m \sigma_i^2\right]^{1/2}}.$$

Observe that $z$ is independent of the warehouse $j$ and the optimal value for $z$, and hence $s$, can be found by setting $\Phi(z) = \frac{b}{b+h}$.

We conclude this section by comparing the inventory position for three different resupply systems. The first is the one that we just analyzed. The second is a system in which there is only one warehouse that satisfies all the demand. The third system is a collection of $m$ independent systems. There is no depot in the third system.

Let us now consider the second system. In this case we assume the material still moves through the depot to the single warehouse. Hence the total warehouse demand per period is the sum of the individual warehouse demands as experienced in the first system we studied. In this depot/single warehouse system, expected holding and backorder costs are incurred $D + A + 1$ periods after a depot order is placed. The demand over this period of time is normally distributed with mean $(D + A + 1)\sum_{i=1}^m \mu_i$ and variance $(D + A + 1)\sum_{i=1}^m \sigma_i^2$. The lowest expected per period cost is found by determining the value of $z$ that satisfies $\Phi(z) = \frac{b}{b+h}$, where, in this second case,

$$z = \frac{s - (D + A + 1)\sum_{i=1}^m \mu_i}{\left[(D + A + 1)\sum_{i=1}^m \sigma_i^2\right]^{1/2}}.$$

In the third system, each warehouse $j$ places its own orders and has its own order up to level $s_j$. To make a comparison, we assume the lead time for each warehouse is $D + A$ periods. Then the optimal value for $z_j$ satisfies

$$z_j = \frac{s_j - (D + A + 1)\mu_j}{(D + A + 1)^{1/2}\sigma_j}.$$

Since $\frac{b}{b+h}$ is the same for all warehouses, $z_j = z$ for all $j$.

Let us now compare the values of $s$ for these three systems. For the first system, which we examined in detail,

$$s = (D + A + 1) \sum_{j=1}^{m} \mu_j + z \left\{ D \sum_{j=1}^{m} \sigma_j^2 + (A + 1) \left( \sum_{j=1}^{m} \sigma_j \right)^2 \right\}^{1/2}.$$

For the second system, in which demand is concentrated at a single location,

$$s = (D + A + 1) \sum_{j=1}^{m} \mu_j + z \left\{ (D + A + 1) \sum_{j=1}^{m} \sigma_j^2 \right\}^{1/2}.$$

For the third, and completely decentralized system,

$$s = \sum_{j=1}^{m} s_j = \sum_{j=1}^{m} \left[ (D + A + 1)\mu_j + z(D + A + 1)^{1/2}\sigma_j \right]$$

$$= (D + A + 1) \sum_{j=1}^{m} \mu_j + z \cdot (D + A + 1)^{1/2} \sum_{j=1}^{m} \sigma_j.$$

We see that the order-up-to level, or inventory position, and hence the safety stock requirements are dependent on the resupply system's structure. The safety stock requirements are clearly greatest for the decentralized system and lowest for the completely centralized system, that is the single warehouse system. The depot and multiple warehouse system requires more safety stock than the single warehouse system and less than the totally decentralized system. The degree of these differences will depend on the system parameters and particularly on the values of $D$ and $A$. When $D$ is significantly greater than $A$, the effect of complete and partial centralization or pooling is the greatest. Let us consider two examples.

First, suppose $m = 10$, $\sigma_j = \sigma = 4$ and $D = 10$ and $A = 5$. Further suppose $z = 1.25$. Then

safety stock for the totally centralized system $= 1.25 \cdot [16 \cdot 10 \cdot 16]^{1/2} = 63.25$,

safety stock for the depot/warehouse system $= 1.25 \left[ 10 \cdot 10 \cdot 16 + 6[10 \cdot 4]^2 \right]^{1/2} = 105.83$,

safety stock for the total decentralized system $= 1.25[16]^{1/2} \cdot 10 \cdot 4 = 200$.

Clearly the cost of decentralization, in terms of safety stock, is substantial. Observe that the depot/warehouse structure requires about 167% of the safety stock level needed in a completely centralized system.

Second, suppose $D = 45$ and $A = 1$ and all other parameters are the same as in the first case. Then

safety stock for the totally centralized system $= 1.25[47 \cdot 10 \cdot 16]^{1/2} = 108.40$,

safety stock for the depot/warehouse system $= 1.25\left[45\cdot 10\cdot 16+2[10\cdot 4]^2\right]^{1/2}$
$= 127.48,$

safety stock for the decentralized system $= 1.25[47]^{1/2}\cdot 10\cdot 4 = 342.78.$

Again we observe that pooling has a very beneficial effect. We also see that when $D$ is much greater than $A$, the depot/warehouse safety stock requirements are closer to those needed in the totally centralized system. In this example, the depot warehouse system requires about 17.6% more safety stock than does the completely centralized system. Both of the systems in which pooling exists require roughly a third of the safety stock required when operating a completely decentralized system.

The reduction in safety stock that results from pooling and operating multi-echelon systems can be substantial. These inventory reductions justify the fixed and operating costs incurred when running depot/warehouse systems. Transportation costs are often reduced when operating depot/warehouse systems, too, since larger shipments over long distances normally result in lower per unit shipping costs.

### 7.2.2  A Continuous Review Three Echelon Pooling Environment

We now study a three echelon system that consists of a set of locations at which demands for parts occur, which we call bases, a set of locations that resupply groups of them, which we call intermediate stocking locations, and a depot that is responsible for resupplying the intermediate stocking locations as needed. This system is an extension of the one discussed in depth in Chapter 5. In the ensuing discussion we will assume that the items are called LRUs to match our earlier discussions, and that demands at bases correspond to removals of failed LRUs from an aircraft. Furthermore, failed LRUs are sent directly to the intermediate stocking facility that is responsible for resupplying the base. That facility is required to send a serviceable unit to the requesting base as soon as one is available. Once the failed unit arrives at the intermediate stocking location, a determination is made as to whether or not it is capable of being repaired there. If it is, the unit is repaired there; otherwise, it is sent to the depot for repair. In the latter case, the depot sends a serviceable LRU to the intermediate stocking location as soon as one is available.

The decision as to where the repair will take place is assumed to depend only on the nature of the failure, that is, the failure mode. Thus we are assuming that the location at which repairs are made depends only on the technical attributes of the failure and not on the workload at a location at the particular moment when the repair is required.

In all cases, shipments of LRUs from the depot to intermediate stocking locations and from intermediate stocking locations to bases are made on a first-come, first-served basis. No prioritization of shipments based on perceived immediate need of the intermediate stocking locations or bases is considered when making allocation decisions.

We also assume that lateral resupply among bases and among intermediate stock locations is not permitted.

While neither the first-come, first-serve nor the lateral resupply assumptions are likely to hold in practice, we will make these assumptions. In fact, if lateral resupply among several bases is the norm, then the collection of these bases should be considered to be a single base in the model. Later in this chapter we will directly address the lateral resupply possibilities.

We observe that these assumptions should provide a conservative solution. That is, actual performance should be at least as good as the model predicts given proper prioritization practices are undertaken when allocating inventories at either the depot or intermediate stocking locations. Performance should also be enhanced if lateral resupply is performed in practice.

Finally, we assume that failures of LRUs at the base level occur according to independent Poisson processes. That is, failures of all LRU types are independent among LRU types and among locations.

### 7.2.2.1  System Operation and Definitions

The structure of the system we are examining is given in Figure 7.3. We assume items are managed independently so we will focus on a single LRU type.

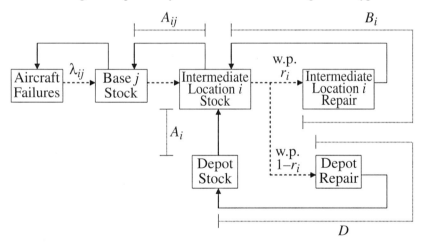

**Fig. 7.3.** A Three-Echelon Resupply System

Removals at base $j$, which are resupplied by intermediate stocking location $i$, occur at a rate of $\lambda_{ij}$ units per day. As mentioned, the removal, or failure, process is a Poisson process. Each failed unit is sent to the appropriate intermediate stocking location immediately when the failure occurs (no batching of failed units is allowed). Thus the arrival process of failed LRUs at intermediate stocking location $i$ is a Poisson process with rate $\sum_{j \in I_i} \lambda_{ij}$, where $I_i$ is the set of bases supplied by intermediate stocking location $i$. The probability that a failed unit arriving at intermediate stocking location $i$ is repaired there is denoted by $r_i$. Hence the arrival process to the repair facility at intermediate stocking location $i$ is a Poisson process with rate $r_i \sum_{j \in I_i} \lambda_{ij}$. Similarly, the arrival process of units requiring repair at the depot is a Poisson process with rate $\sum_i (1 - r_i) \sum_{j \in I_i} \lambda_{ij}$ units per day.

The average depot repair cycle time for units ultimately requiring depot repair is denoted by $D$. This time includes the time required to transport defective LRUs to the depot plus the average time it takes to repair the LRU once the unit is entered into the repair process. Repair times at the depot are assumed to be independent and identically distributed. $D$ is measured in days.

Next, let $B_i$ be the average LRU repair cycle time given the unit is repaired at intermediate stocking point $i$. $B_i$ includes the time to detect a LRU failure at a base, pack, transport it to location $i$, and perform the repair operations at intermediate stocking location $i$. $B_i$ is measured in days.

We let $A_i$ be the average order and ship time for a unit from the depot to intermediate stocking location $i$ (not counting any waiting time for the availability of an LRU at the depot). $A_{ij}$ is the average order and ship time to base $j$ from intermediate stocking location $i$ given stock is available at intermediate stocking location $i$. Both $A_{ij}$ and $A_i$ are measured in days.

Finally, we let $T_{ij}$ represent the average resupply time for an LRU at base $j$ in the group of bases supplied by intermediate stocking location $i$.

The structure we have discussed is of practical importance in cases where $A_{ij}$ is much smaller than $A_i$. As we discussed in the previous section, if they are of roughly comparable values, there is less advantage from an inventory viewpoint to having a three rather than a two echelon resupply system. In fact, the three echelon system could require significantly more inventory.

As we mentioned in the previous section, three echelon systems often exist when $A_{ij}$ is a day or so, $A_i$ is a week or more, and repair resources are expensive and relatively scarce. Although the model we will propose is a stationary model, three echelon systems are also of practical importance when it is uncertain as to which base or bases may have higher activity levels (and failures) for short periods of time. Thus the intermediate stocking point, and its associated repair facility, provides the flexibility needed to respond to highly dynamic requirements.

Although we have described the system assuming repair occurs following the removal of a defective LRU from an aircraft, the model could also represent situations in which there is no repair and all replenishment comes from an outside source. In this case $D$ represents the average procurement lead time, $A_i$ and $A_{ij}$ the average transportation and handling times, and $r_i = 0$.

### 7.2.2.2  Optimization Problem

The model we construct establishes stock levels for each location so that the average total number of outstanding LRU backorders across all bases and all LRUs is minimized subject to an investment constraint on LRUs. While this backorder objective is only a first order approximation to maximizing aircraft availability, it is an accurate approximation and is highly mathematically tractable, as we have already discussed.

We will now develop the mathematical model.

Let $s_{ij}$ represent the LRU stock level at base $j$, $j \epsilon I_i$, where $i$ supplies $j$.

Let $s_i$ represent the LRU stock level at intermediate stocking location $i$, and $s_0$ represent the depot LRU stock level.

As in the two echelon model, we must construct the average resupply time equation which, for example, we use to express the effect of stocking inventory at the intermediate stocking location and the depot on the average base resupply time.

Let

$$T_{ij} = \text{average LRU resupply time for base } j, \ j \epsilon I_i$$
$$= A_{ij} + \text{expected delay due to shortages at } i.$$

But, this expected delay, which we denote by $\delta(s_i)$, is expressed as

$$\delta(s_i) = \frac{\text{average number of outstanding intermediate stocking location } i \text{ backorders}}{\text{intermediate stocking location demand rate}}.$$

If $B_i(s_i)$ represents the average number of backorders outstanding at intermediate stocking location $i$ at a random point in time, then

$$\delta(s_i) = \frac{B_i(s_i)}{\sum_{j \epsilon I_i} \lambda_{ij}} = \frac{B_i(s_i)}{\lambda_i}, \text{ where } \lambda_i = \sum_{j \epsilon I_i} \lambda_{ij}.$$

Our goal is to compute the average number of base $j$ backorders outstanding at a random point in time. To do this, we must establish the probability distribution for the number of units in the resupply system. We let $X_{ij}$ be this random variable for base $j$, $j \epsilon I_i$.

Then

$$E[X_{ij}] = \lambda_{ij} T_{ij}$$
$$= \lambda_{ij} \left( A_{ij} + \frac{B_i(s_i)}{\lambda_i} \right).$$

Furthermore, we compute the variance of the number of units of the LRU in resupply at base $j$, $j \epsilon I_i$, as follows.

Suppose $N_i$ represents the number of backorders existing at the intermediate supply point $i$ at some point in time. Then, based on our assumptions, the probability that $N_{ij}$ of them correspond to failures (and orders) at base $j$, $j \epsilon I_i$, is given by

$$P\left\{ N_{ij} = n_{ij} | N_i = n_i \right\} = \binom{n_i}{n_{ij}} \left( \frac{\lambda_{ij}}{\lambda_i} \right)^{n_{ij}} \left( 1 - \frac{\lambda_{ij}}{\lambda_i} \right)^{n_i - n_{ij}}.$$

The expected value of $N_{ij}$, given $s_i$, is

$$E[N_{ij}|s_i] = E_{N_i}\left[E[N_{ij}|N_i]\right] = \frac{\lambda_{ij}}{\lambda_i}E_{N_i}[N_i]$$

$$= \frac{\lambda_{ij}}{\lambda_i}\mathcal{B}_i(s_i).$$

The variance of $N_{ij}$, given $s_i$, is

$$\text{Var}\,(N_{ij}|s_i) = E\left[N_{ij}^2|s_i\right] - \left[E\left[N_{ij}|s_i\right]\right]^2$$

$$= E\left[N_{ij}^2|s_i\right] - [\mathcal{B}_i(s_i)]^2 \cdot \left[\frac{\lambda_{ij}}{\lambda_i}\right]^2.$$

To determine the variance of $N_{ij}$ given $s_i$ we compute

$$E[N_{ij}^2|s_i] = E_{N_i}\left[E_{N_{ij}}\left[N_{ij}^2|N_i\right]\right]$$

$$= E_{N_i}\left[\text{Var}_{N_{ij}}\left(N_{ij}|N_i\right) + \left(E_{N_{ij}}\left[N_{ij}|N_i\right]\right)^2\right]$$

$$= E_{N_i}\left[N_i\left(\frac{\lambda_{ij}}{\lambda_i}\right)\left(1 - \frac{\lambda_{ij}}{\lambda_i}\right) + \left(\frac{\lambda_{ij}}{\lambda_i}\right)^2 N_i^2\right]$$

$$= \frac{\lambda_{ij}}{\lambda_i}\left(1 - \frac{\lambda_{ij}}{\lambda_i}\right)\mathcal{B}_i(s_i) + \left(\frac{\lambda_{ij}}{\lambda_i}\right)^2 E\left[N_i^2\right]$$

$$= \frac{\lambda_{ij}}{\lambda_i}\left(1 - \frac{\lambda_{ij}}{\lambda_i}\right)\mathcal{B}_i(s_i) + \left(\frac{\lambda_{ij}}{\lambda_i}\right)^2 \left[\text{Var}\,(N_i|s_i) + \mathcal{B}_i(s_i)^2\right],$$

as we did earlier in Chapter 5. Hence

$$\text{Variance}\,(N_{ij}|s_i) = \frac{\lambda_{ij}}{\lambda_i}\left(1 - \frac{\lambda_{ij}}{\lambda_i}\right)\mathcal{B}_i(s_i) + \left(\frac{\lambda_{ij}}{\lambda_i}\right)^2 \text{Var}\,(N_i|s_i).$$

However, $\mathcal{B}_i(s_i)$ depends on the depot stock level, since

$$T_i = r_i B_i + (1 - r_i)(A_i + \text{ depot delay } (s_0)).$$

Then $\delta_i(s_0) = $ expected depot delay given depot stock $s_0 = \frac{\mathcal{B}_0(s_0)}{\lambda_0}$, where $\mathcal{B}_0(s_0)$ represents the average number of outstanding depot backorders at a random point in time and $\lambda_0 = \sum_i(1 - r_i)\sum_{j\epsilon I_i}\lambda_{ij}$. Therefore

$$T_i = \left[A_i + \frac{\mathcal{B}_0(s_0)}{\lambda_0}\right](1 - r_i) + r_i B_i$$

and

$$\lambda_i T_i = r_i\lambda_i B_i + \lambda_i(1 - r_i)A_i + (1 - r_i)\frac{\lambda_i}{\lambda_0}\mathcal{B}_0(s_0).$$

As we just computed for each base, we need to compute the first two moments of the probability distribution for the number of units in resupply for the intermediate stocking location. This calculation is made in exactly the same manner as was done for the random variable $N_{ij}$. The details can be found in the discussion of the two-echelon case.

Hence we have the machinery in place to estimate the first two moments of the probability distributions for the number of units in resupply for both the base and intermediate stocking location. The probability distribution of the number of units in resupply is approximated by a negative binomial distribution with parameters estimated using the computed moments.

The optimization problem we will solve for a single LRU type is as follows:

$$f(b) = \text{minimize} \sum_i \sum_{j \in I_i} \mathcal{B}_{ij}(s_{ij})$$

$$s.t. \quad \sum_i \left\{ \sum_{j \in I_i} s_{ij} + s_i \right\} + s_0 = b \tag{7.9}$$

$$s_{ij}, s_i, s_0 \text{ nonnegative integers,}$$

where $b$ represents the total available stock of the LRU. In fact, rather than solving Problem (7.9) for a single value of $b$, we would solve it for a broad range of values of $b$ so that the function $f(b)$ can be constructed, as we did in Chapter 5.

Suppose we restrict $s_0 \in \mathcal{S}_0$, $s_i \in \mathcal{S}_i$, where $\mathcal{S}_0$ and $\mathcal{S}_i$ represent the sets of values that would be considered for the respective variables. In principle, we would have to look at all possible combinations of stock levels in these sets to obtain an optimal solution to Problem (7.9). This is the case since, as we have seen, $\mathcal{B}_{ij}(s_{ij})$ is a function of $s_{ij}$, $s_i$ and $s_0$ and hence the objective function is not separable. Furthermore, it is not necessarily convex either. Hence an exhaustive search would seem to be necessary.

However, this is not the way we would propose to solve this problem. Suppose we set $a = \sum_i s_i + s_0$ and $\sum_i \sum_{j \in I_i} s_{ij} = b - a$ for some values of $a$ and $b$. The question is how should we select the values of the $s_i$ and $s_0$ variables so that $\sum_i s_i + s_0 = a$. We propose the following method for obtaining the desired partition of the available stock $a$ among the intermediate stocking locations and the depot.

Select the values of the variables $s_i$ and $s_0$ so that the expected delay in responding to base level resupply requests is minimized. That is, solve

$$\min \sum_i \sum_{j \in I_i} \lambda_{ij} T_{ij} = \sum_i \sum_{j \in I_i} \lambda_{ij} \left( A_{ij} + \frac{\mathcal{B}_i(s_i | s_0)}{\lambda_i} \right)$$

$$= \sum_i \sum_{j \in I_i} \lambda_{ij} A_{ij}$$

$$+ \sum_i \left[ \sum_{j \in I_i} \frac{\lambda_{ij}}{\lambda_i} \right] \mathcal{B}_i(s_i | s_0)$$

$$= \text{constant} + \sum_i \mathcal{B}_i(s_i | s_0)$$

subject to $\sum_i s_i + s_0 = a$, where $s_0 \epsilon S_0$, $s_i \epsilon S_i$. Then for all $s_0 \epsilon S_0$, we obtain

$$g(s_0, a) = \min \sum_i \mathcal{B}(s_i | s_0)$$

$$\sum_i s_i = a - s_0$$

$$s_i \epsilon S_i.$$

Since $\mathcal{B}(s_i | s_0)$ is convex in $s_i$ given $s_0$, a marginal analysis method can be used to obtain the optimal allocation of the available stock, $a - s_0$, among the intermediate stocking locations.

Once the values of $g(s_0, a)$ have been obtained for all $s_0 \epsilon S_0$ and for a range of values $a$. We can determine

$$\tilde{g}(a) = \min_{s_0} g(s_0, a). \tag{7.10}$$

The function $\tilde{g}(a)$ may not be a convex function of $a$.

The solution to problem (7.9) is found using knowledge of the optimal values of $s_0$ and $s_i$ for each value of $a$, that is, those values that yield $\tilde{g}(a)$. Observe that problem (7.9) can be written as

$$f(b) = \min_{\substack{c s_0, s_i \\ s_0 \epsilon S_0, s_i \epsilon S_i}} \left\{ \min_{s_{ij}} \sum_{ij} \mathcal{B}_{ij}(s_{ij}) : \sum_i \sum_{j \epsilon I_i} s_{ij} = b - \left( \sum_i s_i + s_0 \right) \right\}.$$
$$s_{ij} = 0, 1, \ldots$$

We find an approximate solution to this problem by solving the following set of problems. For each $a = \sum s_i + s_0$ in an appropriate range, use the allocation found when solving problem (7.10) to compute $T_{ij}$, and hence $\mathcal{B}_{ij}(\cdot)$. Then solve

$$h(a) = \min_{s_{ij}} \left\{ \sum_i \sum_{j \epsilon I_i} \mathcal{B}_{ij}(s_{ij}) : \sum_i \sum_{j \epsilon I_i} s_{ij} = b - a \right\}. \tag{7.11}$$

The solution to problem (7.11) can be found using a marginal analysis algorithm since the objective function is the sum of convex functions, $\mathcal{B}_{ij}(s_{ij})$.

The function $h(a)$ will likely not be a convex function. Hence we would construct the greatest convex minorant of this function, call it $\hat{h}(a)$. These functions would then be used to solve the multi-LRU problem using another marginal analysis algorithm, which is similar to the one employed in finding the solution to the two-echelon, multi-LRU problem that we discussed in detail in Chapter 5.

## 7.3  A Multi-Echelon Periodic Review Pooling Environment

We continue our investigation of lateral resupply and pooling by analyzing a periodic review tactical planning problem for a three-echelon, repairable-parts service network characterized by an uncapacitated repair facility and local opportunities for inventory pooling. As was the case for all of the tactical planning environments examined previously, the central planning problem is to determine the optimal level of total system stock for each part. Once the stock has been acquired, its location in the system, and the resulting service performance of the system, will be managed by employing a real-time execution system such as one of the types discussed in Chapter 10. The more effective this execution system is at managing repair and distribution, the less total stock will be required. Consequently, it is important to capture the possibilities for dynamic optimization when planning total stock levels. Furthermore, because of the large number of parts in such systems, computational efficiency in performing any inventory planning function is a critical concern. Specifically, there is always a balance between computational requirements and modeling complexity. The method we present in this section is a balance between the two concerns. The model we will discuss can be solved in time that is $n \log(n)$, where $n$ is the number of part number-location combinations. In Chapter 8, we extend the ideas presented here to the situation where the depot repair capacity is limited.

### 7.3.1   Multi-Echelon Repairable Parts System with Central Repair

The three-echelon distribution and repair system for repairable parts that we will study in detail is depicted in Figure 7.4. The system consists of a set of inventory pools, each of which contains a number of stocking locations called *bases*; a set of intermediate stocking facilities, each of which resupplies a set of inventory pools; a depot, which resupplies the intermediate stocking facilities; an uncapacitated depot repair facility, at which defective parts are repaired and, once repaired, are sent to the depot stocking location; an external supplier, which provides inventory to replace parts that have been condemned; and third party emergency supply sources. As before, we refer to the recovery, identification, repair, and replacement processes as the *resupply system*. The bases, pools, intermediate stocking facilities, and the depot are referred to as the *distribution system*. The third party emergency supply sources are viewed as a separate system.

The central planning decision to be made is the number of units of each item type to have in the system. We assume that the repair capacity is such that the repair cycle times are independent and identically distributed for all parts of a given item type, and independent across item types. The model developed in this section addresses only the optimal stock level for a single item type for a given repair cycle time distribution, since there will be no constraints in which items interact in our model. We focus on the finite repair capacity case in which interactions do exist and propose methods to allocate this capacity in a multi-item-type environment in Chapter 8. Although the model will set target base stocking levels at each location, that is of secondary concern. Our goal is primarily to establish a desired level of system stock for an item given the possibilities of pooling. Real-time allocation of inventory is the subject of our analysis in Chapter 10.

We begin by assuming that the arborescent resupply network is predetermined and that the cost of a customer shortage is known at each location. We assume that

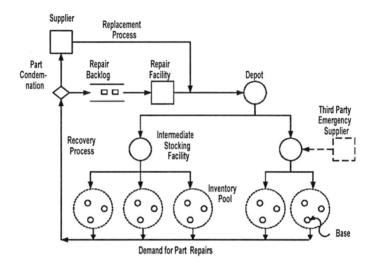

**Fig. 7.4.** Multi-Echelon Distribution Network with Repair and Pooling

the system operates in the following manner in each period. Demand for parts arises randomly at bases and is satisfied out of that base's stock, from another base's stock within the same pool, from a pool within the same subsystem, or from the base's intermediate stocking location, depending on which location has stock on hand. If none do have stock on hand, then stock is obtained immediately from an external emergency supplier at a cost. While bases within the same pool share inventory to satisfy demand, they do so at some cost. The inventory pools are a collection of bases that are within close proximity of each other (e.g., less than two hours travel time from each other). Associated with each demand for a part is a failed unit that enters a recovery process, which includes defect-identification, transport, and a decision of whether to repair or condemn the unit. If the decision is to repair the unit, then the unit enters the depot's repair cycle. Replacement orders for condemned units are placed with an outside supplier.

Once the system is operating, the units that have been acquired are either in the distribution system or in the resupply system; that is, they will either be in a serviceable or reparable status. Optimizing the total system stock requires modeling the operating characteristics of both the distribution system and the resupply system. Our model treats these operational problems as dynamic optimization sub-problems. We assume that transportation times within the distribution system are very short relative to the purchase and repair cycle times experienced in the resupply system. For many high-cost, low-cubic-volume items such as electronic components, air freight is economical and distribution transportation times from an intermediate stocking facility to a base are measured in hours. On the other hand, lead times for repair or re-manufacturing processes are often measured in weeks or months. Consequently, the distribution system can react to changes on a time scale that is much shorter than the repair and procurement system. For simplicity, we assume, as we said earlier, that all emergency transport in the distribution system happens instantaneously at the end of a review period. However, the review period is not assumed to be so short as to negate any concern for positioning stock close to the customer. We also assume that reallocation of stock within a region (i.e. an intermediate stocking facility and its associated inventory pools) occurs instantaneously at the beginning of a review period and occurs at no cost. This assumption is equivalent to the imbalance assumption we made earlier in this chapter. More will be said about this point as we proceed.

In setting total system stock levels, we further assume that the distribution system is managed dynamically to balance the distribution of available stock for best effect. That is, rather than using simple first-come, first-serve allocation policies, the real-time management system allocates available stock to optimize the tradeoff between inventory holding costs and customer shortfalls over the course of a planning horizon. Such a high degree of management attention on the operation of both the distribution and resupply systems can be justified for high-cost, low-demand-rate items as studies by Scudder [214], Scudder and Hausman [215], Pyke [198], Evers [81, 82] and others have demonstrated. This approach requires an integrated implementation of both planning and execution models. In this section we describe the first step of the optimization-based planning and execution process. The second step is discussed in Chapter 10.

### 7.3.2  Linking Resupply and Distribution

Our approach is to develop a cost model based on the steady state probability distribution of the number of units in resupply for a single item. Resupply consists of three processes: recovery and transport, replacement, and repair. Assume that the repair/replace decision is made prior to the unit entering the repair cycle, as illustrated in Figure 7.4. Let $V_B$ denote the number of units in the depot repair cycle system in steady state; and let $V_U$ denote the steady-state number of units of this item on order for replacement from the supplier. We assume that demands occur at bases according to independent Poisson processes. Hence, $V_B$ and $V_U$ are independent random variables. Let $V$ denote the total number of units in resupply in steady state:

$$V = V_B + V_U. \tag{7.12}$$

The stationary probability distribution of $V$ is thus a convolution of two stationary distributions. Since we assume a Poisson demand process and independently distributed repair cycle and procurement lead times, the steady state distribution of $V_B + V_U$ is also a Poisson process. Let $\lambda$ denote the system demand rate for units of the part at the bases and let $r$ denote the probability that a unit will be repaired and $(1 - r)$ the probability that the unit will be condemned. Let $T$ be the average depot repair cycle time and let $U$ denote the average supplier lead time for replacement orders. Then the stationary distribution of $V_B + V_U$ is a Poisson distribution with rate $r\lambda T + (1 - r)\lambda U$.

Let $s > 0$ denote the total planned inventory in the system, both service-able units and units in resupply. The tactical planning model focuses on deter-mining the optimal value of this static decision variable. Let $R$ denote the to-tal distributable physical inventory in the system, that is, $R = (s - V)^+$. Let $X = (V - s)^+$ be the number of units in resupply in excess of planned inventory. The steady state distributions of $R$ and $X$ can be derived from the value of $s$ and the stationary distribution of $V$. We will develop a cost model based on $R$ and $X$ that can be optimized through the appropriate choice of $s$.

The remainder of this section is organized as follows. We first develop a model for the optimal allocation of available stock in a two-echelon distribution system with local opportunities for inventory pooling and external emergency resupply. This analysis yields a cost function which is then used to construct a model for the optimal allocation of stock in a three-echelon distribution system. This analysis, in turn, yields a cost function to describe a model for determining the optimal level of system inventory. We show how to disaggregate the optimal system inventory level into target stock levels (inventory positions) at all locations. We observe that the cost function can be computed in time that is $n \log(n)$ in the number of locations.

### 7.3.3  Optimal Stock Allocation for a Two-Echelon System with Inventory Pools

Let $W$ denote the set of two-echelon subsystems within the overall distribution system. Each subsystem consists of an intermediate stocking facility and a collec-tion of inventory pools supported by this facility. Let $w \in W$ index the individ-ual subsystems. We temporarily focus on a single subsystem and hence suppress the index $w$ from all variables. We will now develop a single-item, single-period model for optimally allocating the physical inventory among this subsystem's stocking locations.

#### 7.3.3.1  Subsystem Structure, the Pooling Assumption, and Penalty Costs

Let $P$ be the set of inventory pools served by a subsystem's intermediate stocking facility and let $B_p$ be the set of bases within pool $p$, $p \in P$. The subsystem is supported by a real-time parts location information system from which knowledge of on-hand stock at each base is available.

The order of events in a review period within this subsystem is as follows. At the beginning of the period, the total subsystem stock available for distribution is known and a decision is made to redistribute this stock among the different facilities in the subsystem. The redistribution of stock takes place before demand is realized. Demand for stock is then realized at the various bases, and lateral resupply or other emergency resupply actions and costs are incurred to satisfy these demands. All demands are assumed to be satisfied by the end of the review period. Three types of emergency shipments are possible, each with different cost consequences:

1. *Allocate in-pool base stock to the needy base:* let the cost of a transfer between bases within the same pool be denoted by $b^p$.
2. *Allocate out-of-pool base stock to the needy base:* let the incremental cost of transfer between pools, within the same subsystem, be denoted by $b^w$. This is also the cost of transferring a unit from the intermediate stocking facility to the pool. When a within subsystem transfer of this type is initiated, the penalty cost of $b^w$ is incurred and is added to the in-pool cost $b^p$ of getting the part to the needy base.
3. *Out-of-subsystem allocation to the needy base:* sometimes the required part is not in stock anywhere within the subsystem and an emergency resupply is required from an external supplier in the region. Let the incremental cost of an emergency shipment of this type to this subsystem be denoted by $b^e$. When this type of transaction is initiated, the per unit cost $b^e$ is incurred and is added to the emergency cost to get the part to the pool and base; that it, it is added to $b^w + b^p$. Note that $b^e$ is also incremental to the regular transport cost to provide a unit to the base from the intermediate stocking facility.

Each of the three identified costs, $b^p$, $b^w$, and $b^e$, is assumed to include not only the incremental transportation cost but also an imputed penalty for the incremental customer waiting time (for the use of a more remote source of supply). The value of $b^w$ also accounts for the reduction in holding costs incurred at another base as a consequence of the lateral resupply transaction. Similarly, $b^e$ reflects the adjustment to the holding costs.

### 7.3.3.2 Stock Allocation Decisions

Let $R_j^p$ be the stock allocation to base $j$, $j \in B_p$, and let $R^p = \sum_{j \in B_p} R_j^p$, the stock allocation to pool $p$. Let $R$ (that is, $R^w$ with the subsystem superscript suppressed) denote the total physical stock level available for allocation in the subsystem at the beginning of a review period. Given $R$, we must choose stock level allocations consistent with this total:

$$\sum_{p \in P} R^p \leq R; \text{ and}$$

$$\sum_{j \in B_p} R_j^p = R^p, \forall p \in P.$$

The difference between $R$ and $\sum_{p \in P} R^p$ is stock that is retained at the subsystem's intermediate stocking facility. As mentioned earlier, redistribution during the review period resulting from these allocation decisions is assumed to take place before demand is realized.

### 7.3.3.3  The Backorder and Inventory Imbalance Assumptions

Denote the net inventory at base $j$ at the beginning of the review period by $I_j^P$, for $j \in B_p$, and let $I^P = \sum_{j \in B_p} I_j^P$, the net inventory of pool $p$. Since we assume lead times are so short within the subsystem, we assume that no stock is in-transit to a base at the beginning of a review period. Recall that we assume that there are no backorders: $I_j^P \geq 0$, for all $p \in P$ and $j \in B_p$. The cost of eliminating backorders through emergency replenishments is captured in the cost function.

A feasible allocation is one satisfying $R_j^P \geq I_j^P, \forall j \in B_p$; otherwise, the allocation will imply costly transshipments to redress imbalances. In an execution model, that is, one that is used for real-time allocation of system stock, we cannot ignore these constraints. However, in a planning model used to set base stock policy parameters and target system inventory levels we do ignore these constraints when review periods are short. As we stated previously, this imbalance assumption is the same assumption that we made earlier in this chapter. Henceforth in this planning model, we ignore the current state of net inventory within and among the pools and assume that material is balanced across locations for a given amount of subsystem inventory. The state of the system is therefore the total distributable inventory of the subsystem, $R$.

### 7.3.3.4  The Allocation Optimization Problem

Assume holding costs do not differ by base. Consequently, how $R^p$ is allocated to the bases does not affect the total holding cost; however, the allocation will affect internal shortage costs. Let $D_j^P$ denote the demand for service parts at base $j \in B_p$ for one review period. Let $C_j^P(R_j^P)$ denote the expected base-to-base lateral resupply cost for base $j \in B_p$, given an initial allocation of $R_j^P$ to base $j$:

$$C_j^P(R_j^P) \equiv b^P E\left[ \left( D_j^P - R_j^P \right)^+ \right]. \tag{7.13}$$

Let $h^P$ denote the incremental holding cost of storing one unit for one review period in pool $p$, over the cost of holding that unit in the subsystem's intermediate stocking facility. We charge holding costs on inventory balances at the end of the review period. Let $D^P$ denote the aggregate demand in pool $p$ for one review period. Let $C^P(R^P)$ be the minimum total expected pool cost for pool $p$ for one review period:

$$C^P(R^P) \equiv h^P E\left[ \left( R^P - D^P \right)^+ \right] + b^w E\left[ \left( D^P - R^P \right)^+ \right] \tag{7.14}$$

$$+ \min_{\substack{s.t. \sum_{j \in B_p} R_j^P = R^P \\ R_j^P \geq 0, \text{ integer}, \forall j \in B_p}} \left\{ \sum_{j \in B_p} C_j^P(R_j^P) \right\}.$$

Two additional costs need to be considered: the cost of emergency shipments from outside the subsystem and the cost of holding inventory held at the intermediate stocking facility. Let $D^w$ denote the random variable for aggregate demand in the subsystem for one review period. Let $h^w$ denote the incremental cost of holding one unit of inventory in the subsystem for one review period, over the cost of holding it in inventory at the depot for one period. It does not include the incremental cost of holding inventory in the individual pools, which are captured by the $h^p$ parameters. Let $C_w(R^w)$ denote the expected cost over one review period for subsystem $w$ assuming we begin the review period with $R^w$ units of distributable inventory in the subsystem and this inventory is allocated optimally. That is:

$$C_w(R^w) \equiv h^w E\left[\left(R^w - D^w\right)^+\right] + b^e E\left[\left(D^w - R^w\right)^+\right] \tag{7.15}$$

$$+ \min_{\substack{s.t. \sum_{p \in P^w} R^p \leq R^w \\ R^p \geq 0, \text{ integer } p \in P^w}} \left\{ \sum_{p \in P^w} C^p(R^p) \right\}.$$

Remember, we assume that the shipments required to achieve the optimal distribution of inventory within the subsystem are performed instantaneously at the beginning of the review period. The expected cost of transporting regular replenishments to a base, which does not depend on the stocking policy, can therefore be ignored. Recall that the cost of lateral resupply to address imbalances at a period's beginning are also ignored in our tactical planning model.

Observe that $C_w(R^w)$ involves a nested optimization of convex newsvendor-style cost functions, which are not difficult to calculate. We use this function to approximate the cost of operating a subsystem rather than using a much more complex model to represent the dynamic behavior of an optimization-based inventory management execution system.

### 7.3.4   Optimal System Inventory

In this section, we consider the set, $W$, of all two-echelon subsystems.

#### 7.3.4.1   The Relevant Cost of Subsystem Inventory

Let $\overline{R}^w$ denote the physical inventory in subsystem $w$ at the beginning of a review period. This includes inventory at each location within the subsystem plus any inventory in transit to the intermediate stocking facility plus any stock that is being allocated for shipment into this subsystem from the depot in the current period. Let $A_w$ denote the lead time to ship units from the depot to the intermediate stocking facility $w$, and let $D^w_{A_w}$ denote the random demand that occurs over this lead time. Thus, $\left(\overline{R}^w - D^w_{A_w}\right)^+$ is the physical inventory available to subsystem $w$ at the beginning of the period that follows the transport lead time, including the current allocation. The relevant cost for the allocation decision is:

$$C^w_{A_w}\left(\overline{R}^w\right) \equiv E\left[C_w\left(\left(\overline{R}^w - D^w_{A_w}\right)^+\right)\right], \tag{7.16}$$

the expected one-review period cost for subsystem $w$ based on the distributable inventory that will be in the subsystem after $A_w$ periods. This expression is correct because we assume that any inventory that is obtained from an external supplier to meet an emergency demand in this subsystem during the lead time $A_w$ is withdrawn from the new allocation, $\overline{R}^w$, and sent to the external supplier that provided the emergency stock.

### 7.3.4.2   The Distributable System Allocation Problem

The variable $R^0$ denotes the total distributable inventory of the system: it equals the sum of the inventory levels in each subsystem, $\sum_{w \in W} \overline{R}^w$, after allocation, plus inventory that is retained at the depot. As in the lower echelon models, we continue to assume that this allocation can be made without regard to the possibility of inventory imbalance in the different subsystems, including units in transit to the subsystems.

Let $h^0$ denote the base cost of having one unit of inventory in the distribution system for one review period, and $D^0$ denote the total system demand that occurs over the next review period, a random variable. Furthermore, let $C^0 \left( R^0 \right)$ denote the expected distribution system cost over one review period of beginning a review period with $R^0$ units of total distributable inventory, assuming this inventory can be allocated optimally among the subsystems and depot. That is:

$$C^0(R^0) \equiv h^0 E \left[ \left( R^0 - D^0 \right)^+ \right] \tag{7.17}$$

$$+ \min_{\substack{s.t.\ \sum_{w \in W} \overline{R}^w \le R^0 \\ \overline{R}^w \ge 0,\ \text{integer } w \in W}} \left\{ \sum_{w \in W} C^w_{A_w} \left( \overline{R}^w \right) \right\}.$$

### 7.3.4.3   The Single-Period System-Wide Cost Function

We have assumed that backorders at bases are satisfied immediately from an outside source. We further assumed that these units are on loan and are repaid with units that complete the resupply process. This assumption ensures that total distributable physical inventory equals planned inventory in excess of units in resupply; that is, $R^0 = (s - V)^+$. Hence, $X \equiv (V - s)^+$ represents the number of units in the resupply system in excess of system stock. The event that $X > 0$ will be rare due to the shortage costs. Nonetheless, we must account for this event in our cost model.

Each unit-loan was charged a cost when it was first incurred, as captured in (7.15). Let $\overline{b}$ denote the per-review-period loan cost, charged for each review-period that the unit-loan is outstanding. Then the single-review-period, system-wide cost is given by $C^0(R^0) + \overline{b}\,X$. We ignore the inventory holding cost of units in repair, replacement, and transit in this model. Since the inventory policy cannot affect the resupply process, this holding cost is irrelevant to determining the economically optimal value of $s$.

Let $G(s)$ denote the expected steady state one-review period system-wide cost given a total system stock level, $s$. Then, using the probability distribution of $V$, we can compute this function as:

$$G(s) \equiv E\left[C^0\left((s-V)^+\right) + \bar{b}\,(V-s)^+\right]. \tag{7.18}$$

The cost function $G(s)$ is the objective function for the central tactical planning model. It is a single-period, convex, newsvendor-style objective that captures tradeoffs among inventory holding costs, lateral resupply costs, and emergency acquisition costs in a dynamically-optimized, three-echelon distribution system with pooling. Furthermore, through the stationary distribution of $V$, this function is sensitive to the design and management parameters of the resupply system.

We are now in a position to describe the optimal system inventory. Let $s^*$ denote the total system inventory level that minimizes this cost. That is:

$$s^* \equiv \arg\min_{s \geq 0} G(s).$$

It should be obvious that $V$ is stochastically decreasing as $T$ and/or $U$ decrease. Thus, system stocks will be nonincreasing as resupply lead times decrease.

### 7.3.4.4  Disaggregating System Inventory Targets

The model developed to this point provides an approach for determining optimal system-wide inventory, $s^*$, for a single part. In practice, it will be desirable to specify target stock levels for each location in the system. This is easily done using the allocation tools already developed. For example, denote the target base stock level (i.e. target inventory position) for subsystem $w$ by $\overline{R}^{w*}$ and let $\left(\overline{R}^{w*}\right)_{w \in W}$ solve

$$\min_{\substack{s.t.\ \sum_{w \in W} \overline{R}^w \leq s^* \\ \overline{R}^w \geq 0, \text{integer},\ w \in W}} \quad \sum_{w \in W} C_{A_w}^w\left(\overline{R}^w\right).$$

The residual, $s^* - \sum_{w \in W} \overline{R}^{w*}$, is the target inventory to be held in reserve at the depot. Similarly, for each subsystem $w$, set $R^* = \overline{R}^{w*} - E\left[D_{A_{w+1}}^w\right]$ and let the target stock levels for the pools, $\left(R^{p*}\right)_{p \in P_w}$, solve

$$\min_{\substack{s.t.\ \sum_{p \in P} R^p \leq R^* \\ R^p \geq 0,\ \text{integer},\ p \in P}} \quad \left\{\sum_{p \in P} C^P(R^P)\right\}.$$

Finally, let the target stock levels for the bases, $\left(R_j^{p*}\right)_{j \in B_p}$, solve

$$\min_{\substack{s.t.\ \sum_{j \in B_p} R_j^P = R^{p*} \\ R_j^P \geq 0,\ \text{integer},\ \forall j \in B_p}} \quad \left\{\sum_{j \in B_p} C_j^p(R_j^P)\right\}.$$

Observe that each of these problems has a convex objective function and hence can be solved using a marginal analysis algorithm.

### 7.3.4.5 Computational Complexity

In this section, we establish a bound on the computational complexity of finding the value of $s^*$, the optimal total system stock.

The algorithmic approach is to develop piecewise linear approximations to each cost function. Let $\widetilde{C}_j^p(\cdot)$, $\widetilde{C}^p(\cdot)$, $\widetilde{C}_w(\cdot)$, $\widetilde{C}_{A_w}^w(\cdot)$, $\widetilde{C}^0(\cdot)$, and $\widetilde{G}(\cdot)$ denote the piecewise linear approximations to $C_j^p(\cdot)$, $C^p(\cdot)$, $C_w(\cdot)$, $C_{A_w}^w(\cdot)$, $C^0(\cdot)$, and $G(\cdot)$, respectively, for $j \in B_p$, $p \in P_w$, and $w \in W$. Let

$$r_j^p = \left\{ r_{j0}^p, r_{j1}^p, \dots, r_{jn(p,j)}^p \right\}$$

denote the grid for the breakpoints of $\widetilde{C}_j^p(\cdot)$, where $n(p, j)$ denotes the number of points, less one, in the grid. We require $r_{j0}^p = 0$ and $r_{jn}^p > r_{jn-1}^p$ for $n = 1, 2, \dots, n(p, j)$. Let

$$c_j^p = \left\{ c_{j0}^p, c_{j1}^p, \dots, c_{jn(p,j)}^p \right\}$$

denote the breakpoints of $\widetilde{C}_j^p(\cdot)$: i.e., $c_{jn}^p = \widetilde{C}_j^p\left( r_{jn}^p \right)$ for $n = 1, 2, \dots, n(p, j)$. Similarly, define pairs of vectors $(r^p, c^p)$, $(r_w, c_w)$, $\left( r_{A_w}^w, c_{A_w}^w \right)$, $(r^0, c^0)$, and $(r, c)$ to denote the grids and breakpoints of $\widetilde{C}^p(\cdot)$, $\widetilde{C}_w(\cdot)$, $\widetilde{C}_{A_w}^w(\cdot)$, $\widetilde{C}^0(\cdot)$, and $\widetilde{G}(\cdot)$, respectively. Let $n_w(p)$, $n_w(0)$, $n_A(w)$, $n_A(0)$, $n(0)$, respectively, denote the number of points in each respective grid, less the origin. Let $\bar{n}$ denote an upper bound on the number of grid points in any of these approximations.

The piecewise linear approximations are computed by solving equations (7.13), (7.14), (7.15), (7.16), (7.17), and (7.18) using previously computed piecewise linear approximations to cost functions on the right hand side of these equations wherever appropriate. We assume constant time algorithms exist to compute the probability distributions and expectations required in each equation. Let $\overline{M}_B$ denote an upper bound on the number of locations that must be considered in any of the pooling allocation optimizations (7.14):

$$\overline{M}_B = \max_{w \in W} \max_{p \in P_w} \left| B_p \right|.$$

Similarly, let $\overline{M}_P$ denote an upper bound on the number of locations that must be considered in any of the subsystem allocation optimizations (7.15):

$$\overline{M}_W = \max_{w \in W} |P_w|.$$

Let $\overline{M}_P = |W|$, the number of subsystems that must be considered in the system-wide optimization (7.17). Let $\overline{M}$ denote an upper bound on the number of locations that must be considered in any of the optimizations:

$$\overline{M} \equiv \max \left\{ \overline{M}_B, \overline{M}_P, \overline{M}_W \right\}.$$

Let $\overline{N}$ denote the total number of locations to consider:

$$\overline{N} = 1 + |W| + \sum_{w \in W} |P_w| + \sum_{w \in W} \sum_{p \in P_w} |B_p| .$$

**Proposition 1.** *Assuming constant time algorithms exist for computing the probability distributions required, the number of calculations required to compute $\widetilde{G}(\cdot)$ is $O\left(\left(1 + \frac{3}{4} \log_2(\overline{M})\right)\overline{N}\bar{n}\right)$.*

*Proof.* $O\left(\overline{N}\bar{n}\right)$ is a simple bound on the number of calculations to evaluate all the gridpoints, excluding optimizations. By Proposition 2 in the appendix, the number of calculations to perform the optimization in (7.17) is $O\left((1 + \log_2\left(\overline{M}_W\right))\,\overline{M}_W\bar{n}\right)$. Similarly, the number of calculations to perform each optimization of the form (7.15) is at most $O\left((1 + \log_2(\overline{M}_P))\overline{M}_P\bar{n}\right)$. There are $\overline{M}_W$ optimizations of that form. Likewise, the number of calculations to perform each optimization of the form (7.14) is at most

$$O\left(\left(1 + \log_2\left(\overline{M}_B\right)\right)\overline{M}_B\bar{n}\right)$$

and there are at most $\overline{M}_W\,\overline{M}_P$ optimizations of that form. Assembling these facts, we have that the number of calculations required to compute $\widetilde{G}(\cdot)$ is:

$$O\left( \begin{array}{c} \overline{N}\bar{n} + \left(1 + \log_2\left(\overline{M}_W\right)\right)\overline{M}_W\bar{n} + \overline{M}_W\left(1 + \log_2\left(\overline{M}_P\right)\right)\overline{M}_P\bar{n} \\ + \overline{M}_W\overline{M}_P\left(1 + \log_2\left(\overline{M}_B\right)\right)\overline{M}_B\bar{n} \end{array} \right)$$

$$\leq O\left(\overline{N}\bar{n} + 3\overline{M}_W\overline{M}_P\overline{M}_B\left(1 + \log_2\left(\overline{M}\right)\right)\bar{n}\right).$$

Noting that $\overline{M}_W\overline{M}_P\overline{M}_B$ is of the same order of magnitude as $\overline{N}$ , the result follows.   □

*Remark 1.* Under the further assumption that $\overline{M} = \overline{M}_W = \overline{M}_P = \overline{M}_B$ and that $\overline{M} \simeq \overline{N}^{1/3}$, then the bound on the number of calculations is

$$O\left(\left(1 + \frac{1}{4}\log_2(\overline{N})\right)\overline{N}\bar{n}\right).$$

This is the source of our claim that the optimization of total system inventory can be performed in time that is $n \log(n)$ in the number of locations.

## 7.4   Appendix: The Allocation Optimization

We are given a set $M = \{1, 2, \ldots, \overline{M}\}$ of locations and an augmented set $M_0 = \{0\} \cup M$ that includes one location at a higher level. For each location $m \in M_0$, we are given a set of integer gridpoints $r^m = \left\{r_0^m, r_1^m, \ldots, r_{n(m)}^m\right\}$ indexed by a set $N_m = \{0, 1, \ldots, n(m)\}$, satisfying $r_0^m = 0$ and $r_n^m > r_{n-1}^m$, for all $n > 0$. At each gridpoint, $r_n^m$, for $m \in M$ and $n \in N_m$, we are given a function evaluation,

$c_n^m$, of a convex function. We define a piecewise linear approximation to each original convex function as follows. For each gridpoint, we compute a slope, $\widehat{c}_n{}^m$, according to the following rule:

$$\widehat{c}_n^m = \begin{cases} \dfrac{c_{n+1}^m - c_n^m}{r_{n+1}^m - r_n^m}, & n < n(m); \\[2ex] \dfrac{c_n^m - c_{n-1}^m}{r_n^m - r_{n-1}^m}, & n = n(m). \end{cases} \qquad (7.19)$$

By convexity of the original function, we have $\widehat{c}_n^m \geq \widehat{c}_{n-1}^m$ for all $n > 0$. The piecewise linear approximation function for location $m \in M$ is given by $\widetilde{C}_m(r)$ :

$$\widetilde{C}_m(r) \equiv c_0^m + \sum_{n=0}^{n(m)-1} \left\{ 1_{\{r \geq r_n^m\}} \left( r \wedge r_{n+1}^m - r_n^m \right) \widehat{c}_n^m \right\} \qquad (7.20)$$

$$+ 1_{\{r \geq r_{n(m)}^m\}} \left( r - r_{n(m)}^m \right) \widehat{c}_{n(m)}^m. \qquad (7.21)$$

In addition, we are given a convex function $f(\cdot)$ defined on $\mathcal{R}^+$. The allocation optimization is to find function evaluations, $c_n^0$, for all $n \in N_0$, satisfying

$$c_n^0 = f(r_n^0) + \min_{\substack{r_m \geq 0, \\ r_m \text{ integer}, \ \forall m \in M; \\ \sum_{m \in M} r_m = r_n^0}} \sum_{m \in M} \widetilde{C}_m(r_m). \qquad (7.22)$$

The following marginal analysis algorithm can be used to solve the allocation optimization:

### Definition 4 (*Algorithm* AllocOpt).

1. For each $m \in M$, and each $n \in N_m$, compute $\widehat{c}_n^m$ using (7.19).
2. For each $m \in M$, set $n^*(m) \leftarrow 0$ and $r^*(m) = 0$.
3. Set $m^* = \arg\min_{m \in M} \left\{ \widehat{c}_0^m \right\}$.
4. Set $z \leftarrow \sum_{m \in M} c_0^m$.
5. Set $c_0^0 \leftarrow z$.
6. Set $n \leftarrow 1$.
7. While $n \leq n(0)$, do:
   a) Set $u \leftarrow r_n^0 - r_{n-1}^0$.
   b) While $u > 0$, do:
      i. If $n^*(m^*) = n(m^*)$ then set $x \leftarrow u$; else set

$$x \leftarrow u \wedge \left( r_{n^*(m^*)+1}^{m^*} - r^*(m^*) \right).$$

      ii. Set $z \leftarrow z + x \cdot \widehat{c}_{n^*(m^*)}^m$.
      iii. Set $r^*(m^*) \leftarrow r^*(m^*) + x$.
      iv. If $n^*(m^*) < n(m^*)$ and $r^*(m^*) = r_{n^*(m^*)+1}^{m^*}$, then set $n^*(m^*) \leftarrow n^*(m^*) + 1$.

> v. Set $m^* = \arg\min_{m \in M} \left\{ \widehat{c}^{\,m}_{n^*(m)} \right\}$.
>
> vi. Set $u \leftarrow u - x$.
>
> c) Set $c^0_n \leftarrow z$.
>
> d) Set $n \leftarrow n + 1$.

8. For $n = 0, 1, \ldots, n(0)$, set $c^0_n \leftarrow c^0_n + f(r^0_n)$.

**Proposition 2.** *Algorithm **AllocOpt** terminates with a set $c^0 = \left( c^0_n \right)_{n \in N_0}$ satisfying (7.22) for each $n \in N_0$. Assuming a constant time algorithm exists to compute $f(r)$ for any $r \in \mathcal{R}^+$, algorithm **AllocOpt** requires*

$$O\left( \left( 1 + \log_2 \left( \overline{M} \right) \right) \sum_{m \in M_0} n(m) \right)$$

*calculations.*

*Proof.* Observe that $u$, $n$, and $r^*(m)$, for all $m$, are integers throughout the algorithm. Convexity of the piecewise linear functions (7.20) ensures that a marginal analysis algorithm of the form **AllocOpt** can be used to solve (7.22). The outer loop, step 7, is performed at most $n(0)$ times. The inner loop, (7b), is performed at most $\sum_{m \in M_0} n(m)$ times. This follows because on each loop either $n^*(m)$ is incremented by one for some $m$, or $u$ is set to zero and the loop is terminated. The maximum number of times $n^*(m)$ can be incremented for any $m$ is $n(m)$. The main optimization step, step 7(b)v, requires at most $\log_2 \left( \overline{M} \right)$ comparisons, provided the vector $\widehat{c} = \left( \widehat{c}^{\,m}_{n^*(m)} \right)_{m \in M}$ is maintained as a heap. The number of other calculations, as in steps (1) and (8), is proportional to $\sum_{m \in M_0} n(m)$.   □

*Remark 2.* Equation (7.22) requires a minimization subject to the constraint $\sum_{m \in M} r_m = r^0_n$. It is trivial to extend algorithm **AllocOpt** to constraints of the form $\sum_{m \in M} r_m \leq r^0_n$. One simply modifies the inner loop, step 7b, to read: "While $u > 0$ and $\widehat{c}^{\,m}_{n^*(m^*)} \leq 0$, do … "

## 7.5  Problem Set, Chapter 7

**7.1.** Verify that the equations for the stationary state transitions and the solution to this set of equations, as given in equations (7.3-7.5), are correct.

**7.2.** Extend the analysis provided in Section 7.1.1 in the following ways. First, suppose the system consists of a depot and 3 bases, where the demand rates are unequal. When a demand arises at a base that is out of stock, the base that is selected to resupply it is chosen randomly. All other assumptions made in Section 7.1.1 remain. Develop the lateral resupply model in this case.

Second, again assume the system consists of a depot and 3 bases, where demand rates are unequal. Now assume when a lateral resupply request occurs, the base having the maximum number of days of supply on-hand provides the stock to satisfy this request. Thus the inventory level at each base must be taken into account when constructing the steady state transition rates and equations. Develop these equations.

**7.3.** When developing the probability distributions in Section 7.1.2.1, we assumed that equation (7.8) provided an accurate approximation of the steady state probability that a base has a specified number of units in a backorder status at the depot. Suppose there is a single pool in the system consisting of 5 identical bases. Explore the accuracy of this approximation. Accomplish this task by constructing a simulation model, and then conducting an experiment to test the validity of the approximation for a range of values of $\lambda$ and $A$.

**7.4.** The objective function in the optimization model described in Section 7.1.2.2 contains backorder and lateral resupply costs. The constraints limit inventory investment and ensure that minimum stock levels are maintained. Suppose the model is altered as follows. Suppose the objective is to minimize the sum of holding and lateral resupply costs subject to fill rate constraints at each base and within each pool. Construct a model for this problem and provide an algorithm for finding the base and depot stock levels.

**7.5.** Extend the single-item model presented in Section 7.1.2.2 to a multi-item model. Replace the budget constraint in the single item model with a constraint that limits total investment for all items at all locations. Provide an algorithm that could be used to obtain the desired stock levels at minimum total cost.

**7.6.** Show that the function $C_i(s_i|s_0)$ discussed in Section 7.1.2.2 is convex.

**7.7.** An algorithm for finding depot and base stock levels is outlined in the final paragraph of Section 7.1.2.2. Provide a formal statement of this algorithm. Implement your algorithm using the following data: $P = 2, c = 100, b = 1, a_1 = .2, a_2 = .1, s_i^l = 0$ for $i = 1, 2, s_0^l = 2, B = 1000, A_i = 2$ for $i = 1, 2, D = 3, n_1 = 2, n_2 = 3, \lambda_1 = .1, \lambda_2 = .2.$

**7.8.** The analysis presented in Section 7.2.1 is based on the "no imbalance" assumption. Is this assumption a reasonable one to make when the demand process is a Poisson process and the demand rates are low? What impact does making this assumption have on expected costs? Answer these questions by constructing a simulation and conducting an experiment.

**7.9.** The pooling environment described in Section 7.2.1 is based on the assumption that demands are independent from period to period. As in Section 7.2.1, let $d_{jt}$ be a random variable describing demand at warehouse $j$ in period $t$. However, let

$$d_{jt} = \mu_j L_t + e_{jt}$$

where $\mu_j$ is the expected long term average demand rate at warehouse $j$, $L_t$ is an index random variable for period $t$ that is normally distributed with $E[L_t] = 1$

and Var $[L_t] = \sigma_L^2$, and $e_{jt}$ is the forecast error random variable for warehouse $j$ in period $t$, which has a normal distribution and $E[e_{jt}] = 0$ and Var $[e_{jt}] = \sigma_j^2$.

Suppose the demand correlation that exists among the warehouses and periods is the following autoregressive process of order one:

$$L_t = \sigma_L V_t + 1$$

where

$$V_t = a V_{t-1} + e_{V_t}$$

and the error terms $e_{V_t}$ are independent and identically distributed normal random variables with $E[e_{V_t}] = 0$ and Var $[e_{V_t}] = 1 - a^2$. What is the interpretation of $\sigma_L^2$?

Demonstrate that $V_t$ is normally distributed with $E[V_t] = 0$ and Var $[V_t] = 1$. Thus show that $L_t$ also has a normal distribution with $E[L_t] = 1$ and Var $[L_t] = \sigma_L^2$.

Suppose $d_t = \sum_j d_{jt} = (\sum_{j=1}^m \mu_j) L_t + \sum_{j=1}^m e_{jt}$. Then show that $E[d_t] = \sum_{j=1}^m \mu_j$, and Var $[d_t] = (\sum_{j=1}^m \mu_j)^2 \sigma_L^2 + \sum_{j=1}^m \sigma_j^2$.

Show next that the random variables $d_{jt}$ are not independent by showing that Var $[\sum_{j=1}^m d_{jt}] \neq \sum_{j=1}^m$ Var $[d_{jt}]$.

Next, recall that $D$ is the depot lead time. As earlier in Section 7.2.1.2, let $Y_0 = \sum_{t=1}^D d_t$. Show that $Y_0$ has a normal distribution, with $E[Y_0] = \sum_{t=1}^D \sum_{j=1}^m \mu_j$, and Var $[Y_0] = \sum_{t=1}^D$ Var $[d_t] + 2 \sum\sum_{t>u}$ Cov$(d_t, d_u)$.

Remember when $t > u$,

$$\text{Cov}(d_t, d_u) = E\left(\left(\sum_{j=1}^m \mu_j\right) L_t + \sum_{j=1}^m e_{jt}\right)\left(\left(\sum_{j=1}^m \mu_j\right) L_u + \sum_{j=1}^m e_{ju}\right)\right) - \left(\sum_{j=1}^m \mu_j\right)^2$$

$$= \left(\sum_{j=1}^m \mu_j\right)^2 E[L_t L_u] - \left(\sum_{j=1}^m \mu_j\right)^2.$$

Show that $E[L_t L_u] = \sigma_L^2 E[V_t V_u] + 1$. But $E[V_t V_u] = a^{t-u}$. Using these results, show that

$$\text{Var}[Y_0] = \sum_{t=1}^D \left(\left(\sum_{j=1}^m \mu_j\right)^2 \sigma_L^2 + \sum_{j=1}^m \sigma_j^2\right) + 2\sigma_L^2 \sum\sum_{t>u} \left(\sum_{j=1}^m \mu_j\right)^2 a^{t-u}.$$

Recall that in Section 7.2.1.2 we defined $Y_j = \sum_{t=D+1}^{D+A+1} d_{jt}$. Thus $E[Y_j] = \sum_{t=D+1}^{D+A+1} E[\mu_j L_t + e_{jt}] = \sum_{t=D+1}^{D+A+1} \mu_j = (A+1)\mu_j$. Show that

$$\text{Var}[Y_j] = \sum_{t=D+1}^{D+A+1} (\mu_j^2 \sigma_L^2 + \sigma_j^2) + 2\sigma_L^2 \sum\sum_{\substack{t>u \\ t=D+1,\dots,D+A+1 \\ u=D+1,\dots,D+A}} \mu_j^2 a^{t-u}.$$

Suppose the coefficient of variation of the demand processes is the same at each warehouse. Let $c$ be this common value, that is, $c = \sqrt{\mu_j^2 \sigma_L^2 + \sigma_j^2} / \mu_j$ which in turn implies that $\sigma_j^2 = b\mu_j$ for some $b > 0$. Thus

$$c = \sqrt{\sigma_L^2 + b} \quad \text{or} \quad b = c^2 - \sigma_L^2.$$

Using these observations, derive results similar to those found in Section 7.2.1.2. These results were shown in Erkip, Hausman and Nahmias [79].

**7.10.** The model developed in Section 7.2.2 was based on the assumption that all depot stock was allocated to the warehouses in each period. Prove that when the holding costs are the same at the depot and warehouses and that the imbalance assumption holds, it is optimal to allocate all the depot stock to the warehouses in each period.

**7.11.** Provide an outline of how the three echelon model and algorithm presented in Section 7.2.2 can be extended to an N echelon environment.

**7.12.** Suppose we have a two-echelon, reparable parts service network. The upper echelon, called the depot, resupplies the lower echelon locations, which we call bases. Each demand arises at a base and is the result of a part failure. Part failures at bases arise according to a Poisson process with rate $\lambda_{ij}$ for part type $i$ at base $j$. Once a failure occurs, it is sent to the depot for repair, and the depot provides inventory to replenish the stock. All failed units arriving at the depot are repaired there within an uncapacitated repair facility. Furthermore, bases are grouped into pools. If a demand arises at a base within a pool, and that base does not have stock on hand, the unit will be laterally resupplied to the needy base. If no stock is on hand at any base in the pool, then a third party will "loan" the needy base inventory to meet its current demand. When the depot provides the replenishment stock corresponding to a demand at a base, that unit of replenishment stock goes to the base at which the failure occurred if the unit needed to satisfy the demand came from that base's inventory; otherwise, it goes either to the base that laterally resupplied the base, or to the external source that loaned the base a unit of the item.

Construct a periodic review tactical planning model for this situation. Use the notation found in Section 7.3. State any assumptions that you make. Also, provide an algorithm for obtaining the target stock level for each item at each location.

# 8

# Capacity-limited Systems

The analyses presented in the preceding chapters were based on the premise that the resupply systems either had a reliable source of material (constant lead times) or that repair cycle times were independent and identically distributed. Palm's theorem and its extensions were applied repeatedly to develop required probability distributions. These distributions provided the basis for the tactical planning models that were employed for finding the optimal system stock levels.

We will continue to construct tactical level planning models in this chapter. However, these models will be based on the assumption that resupply is limited by a capacitated resource. The presence of capacity alters the method of analysis substantially.

Many models related to the tactical planning problem can be and have been constructed when capacities exist. One type of model is based on a continuous-time view. These models are often queuing-like models. These types of models have a long history and are described in Buzacott and Shanthikumar [34], a series of papers by Gross and various colleagues [104, 105, 108, 107, 110, 113], and Nahmias [186]. Another type of tactical planning model is a variation of the periodic review model presented in Chapter 7. These periodic review models are often based on ideas presented by Prabhu [195, 196], Glasserman [95] , Glasserman and Tayur [96, 97, 98], Tayur [247], and Roundy and Muckstadt [182, 206].

In the remainder of this chapter we will examine both queuing and periodic review based tactical planning models. We will first focus on the periodic review model. However, we begin by presenting some key results that provide the foundation for the periodic review planning models. Specifically, we will develop the concept of a shortfall random variable and will discuss how its probability distribution can be constructed both in an exact and in an approximate manner.

## 8.1 The Shortfall Distribution

Let us introduce several important ideas by examining a simple capacity limited production facility. Production decisions are made in each period of an infinite planning horizon. There is only one item being produced and demands for this item are independent and identically distributed from period to period.

Assume that the system operates in the following manner. At the start of a period we observe the demand for the single item. Based on that period's demand and the item's current net inventory level, a production quantity is determined. Production then takes place. The production quantity is limited to a maximum of $c$ units, the facility's per period production capacity. Whenever the demand exceeds $c$ plus the on-hand stock, we assume the unsatisfied demand is backordered. On the other hand, if the demand is less than $c$ plus the on-hand stock, then all customer demands are satisfied. After this production occurs, shipments are made to customers. The final act in a period is to charge holding and backorder costs of $h$ and $b$ dollars per unit held or backordered, respectively.

The optimal policy for managing this system is a modified version of the (s−1,s) policy. Suppose $s$ is our target inventory level. The modified (s−1,s) policy states that we should produce to have $s$ units on-hand after satisfying demand if capacity permits; otherwise, produce $c$ units, the system's production capacity. As long as there is positive probability that demand in a period can exceed $c$ units, then there is a positive probability that the production capacity in a period is not sufficient to raise the period ending inventory to $s$. The amount by which $s$ exceeds the actual period ending inventory level is called the shortfall. The shortfall is a random variable. Our goal is to show how to compute its distribution.

The proof that the modified (s−1,s) policy is optimal can be found in Janakiraman and Muckstadt [140] and elsewhere. There are a number of authors who have addressed problems relating to the system we have described and its multi-echelon generalizations (e.g. see Glasserman and Tayur [96, 98]). We will first show some general properties of this shortfall random variable and then focus on the case where demand is described by a discrete demand distribution. We will then examine the case in which per period demand is approximated by a continuous random variable. In the latter case, we will show how an approximate probability distribution for the shortfall random variable can be computed.

### 8.1.1  General Properties

Let $V_n$ represent the random variable for the shortfall in period $n$. We assume that the expected per period demand is strictly less than $c$. If this is not the case, then the backorder quantity will grow without bound as $n \to \infty$ with probability 1. Suppose $D_n$ measures the demand in period $n$. Since $E[D_n] < c$ for all $n$, and the random variables $D_n$ are independent and identically distributed, a stationary distribution exists for the shortfall process. Let $V$ represent this random variable. Thus, in steady state, $V = s - I$, where $I$ is the period ending net inventory level.

Recall the sequence of events that occur in period $n$. The initial net inventory $I_n$ is equal to $s - V_{n-1}$. Demand is observed, that is, $D_n$ is observed. A production quantity is then determined which equals $\min\{c, V_{n-1} + D_n\}$. Note that if $V_{n-1} > s$, then $(V_{n-1} - s)$ backorders exist at the beginning of period $n$. At the end of period $n$, the net inventory is $s - V_n$. If this quantity is positive, then there is stock on hand and is charged a holding cost of $h$ dollars per unit; if negative, then there are backorders which are charged at a cost of $b$ dollars per unit backordered. Figure 8.1 illustrates the evolution of the net inventory random variable.

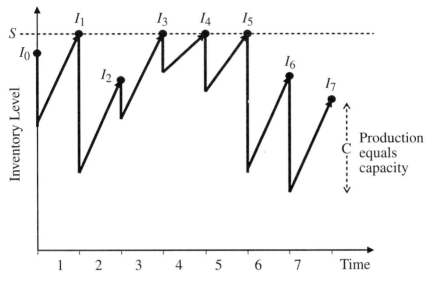

**Fig. 8.1.** System Evolution of Net Inventory

Observe that

$$V_n = \left[V_{n-1} + D_n - c\right]^+ . \tag{8.1}$$

That is, if the capacity is large enough to satisfy both the entering shortfall plus the current period's demand, then $V_n = 0$; otherwise $V_n$ equals the difference between the total requirement ($V_{n-1} + D_n$) and the production capacity ($c$).

We observe that $V_n$ is independent of the target stock level $s$. Equation (8.1) describes the period-to-period dynamic behavior of the shortfall random variable. This behavior of $V_n$ is illustrated in Figure 8.2.

### 8.1.2 Discrete Demand Case

Assume $V_0 = 0$. Observe from Equation (8.1) that $V_n$ depends only on $V_{n-1}$ and $D_n$. That is, it does not depend on $V_0, \ldots, V_{n-2}$. Hence we can model the transitions of the shortfall process as a Markov chain. Specifically the transition probabilities for this chain are as follows:

$$p_{ij} \equiv P\{V_n = j \mid V_{n-1} = i\} = \begin{cases} P\{D \le c - i\}, & j = 0 \text{ and } i \le c \\ P\{D = c + (j - i)\}, & j > 0, i \le c + j \\ 0, & \text{otherwise} \end{cases}$$

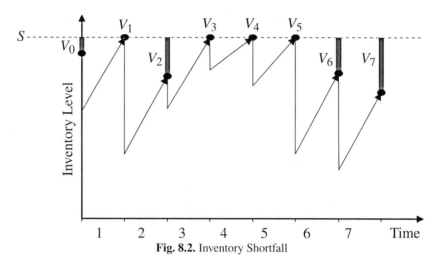

**Fig. 8.2.** Inventory Shortfall

Let $\mathcal{P} = [p_{ij}]$ be the matrix of transition probabilities. Since we assume that $E[D] < c$, a steady state distribution exists for the random variable, $V$, since the chain is ergodic. Let the stationary distribution that $V = i$ be denoted by $\pi_i$, and $\pi$ the vector whose $i^{\text{th}}$ component is $\pi_i$. Then $\pi$ solves

$$\pi \mathcal{P} = \pi$$
$$\sum \pi_i = 1$$
$$\pi_i \geq 0.$$

For practical situations, the matrix $\mathcal{P}$ and the corresponding vector $\pi$ can be truncated to yield a finite system of equations. Some testing needs to be done to insure that accuracy is not sacrificed. The truncation process will depend on the difference between $c$ and $E(D)$ and the variance of the demand process.

Suppose $c$ varies from period-to-period. Specifically, suppose $c$ is a random variable that is independent and identically distributed from period to period. We assume that $E[D] < E[c]$. In this case, we can again represent the transitions of the shortfall process as a Markov chain. The process remains ergodic so that a stationary distribution will exist. In this case, the transition probabilities are given by

$$p_{ij} \equiv P\{V_n = j | V_{n-1} = i\} =$$
$$\begin{cases} \sum_{a \geq i} P\{D \leq a - i\} \cdot P\{c = a\}; & j = 0, i \leq c \\ \sum_{a \geq i-j} P\{D = a + (j - i)\} P\{c = a\}; & j > 0, i \leq c + j \\ 0; & \text{otherwise.} \end{cases}$$

In many cases the capacity is a random variable. If capacity can vary significantly from period to period, then the target stock level can increase substantially.

Let us now illustrate how the steady state shortfall distribution of $V$ behaves for different levels of demand variation and available capacity. We assume that the expected demand per period is 100 units in all cases. We further assume the capacity $c$ does not vary from period to period. Table 8.1 shows how both the expected value of $V$ and standard deviation of $V$ change for 20 combinations of the per period variance of demand and the amount of available per period capacity. These data show how sensitive the mean and standard deviations of $V$ are to changes in these values, and hence how inventory requirements will also depend on these values.

| | | *Capacity* **Per Period** *(Utilization)* | | | |
| --- | --- | --- | --- | --- | --- |
| | | *120* *(83.3%)* | *110* *(90.9%)* | *105* *(95.2%)* | *101* *(99.0%)* |
| *Demand Variance Per Period (Coefficient of Variation)* | *101* *(0.10)* | E(V) = 0.11 StDev 0.93 | 1.36 3.83 | 5.51 9.44 | 45.04 50.59 |
| | *200* *(0.14)* | E(V) = 0.76 StDev 3.22 | 4.26 9.06 | 13.32 19.83 | 92.47 100.72 |
| | *500* *(0.22)* | E(V) = 4.48 StDev 11.48 | 15.35 25.48 | 39.41 51.56 | 238.60 252.68 |
| | *1000* *(0.32)* | E(V) = 12.96 StDev 26.18 | 36.34 53.26 | 85.45 104.63 | 484.69 505.96 |
| | *2000* *(0.45)* | E(V) = 32.91 StDev 56.36 | 81.39 109.17 | 180.57 210.92 | 979.88 1,012.55 |

**Table 8.1.** Expected Shortfall and Standard Deviation of Shortfall for Various Combinations of Capacity Utilization and Demand Variation

Additionally, Figures 8.3-8.8 provide the probability distributions for $V$ for certain cases. Figures 8.3-8.5 contain plots of the distribution of $V$ when the variance of the demand is set to 101 and when the capacity utilization rate assumes three different values; .833, .952, and .99. Similarly, Figures 8.6-8.8 show the probability distributions for $V$ when the variance of demand per period is 2000, again for the same three different amounts of available capacity. Note how the shapes of the distributions are affected by the variance of the demand process and the utilization rate. Furthermore note how the tail of the distributions behave.

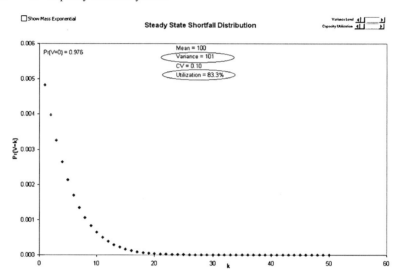

**Fig. 8.3.** Steady State Shortfall Distribution: Exact

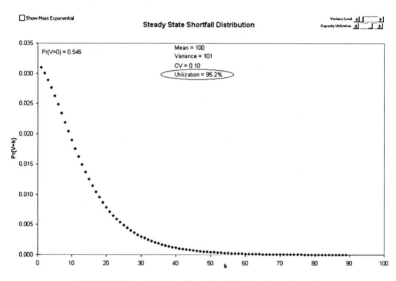

**Fig. 8.4.** Steady State Shortfall Distribution: Exact

### 8.1.3  Continuous Demand Case

When the demand per period is sufficiently large the calculation of the stationary probabilities for $V$ using the Markov chain approach is impractical. That is, the time required to compute the probabilities is substantial in this case. Furthermore, if the demand process is represented by a continuous distribution, then $V$ will also be a continuous random variable. Thus it is of interest to calculate the distribution of $V$ using a continuous approximation to the demand process. We will

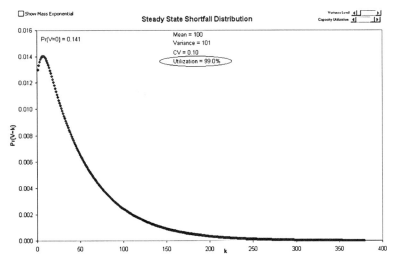

**Fig. 8.5.** Steady State Shortfall Distribution: Exact

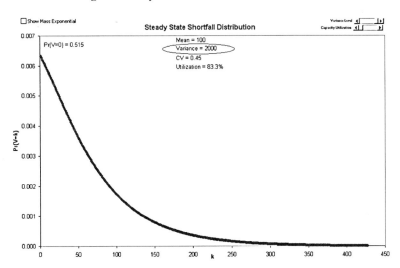

**Fig. 8.6.** Steady State Shortfall Distribution: Exact

also approximate the distribution of $V$. We will see that the computation of this approximate distribution can be made quite easily. Let us now see how this can be done.

Let $v > 0$. Then

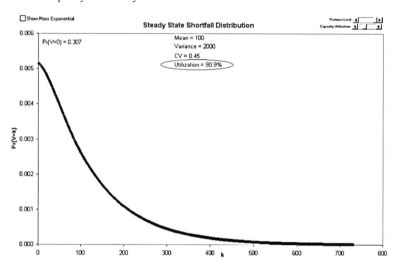

**Fig. 8.7.** Steady State Shortfall Distribution: Exact

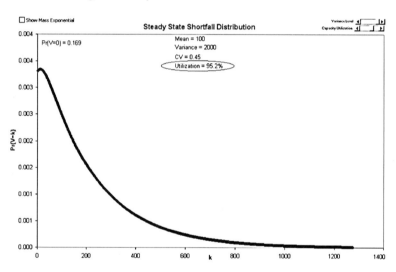

**Fig. 8.8.** Steady State Shortfall Distribution: Exact

$$
\begin{aligned}
P\left\{V_n > v\right\} &= P\left\{[V_{n-1} + D_n - c]^+ > v\right\} \\
&= P\left\{V_{n-1} + D_n - c > v\right\} \\
&= P\left\{D_n > v + c\right\} \\
&\quad + E_D\left[1(d \le v + c) \cdot P\left[V_{n-1} > v + c - d \,|\, D_n = d\right]\right],
\end{aligned}
\tag{8.2}
$$

where $1(A) = 1$ if true and 0 otherwise. Furthermore, let $F_V(\cdot)$ and $\overline{F}_V(\cdot)$ denote the cumulative and complementary cumulative distribution functions for the random variable $V$, and let $f_D(\cdot), \overline{F}_D(\cdot)$ represent the density and complementary cumulative distribution functions for $D$. If $P[D = 0] = q$, then equation (8.2) can be written as

$$\overline{F}_{V_n}(v) = \overline{F}_{D_n}(v+c) + q\overline{F}_{V_{n-1}}(v+c)$$
$$+ \int_0^{v+c} \overline{F}_{V_{n-1}}(v+c-x)f_D(x)\,dx.$$

The random variable $V$ is sometimes assumed to have a mass exponential distribution. The probability that $V = 0$ is positive and the remainder of the distribution is exponential in this case. One reason this approximation is made is due to a theorem proven by Glasserman [95]. Roughly speaking, he shows that the tail of the shortfall distribution behaves like an exponential function. Specifically, his theorem is as follows.

**Theorem 11.** *Assume that* $E[e^{\alpha D}] < \infty$ *for all* $\alpha < \delta$, $\delta > 0$ *and that* $P[D > c] > 0$. *Then there exist* $\beta$ *and* $\alpha$ *such that* $\frac{P\{V > v\}}{\beta e^{-\alpha v}} \to 1$ *as* $v \to \infty$. *Also,* $\alpha$ *is the unique, strictly positive solution to* $E[e^{-\alpha(c-D)}] = 1$. *When* $D$ *is normally distributed, Glasserman states that* $\beta \approx e^{-2(.583)\frac{(c-E(D))}{\sigma}}$.

Let us use a mass exponential distribution to approximate the distribution of $V$. We will discuss the quality of this approximation subsequently. Additional discussion can be found in Roundy and Muckstadt [182, 206].

Let $p_0$ approximate $P\{V = 0\}$, the probability that there is no shortfall, and $\overline{p}_0 = 1 - p_0$. Assuming $V$ has a mass exponential distribution.

$$\overline{F}_V(v) = \begin{cases} \overline{p}_0 e^{-\gamma v}, & v \geq 0 \\ 0, & \text{otherwise} \end{cases}. \tag{8.3}$$

The parameters $p_0$ and $\gamma$ depend on $A = c - E(D)$ and the distribution of demand.

First, substitute (8.3) into (8.2) to obtain

$$\overline{p}_0 e^{-\gamma V} = \overline{F}_D(v+c) + E[1(D \leq v+c) \cdot \overline{p}_0 e^{-\gamma(v+c-D)}]$$

Multiplying both sides by $e^{\gamma v}$ yields

$$\overline{p}_0 = e^{\gamma v}\overline{F}_D(v+c) + \overline{p}_0 e^{-\gamma c} E[1(D \leq v+c)e^{\gamma D}]. \tag{8.4}$$

Letting $v \to \infty$ results in

$$\overline{p}_0 = \overline{p}_0 e^{-\gamma c} \lim_{v \to \infty} E[1(D \leq v+c)e^{\gamma D}]$$
$$= \overline{p}_0 e^{-\gamma c} E[e^{\gamma D}]$$

since, for a stable solution to exist, $\lim_{v \to \infty} e^{\gamma V}\overline{F}_D(v+c) = 0$. Hence the parameter $\gamma$ must solve

$$e^{\gamma c} = E[e^{\gamma D}].$$

Now when $v = 0$ we see from (8.4) that

$$\overline{p}_0 = \overline{F}_D(c) + \overline{p}_0 e^{-\gamma c} E(1(D \le c) e^{\gamma D})$$

or

$$\overline{p}_0 = \frac{\overline{F}_D(c)}{1 - e^{-\gamma c} E(1(D \le c) e^{\gamma D})}$$

and

$$p_0 = 1 - \frac{\overline{F}_D(c)}{1 - e^{-\gamma c} E(1(D \le c) e^{\gamma D})}.$$

An important special case occurs when demand is normally distributed, which is a reasonable approximation when the demand process is a Poisson process with a large mean. In this case

$$\overline{p}_0 = \frac{\overline{F}_D(c)}{1 - \overline{F}_D(c)}$$

and

$$\gamma = \frac{2(c - E(D))}{\sigma^2},$$

where $\sigma^2$ is the variance of the per period demand.

Let us now turn to evaluating a key performance measure that depends on the system's shortfall distribution, the fill rate. As before $V_n$ and $V_{n-1}$ represent the shortfall random variables at the end of periods $n$ and $n - 1$, respectively. Suppose the stock level for the system is denoted by $s$. Observe that the number of backorders that remain at the end of period $n$ corresponding to demands occurring in periods $n - 1$ and earlier is given by

$$\left[V_{n-1} - s - c\right]^+.$$

Then the number of demands arising in period $n$ that are backlogged at the end of period $n$ is given by

$$[V_n - s]^+ - \left[V_{n-1} - s - c\right]^+.$$

Therefore, in steady state,

$$\eta(s) = E\left[(V - s)^+ - (V - s - c)^+\right]$$

represents the expected number of units that are demanded in a period that are backordered at the period's end. The fill rate is given by

$$a = \mathcal{FR}(s) = 1 - \frac{\eta(s)}{E(D)}.$$

We observe that

$$\eta(s) \approx \frac{\overline{p}_0}{\gamma} e^{-\gamma s}(1 - e^{-\gamma c})$$

and that

$$s \approx \left[ -\frac{1}{\gamma} \ell n \left\{ \frac{\eta(s) \cdot \gamma}{\overline{p}_0(1 - e^{-\gamma c})} \right\} \right]^+ = \left[ -\frac{1}{\gamma} \ell n \left\{ \frac{(1-a)\gamma E(D)}{\overline{p}_0(1 - e^{-\gamma c})} \right\} \right]^+$$

as discussed in Roundy and Muckstadt [182, 206].

An interesting question arises as to how well the mass exponential approximation matches the actual distribution function for $V$. To illustrate how well the approximation represents the distribution for $V$, consider again the example environment presented in the previous section. Figure 8.9 through Figure 8.14 contain graphs of both the exact and approximate distributions. As can be seen, the quality of the approximation varies from case to case. Roundy and Muckstadt [182, 206] performed an extensive analysis of the accuracy of the mass exponential approximation. They showed in many cases that this approximation is poor and that using it to set stock levels can result in excessive amounts of inventory. Nonetheless, the approximation is appealing because of its simplicity. Roundy and Muckstadt [182, 206] provide an extensive discussion of the mass exponential for various demand distributions.

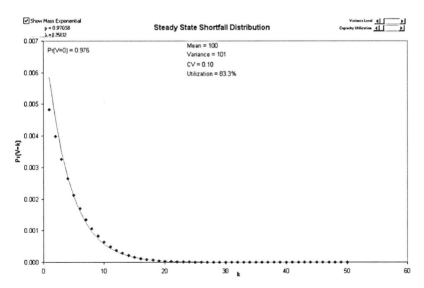

**Fig. 8.9.** Steady State Shortfall Distribution: A Comparison of Exact and Approximate Distributions

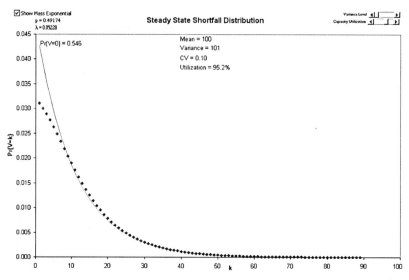

**Fig. 8.10.** Steady State Shortfall Distribution: A Comparison of Exact and Approximate Distributions

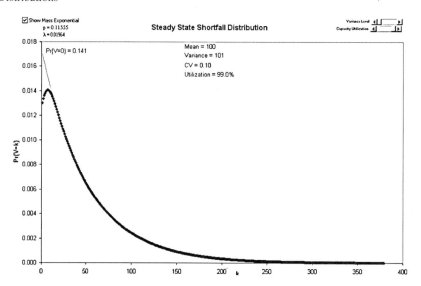

**Fig. 8.11.** Steady State Shortfall Distribution: A Comparison of Exact and Approximate Distributions

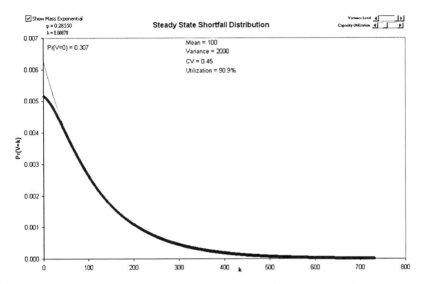

**Fig. 8.12.** Steady State Shortfall Distribution: A Comparison of Exact and Approximate Distributions

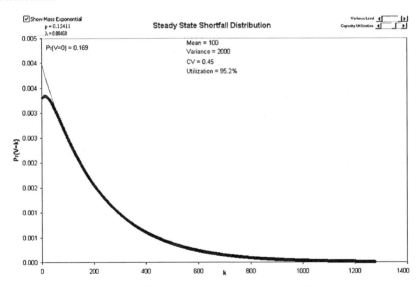

**Fig. 8.13.** Steady State Shortfall Distribution: A Comparison of Exact and Approximate Distributions

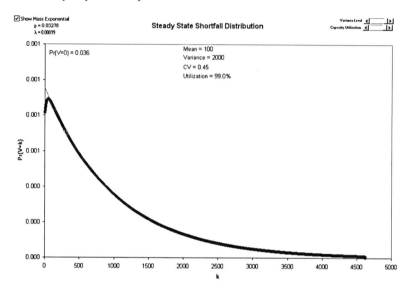

**Fig. 8.14.** Steady State Shortfall Distribution: A Comparison of Exact and Approximate Distributions

Because the approximation is often inaccurate, Roundy and Muckstadt [182, 206] proposed and tested another approximation, which is also based on equation (8.2). We will next briefly describe the alternative approximation.

Suppose we alter equation (8.3) slightly and define

$$\overline{F}_V^0(v) = \begin{cases} \overline{p}_1 e^{-\gamma v}, & v \geq 0 \\ 0, & \text{otherwise.} \end{cases}$$

We will define the parameter $\overline{p}_1$ subsequently. The new approximation is

$$\overline{F}_V^1(v) = \overline{F}_D(v + c) + E_D \left[ 1(D \leq v + c) \cdot \overline{F}_V^0(v + c - d | D = d) \right]$$

$$= \overline{F}_D(v + c) + E_D \left[ 1(D \leq v + c) \cdot \overline{p}_1 e^{-\gamma(v+c-d)} | D = d \right]$$

$$= \overline{F}_D(v + c) + \overline{p}_1 \int_{-\infty}^{v+c} e^{-\gamma(v+c-x)} f_D(x) \, dx$$

$$= \overline{F}_D(v + c) + \overline{p}_1 e^{-\gamma(v+c)} \int_{-\infty}^{v+c} e^{\gamma x} f_D(x) \, dx$$

Our goal is to improve upon our estimate of the probability of the shortfall being positive, that is, $\overline{p}_1$. Suppose we substitute the above expression for $\overline{F}_V^1(v)$ for $\overline{F}_V^0(v)$ and set $v = 0$. Then we have

$$\overline{F}_V^1(0) = \overline{F}_D(c) + \int_{-\infty}^c \overline{F}_V^1(c-x) f_D(x) \, dx$$

$$= \overline{F}_D(c) + \int_{-\infty}^c \left[ \overline{F}_D(2c-x) + \overline{p}_1 e^{-\gamma(2c-x)} \int_{-\infty}^{2c-x} e^{\gamma y} f_D(y) \, dy \right] f_D(x) \, dx$$

$$= \overline{F}_D(c) + \int_{-\infty}^c \overline{F}_D(2c-x) f_D(x) \, dx$$

$$+ \overline{p}_1 \int_{-\infty}^c e^{-\gamma(2c-x)} \left[ \int_{-\infty}^{2c-x} e^{\gamma y} f_D(y) \, dy \right] f_D(x) \, dx.$$

Also, from above, we know that

$$\overline{F}_V^1(0) = \overline{F}_D(c) + \overline{p}_1 e^{-\gamma c} \int_{-\infty}^c e^{\gamma x} f_D(x) \, dx.$$

Combining the above we obtain

$$\overline{p}_1 = \frac{\int_{-\infty}^c \overline{F}_D(2c-x) f_D(x) \, dx}{e^{-\gamma c} \int_{-\infty}^c e^{\gamma x} f_D(x) \, dx - \int_{-\infty}^c e^{-\gamma(2c-x)} \left[ \int_{-\infty}^{2c-x} e^{\gamma y} f_D(y) \, dy \right] f_D(x) \, dx}.$$

The process could be continued to compute $\overline{F}_V^2(v)$ in which we would use $\overline{F}_V^1(v)$ as we did $\overline{F}_V^0(v)$ above.

Roundy and Muckstadt [182, 206] show that this new approximation is quite accurate for a wide range of standard continuous probability models. Unfortunately, the computations do require numerical integration.

## 8.2 Capacity-limited Multi-Echelon Repair System

In Chapter 7 we studied a three echelon resupply system for reparable parts. That system consisted of a lowest echelon in which there were a set of inventory pools, each of which contained a set of bases; a set of intermediate stocking facilities, each of which resupplied a collection of inventory pools; a depot, which resupplied intermediate stocking facilities (the depot consists of a depot repair facility and a depot stocking facility); an external supplier, which supplies parts that have been condemned; and a set of third-party emergency supply sources. In Chapter 7 we assumed that the repair cycle times for all parts entering the depot repair process were independent and identically distributed. In this section we will revisit this same problem, but will now assume that the depot has a capacity on the number of units that can be repaired per period. We begin by assuming that we have only one part type in the system. We will then extend our analysis to the case where there are multiple part types.

### 8.2.1  Single Item System

The tactical planning problem we will study has as its goal the determination of the target stock level for the system.

Recall that the system that we are studying operates in the following manner. In each period, demand for units arises at each base. If the base has stock on hand, the demand is satisfied from that base's stock. If a unit of demand can not be satisfied from that base's stock, then a request is placed for a unit of stock on other members of the pool. If one of them has a unit available, then a lateral resupply transaction occurs. If no such stock is available, then a request is made on the remainder of the subsystem of which the base is a member. That is, the other pools within the collection of pools supported by the same intermediate stocking facility and the intermediate stocking facility are requested to satisfy the demand. If there is no stock anywhere in that subsystem, we assume an external supplier provides the needed unit. In this case, this external supplier will be compensated and will receive a replacement for the "borrowed" part in a subsequent time period. Costs are incurred each period for lateral and emergency resupply transaction.

Each part that is demanded must be resupplied to the system either through a repair process or from an external supplier (e.g. the manufacturer). When a unit fails, there is a probability $r$ that it will enter the depot repair resupply process and $1 - r$ that it will be resupplied from the external supplier. If it is repaired, we assume the unit is shipped from the base to the depot, which takes $D$ periods. The unit then enters the depot repair queue where it remains until it enters the repair process. The repair process is completed in one period. When the failed unit is condemned, a replacement unit is ordered from the supplier.

Corresponding to each demand at the bases, which occur according to independent Poisson processes, there is a resupply request placed on the depot. The depot sends replenishment stock to the appropriate intermediate stocking facility, which, in turn, resupplies the base or external emergency supply location that provided the unit of stock needed to satisfy a demand. Allocation of stock is always made based on a greatest need basis rather than a first-come-first-serve basis, as discussed in Chapter 7.

Figure 8.15 shows the structure of this system.

Recall that in Chapter 7 we developed a tactical planning model based on this system's operation and also on an assumption which we called the imbalance assumption. A key component of that model was the development of the probability distribution of the number of units in resupply, which we denoted by the random variable $V$. We now see how to construct this distribution when there is limited repair capacity each period. Once this distribution is determined, we would employ exactly the same method as described in Chapter 7 to find the target system stock level.

When repair capacity is limited, there are three separate segments to the resupply system. First, there is the transport process associated with the movement of reparable units to the depot. Second, there are units in the repair queue and repair process. Third, there are units on order from the external supplier.

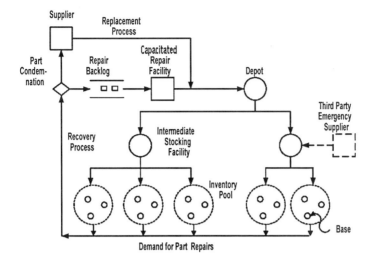

**Fig. 8.15.** Structure of the Resupply System

Recall that we assume the demand process at each base is an independent and identically distributed Poisson process from period to period. Furthermore, demands at one base are independent from those occurring at other bases.

Let $V^T$ be the random variable describing the number of units on the way to the depot from the bases that require repair, which is measured at the end of a period. Let $V^R$ be the random variable describing the number of units awaiting repair at the end of a period. Finally, let $V^U$ denote the random variable representing the number of units on order from the external supplier at the end of a period.

Then $V = V^R + V^T + V^U$. Based on our assumptions, these three random variables are independent. Consequently, the distribution of $V$ is the convolution of the three distributions. As was the case in Chapter 7, $V^T$ and $V^U$ have Poisson distributions. Hence $V^T + V^U$ is Poisson distributed. Thus it remains for us to find the distribution for $V^R$. This distribution is easily determined using the methods described in the previous section, as we will now see.

The units arrive to the repair queue as a consequence of the failures of units each period at the bases. The number of arrivals per period is described by a Poisson distribution with rate $r \cdot \sum_j \lambda_j$, where $\lambda_j$ is the demand rate at base $j$. The number of units remaining in the queue awaiting repair depends on this arrival process as well as the per period repair capacity, which we continue to denote by $c$.

The number of units in the repair queue at the end of period $n$ we denote by $V_n^R$. As we saw earlier,

$$V_n^R = \left[ V_{n-1}^R + D_n - c \right]^+,$$

where now $D_n$ measures the number of units arriving to the depot repair facility in period $n$. We assume that

$$E[D_n] = r \cdot \sum_j \lambda_j < c$$

so that a stationary distribution for $V^R$ will exist. As we discussed in the last section, the period to period number of units awaiting repair at the end of a period can be represented by a Markov chain. The transition probabilities are those given in Section 8.1.2 of this chapter.

Assume that there is a maximum queue size that is permitted at the end of a period. This maximum can be assured by running overtime to eliminate any excess. In this situation, the Markov chain will have a finite number of states. This chain is clearly ergodic so a stationary distribution will exist for $V^R$. The random variable $V^R$ is analogous to the shortfall random variable. If the stock level is zero, then the shortfall and the number of units in the repair queue are identical.

Thus we can now compute the distribution for $V$ since we have shown how the distributions of the three independent random variables can be determined.

Now that we know the probability distribution for $V$, we can employ exactly the same method of analysis presented in section 7.3 to compute the optimal total system stock for the items.

### 8.2.2  Multiple Item System

Suppose the system we discussed in the previous section exists for many items rather than just a single item. Suppose further that demands at the bases are independent among items in addition to the other assumptions we have made concerning the demand process. Finally, suppose that the only way in which they interact is through the depot repair system. That is, these items share repair capacity. We assume the repair capacity required to repair any item type is the same.

The shortfall distribution $V^R$ that we just derived pertains to the total number of units in the repair queue when the repair capacity is shared. Suppose we compute $V$ using this distribution, where $V$ now measures the total number of units in the resupply system. That is, $V = \sum_i (V_i^T + V_i^U) + V^R$, where $V_i^T$ and $V_i^U$ are, respectively, the number of units in the transport system to the depot for item $i$ and on order with the supplier for item $i$. Since all random variables are assumed to be independent with Poisson distributions, $\sum_i (V_i^T + V_i^U)$ also has a Poisson distribution. $V^R$ has a distribution computed as discussed in the last section. Thus we can compute the distribution of the random variable $V$. We now show how we can use this distribution to obtain stock levels for each item.

Suppose there are $n$ item types in the resupply system that share a common repair resource. Furthermore, suppose the total system stock level for item $i$ is $s_i$ units. Then the total system stock over all items is $s = \sum_i s_i$. Now $V$ measures the total number of units of all item types that are in the resupply system just after all repairs in the period have been completed and all shipments from the suppliers have been received.

Suppose we let $V_i$ be a random variable representing the number of units of item $i$ that are in the resupply system in the steady state. Additionally, let $V_i^R$ be a random variable that measures the number of units of item $i$ that are in the depot repair queue at the end of a period. Suppose $\overline{\lambda}_i = r_i \sum_j \lambda_{ij}$, where $\lambda_{ij}$ is the per period demand rate for item $i$ at base $j$, and $\overline{\lambda} = \sum_i \overline{\lambda}_i$.

Suppose there is no prioritization of repair at the depot. That is, units enter the repair process on a first come, first served basis. Then

$$P\{V_i^R = v_i \mid V^R = v\} = \binom{v}{v_i} \left[ \frac{\overline{\lambda}_i}{\overline{\lambda}} \right]^{v_i} (1 - \frac{\overline{\lambda}_i}{\overline{\lambda}})^{v - v_i}$$

and

$$P\{V_i^R = v_i\} = \sum_{v \geq v_i} P\{V_i^R = v_i \mid V^R = v\} P\{V^R = v\}.$$

In the previous section we discussed how to determine the probability distribution for the random variable $V^R$. Thus we may readily compute $P\{V_i^R = v_i\}$.

Recall that the random variables $V_i^T$ and $V_i^U$ each have Poisson distributions. Furthermore, note that $V_i^T$, $V_i^U$, and $V_i^R$ are independent. Hence the random variable $V_i = V_i^T + V_i^U + V_i^R$ can be calculated as the convolution of the three known distributions.

Now that we have established the distribution of $V_i$, we can find the values for $s_i$, and hence $s$. Recall in Section 7.3 we showed how to find the optimal total system stock level for an item. By replacing the random variable $V$ in that section with $V_i$, we can use the methodology presented in Section 7.3 to find the optimal values of $s_i$, given the imbalance assumptions that were invoked. Hence the multiple item problem decomposes into solving a series of single item problems.

## 8.3   A Continuous-Time Capacity-limited System

Let us now consider a continuous-time tactical planning model of a simple system consisting of a single location, which we call the depot, at which repairs are conducted and stock is held for $n$ item types. We assume material flows in the system as shown in Figure 8.3.

Reparable units of the $n$ items arrive to the system according to a Poisson process with rate $\lambda_i$, $i = 1, \ldots, n$, measured in units per hour. Once the reparable units arrive, they are immediately entered into the repair queue. The repair actions are conducted on a first-come, first-serve basis. Each arriving reparable unit triggers a replenishment order for a unit of the same type. If one is on hand, it is shipped immediately; otherwise, the replenishment request is backordered.

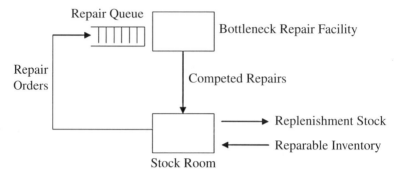

**Fig. 8.16.** Resupply System

We assume that once a unit of any item type enters the bottleneck repair work center, its repair time is exponentially distributed with an expected repair time of $1/\mu$ hours. Repair times are assumed to be independent of each other as well. However, the repair system works on only one unit at a time. We also assume that there are no changeover or set up times so that batching of repair is not an issue.

The goal is to find stock levels for each of the items, which we denote by $s_i$, so as to minimize the expected holding and backorder costs incurred per unit of time, in this case, per hour. We charge holding costs proportional to the expected on hand serviceable stock. Let $h_i$ be the holding cost rate for item $i$. We also incur backorder costs for item $i$ at a rate of $b_i$.

Let us now introduce notation that we will use as we proceed. Recall that $\lambda_i$ is the demand rate for replenishment stock for item $i$. Let $\lambda = \sum \lambda_i$ be the total system demand rate, which is the arrival rate to the bottleneck repair facility. Furthermore, let

$$p_i = \frac{\lambda_i}{\lambda}, q_i = 1 - p_i,$$

$$p = \frac{\lambda}{\mu}, \text{ the system utilization rate,}$$

$N_i =$ steady state number of units of item type $i$ in the repair system, and

$$N = \sum_{i=1}^{n} N_i.$$

### 8.3.1  Basic Repair System Model

We first examine a queueing model, for which we will be able to derive exact analytic results. Specifically, since we have assumed that the arrival process for orders is a Poisson process and that production times are exponentially distributed, the system is a $M/M/1$ queueing system. As such, we know that $N$, the number of units in the production system is geometrically distributed. (This result can be derived using the standard birth-death equations.)

Furthermore, it is also true that $N_i$ is geometrically distributed, that is

$$P[N_i = j] = (1 - \eta_i)\eta_i^j, \text{ where } \eta_i = \frac{\lambda_i}{\mu - \lambda + \lambda_i},$$

which we now prove.

Recall that the generating function for a geometrically distributed random variable is

$$E[s^N] = \sum_{j=0}^{\infty} s^j P[N = j]$$

$$= \sum_{j=0}^{\infty} s^j (1 - \rho)\rho^j$$

$$= (1 - \rho) \sum_{j=0}^{\infty} (s\rho)^j$$

$$= \frac{(1 - \rho)}{1 - \rho s}, \text{ for sufficiently small } s,$$

where $\rho$ also is the probability that one or more units are in the queueing system. Let's construct the generating function for $N_i$

$$E[s^{N_i}] = \sum_{j=0}^{\infty} E[s^{N_i} | N = j]P[N = j]$$

$$= \sum_{j=0}^{\infty} E[s^{N_i} | N = j](1 - \rho)\rho^j.$$

But

$$E[s^{N_i} | N = j] = \sum_{k=0}^{j} s^k P[N_i = k | N = j]$$

and, as we discussed in Chapter 3, we know that

$$P[N_i = k | N = j] = \binom{j}{k}\left(\frac{\lambda_i}{\lambda}\right)^k \left(1 - \frac{\lambda_i}{\lambda}\right)^{j-k}, \ j \geq k.$$

Therefore

$$E[s^{N_i}] = \sum_{j=0}^{\infty} \sum_{k=0}^{j} s^k \binom{j}{k} \left(\frac{\lambda_i}{\lambda}\right)^k \left(1 - \frac{\lambda_i}{\lambda}\right)^{j-k} (1-\rho)\rho^j$$

$$= \sum_{j=0}^{\infty} \left[ (1-\rho)(\rho q_i)^j \sum_{k=0}^{j} \binom{j}{k} \left(\frac{sp_i}{q_i}\right)^k \right], \left(p_i = \frac{\lambda_i}{\lambda}; q_i = 1 - p_i\right),$$

$$= \sum_{j=0}^{\infty} (1-\rho)(\rho q_i)^j \left(1 + \frac{sp_i}{q_i}\right)^j$$

$$= \sum_{j=0}^{\infty} (1-\rho)(\rho q_i + \rho s p_i)^j$$

$$= (1-\rho)\frac{1}{1 - (q_i\rho + p_i\rho s)}.$$

Since

$$1 - \rho = 1 - (1 - p_i + p_i)\rho = 1 - q_i\rho - p_i\rho,$$

$$E[s^{N_i}] = \frac{1 - q_i\rho - p_i\rho}{1 - (q_i\rho + p_i\rho s)}$$

$$= \frac{1 - \frac{p_i\rho}{1-q_i\rho}}{1 - \frac{p_i\rho s}{1-q_i\rho}}$$

$$= \frac{1 - \eta_i}{1 - \eta_i s},$$

where

$$\eta_i = \frac{p_i\rho}{1 - q_i\rho}$$

$$= \frac{\frac{\lambda_i}{\lambda} \cdot \frac{\lambda}{\mu}}{1 - (1 - \frac{\lambda_i}{\lambda})\frac{\lambda}{\mu}} = \frac{\lambda_i}{\mu - \lambda + \lambda_i}.$$

Hence $N_i$ is geometrically distributed, that is

$$P[N_i = j] \equiv p_i(j) = (1 - \eta_i)\eta_i^j.$$

Thus the random variable measuring the number of units of item type $i$ in the repair system at a random point in time is geometrically distributed. If $s_i$ is the stock level for item $i$, the on-hand serviceable stock is $[s_i - j]^+$ when $j$ units are in the repair system.

We next show how to find the stock levels $s_i$ such that we minimize the *expected holding and backorder costs* for the $n$ products. Our objective is to find the values of $s_i \geq 0$ and integer that

$$\min \sum_{i=1}^{n} \left[ h_i \sum_{j=0}^{s_i} (s_i - j) p_i(j) + b \sum_{j=s_i}^{\infty} (j - s_i) p_i(j) \right].$$

Note this problem can be rewritten as

$$\sum_{i=1}^{n} \left\{ \min_{s_i=0,1,\dots} h_i \sum_{j=0}^{s_i} (s_i - j) p_i(j) + b \sum_{j=s_i+1}^{\infty} (j - s_i) p_i(j) \right\},$$

since there are no constraints among the items. The single product optimization problems have the form of the classic newsvendor problem. Therefore, the optimal $s_i$ is the smallest integer (nonnegative) for which

$$\sum_{j=s_i+1}^{\infty} p_i(j) \le \frac{h_i}{h_i + b}.$$

But in this case

$$\sum_{j=s_i+1}^{\infty} p_i(j) = \sum_{j=s_i+1}^{\infty} (1 - \eta_i) \eta_i^{j}$$

$$= (1 - \eta_i) \eta_i^{s_i+1} \sum_{j=s_i+1}^{\infty} \eta_i^{j-(s_i+1)}$$

$$= \frac{1 - \eta_i}{1 - \eta_i} \eta_i^{s_i+1} = \eta_i^{s_i+1} \le \frac{h_i}{h_i + b}.$$

Clearly $s_i$ depends on $\mu$. But how does $s_i$ change as a function of the production rate, $\mu$? The optimality condition we established for the stock level for item $i$ is to find the smallest nonnegative integer $s_i$ for which

$$\eta_i^{s_i+1} \le \frac{h_i}{h_i + b}.$$

Hence

$$(s_i + 1)\ell n \eta_i \le \ell n h_i - \ell n (h_i + b)$$

or

$$s_i \ge \frac{\ell n h_i - \ell n (h_i + b)}{\ell n \eta_i} - 1.$$

Let

$$f(\mu) = \frac{\ell n h_i - \ell n (h_i + b)}{\ell n \frac{\lambda_i}{\mu - \lambda + \lambda_i}} = \frac{\ell n h_i - \ell n (h_i + b)}{\ell n \lambda_i - \ell n (\mu - \lambda + \lambda_i)}.$$

Then

$$f'(\mu) = \frac{\ell n h_i - \ell n(h_i + b)}{[\ell n \lambda_i - \ell n(\mu - \lambda + \lambda_i)]^2} \cdot \frac{1}{\mu - \lambda + \lambda_i}$$

$$= \ell n\left(\frac{h_i}{h_i + b}\right)\left\{\frac{1}{(\ell n \lambda_i - \ell n(\mu - \lambda + \lambda_i))^2} \cdot \frac{1}{\mu - \lambda + \lambda_i}\right\} < 0,$$

since

$$\ell n\left[\frac{h_i}{(h_i + b)}\right] < 0.$$

Thus $s_i$ is a nonincreasing function of $\mu$, as would be expected.

Let us consider the following example. Suppose $n = 5$ and

$$\lambda = 950/\text{wk}$$
$$\mu = 1000/\text{wk}$$
$$\lambda_1 = 450/\text{wk}$$
$$h_1 = \$1/\text{wk}$$
$$b = \$10.$$

Then

$$\eta_1 = 450/(1000 - 950 + 450) = .9$$

$$s_i = \left[\frac{\ell n 1 - \ell n 11}{\ell n .9} - 1\right]^+$$

$$= \lceil 21.7 \rceil^+ = 22 \text{ units}$$

Observe that, in general,

$$E(\text{on} - \text{hand}) = \sum_{j=0}^{s_i}(s_i - j)(1 - \eta_i)\eta_i^j$$

$$= s_i(1 - \eta_i)\sum_{j=0}^{s_i}\eta_i^j - \sum_{j=0}^{s_i}j(1 - \eta_i)\eta_i^j$$

$$= s_i(1 - \eta_i)\frac{1 - \eta_i^{s_i+1}}{1 - \eta_i} - \eta_i\sum_{j=0}^{s_i}j(1 - \eta_i)\eta_i^{j-1}$$

$$= s_i(1 - \eta_i^{s_i+1}) - \eta_i(1 - \eta_i)\sum_{j=1}^{s_i}j\eta_i^{j-1}$$

$$= s_i(1 - \eta_i^{s_i+1}) - \eta_i(1 - \eta_i)\frac{1 - \eta_i^{s_i} - s_i(1 - \eta_i)\eta_i^{s_i}}{(1 - \eta_i)^2}$$

$$= \frac{s_i(1 - \eta_i^{s_i+1})(1 - \eta_i) - \eta_i[1 - \eta_i^{s_i}] + s_i\eta_i(1 - \eta_i)\eta_i^{s_i}}{(1 - \eta_i)}$$

$$= s_i - \eta_i\frac{1 - \eta_i^{s_i}}{1 - \eta_i}.$$

In this case,

$$E(\text{on} - \text{hand}) = 22 - .9 \left[ \frac{1 - (.9)^{22}}{1 - .09} \right] = 13.89 \text{ units.}$$

We can also show that

$$E(\text{Backorder}) = \frac{\eta_i^{s_i+1}}{1 - \eta_i} = .8863 \text{ in this case, and the}$$

$$E(\text{in repair}) = \frac{\lambda_i}{\mu - \lambda} = \frac{\eta_i}{1 - \eta_i} = 9 \text{ in this case.}$$

## 8.3.2   Compound Poisson Arrival Process (GI/M/1)

We now assume that repair orders for a single item type arrive to the repair system according to a Poisson process with rate $\lambda$; but, the number of units requested per arrival is a random variable. That is, the demand process is a compound Poisson process. Let us consider a specific example where the order quantity is geometrically distributed. Let $X$ be the random variable describing the order size, that is,

$$P\{X = k\} = \begin{cases} (1 - \alpha)\alpha^{(k-1)}; & k \geq 1 \\ 0; & \text{otherwise} \end{cases}$$

and

$$E[X] = \frac{1}{1 - \alpha} \left( \text{or } \alpha = 1 - \frac{1}{E(X)} \right).$$

Although we will not develop the details of the analysis, one can show that the probability distribution of the number of units in the repair system, which we denote by $N$, is

$$P\{N = n\} = \begin{cases} 1 - \rho; & n = 0 \\ (1 - \rho)(1 - \alpha)\rho\gamma^{n-1}; & n > 0 \end{cases}$$

where $\gamma = \alpha + (1 - \alpha)\rho$ and $\rho = \frac{\lambda \cdot E[X]}{\mu}$.

As before, let us assume that the holding and backorder cost rates are $h$ and $b$, respectively. Then the model is

$$\min f(s) = \min_{s=0,1,...} \left( h \sum_{j=0}^{s} (s - j)p(j) + b \sum_{j=s+1}^{\infty} (j - s)p(j) \right).$$

Since $f(s)$ is convex in $s$, we use a first difference approach to find $s^*$. Due to this convexity, the goal is to find the smallest $s$ for which

$$f(s) - f(s + 1) \leq 0, \text{ or}$$

$$h\left(\sum_{j=0}^{s}(s-j)p(j) - \sum_{j=0}^{s}(s+1-j)p(j)\right)$$

$$+ b\left(\sum_{j=s+1}^{\infty}(j-s)p(j) - \sum_{j=s+1}^{\infty}(j-(s+1))p(j)\right)$$

$$= -h\sum_{j=0}^{s}p(j) + b\sum_{j=s+1}^{\infty}p(j)$$

$$= -h(1 - \sum_{j=s+1}^{\infty}p(j)) + b\sum_{j=s+1}^{\infty}p(j)$$

$$= (b+h)\sum_{j=s+1}^{\infty}p(j) - h \le 0$$

$$\text{or } \sum_{j=s+1}^{\infty}p(j) \le \frac{h}{b+h}.$$

Since

$$p(j) = \rho(1-\rho)(1-\alpha)\gamma^{j-1}, \, j > 0,$$

$$(1-\alpha)\rho(1-\rho)\sum_{j=s+1}^{\infty}\gamma^{j-1} = (1-\alpha)\rho(1-\rho)\gamma^{s}\sum_{j=0}^{\infty}\gamma^{j}$$

$$= \frac{(1-\alpha)\rho(1-\rho)\gamma^{s}}{1-(\alpha+(1-\alpha)\rho)}$$

$$= \frac{(1-\alpha)\rho(1-\rho)\gamma^{s}}{(1-\alpha)-(1-\alpha)\rho} = \frac{(1-\alpha)\rho(1-\rho)\gamma^{s}}{(1-\alpha)(1-\rho)}$$

$$= \rho\gamma^{s}.$$

Thus, the goal is to find the smallest $s$ for which

$$\gamma^{s} \le \frac{h}{\rho(b+h)}$$

or

$$s^* = \left\lceil \frac{\ell nh - \ell n\rho - \ell n(b+h)}{\ell n\gamma} \right\rceil^{+}.$$

We also have

$$E(\text{on} - \text{hand}) = s + \frac{\rho\gamma^{s}}{1-\gamma} - \frac{\rho}{(1-\gamma)}$$

and

$$E[\text{backorders}] = \frac{\rho\gamma^{s}}{1-\gamma}.$$

For example, suppose

$$h = \$/\text{wk}$$
$$b = \$10$$
$$\lambda = 450/\text{wk}$$
$$E[X] = 2 \text{ units}$$
$$\mu = 1000/\text{wk}$$

Then we have

$$\rho = \frac{\lambda E[X]}{\mu} = .9$$
$$\alpha = 1 - \frac{1}{2} = \frac{1}{2}$$
$$\gamma = \frac{1}{2} + \frac{1}{2}(.9) = .95.$$

Consequently

$$s^* = \left\lceil \frac{\ell n 1 - \ell n(.9) - \ell n 11}{\ell n(.95)} \right\rceil^+ = \lceil 44.7 \rceil^+ = 45 \text{ units,}$$
$$E(\text{on} - \text{hand}) = 28.79,$$
$$E(\text{backorders}) = 1.79,$$

and
$$f(s^*) = 1 \cdot (28.79) + 10(1.79) = 46.69/\text{wk}.$$

Now compare this with the alternative system in which

$$\lambda = 900 \text{ and } E[X] = 1, \text{ and}$$

we have an $M/M/1$ environment. How do the stock levels compare? In this case $\alpha = 0$, $\gamma = \rho$ and, as we saw earlier, $s^*$ is the largest integer for which

$$\rho^{s+1} \le \frac{h}{h+b}.$$

($\rho = \eta$ in the earlier model) or

$$s^* = \left\lceil \frac{\ell n h - \ell n(n+b) - \ell n \rho}{\ell n \rho} \right\rceil^+.$$

When $\rho = .9$, $h = 1$, and $b = 10$,

$$s^* = \left\lceil \frac{\ell n 1 - \ell n 11 - \ell n(.9)}{\ell n(.9)} \right\rceil^+ = \lceil 21.7 \rceil^+ = 22.$$

Thus as the coefficient of variation of the demand process increases, we see that the optimal stock level goes up, and up by a factor of 2 in this case. Furthermore,

$$f(s^*) = \$22.75/\text{wk}.$$

If the order size distribution changes, the stock level requirement would also change. The more lumpy the demand, the greater the inventory requirement will be. Hence it is desirable to keep the relative variation in the demand process as low as possible to keep inventory levels as low as possible.

### 8.3.3  A Budget Constrained System

Suppose that rather than charging inventory carrying costs, we impose a constraint on the amount of inventory held. For example, suppose we limit the amount of floor space that is available to hold inventory, call it $C$. Let $a_i$ be the amount of space required to hold one unit of item $i$. Furthermore, assume that our goal is to maximize the expected fill rate subject to this floor space constraint. The demand process for item $i$ is again assumed to be a Poisson process with rate $\lambda_i$.

Then the problem can be stated as

$$\max \sum_{i=1}^{n} \frac{\lambda_i}{\lambda} \sum_{j=0}^{s_i-1} (1 - \eta_i)\eta_i^j$$

$$\text{subject to } \sum_{i=1}^{n} a_i s_i \leq C, \quad s_i = 0, 1, \dots . \tag{8.5}$$

We could solve this knapsack like problem using a variety of methods. Rather than considering this general problem, we will more closely examine the case where $a_i = 1$ for all $i$. Consequently, this problem can be solved to optimality using a greedy marginal analysis algorithm.

Let

$$f_i(s_i) = \frac{\lambda_i}{\lambda} \sum_{j=0}^{s_i-1} (1 - \eta_i)\eta_i^j.$$

The effect on the fill rate of increasing the stock level of item $i$ from $s_i$ to $s_i + 1$ is

$$f_i(s_i + 1) - f_i(s_i) = \frac{\lambda_i}{\lambda}(1 - \eta_i)\eta_i^{s_i}. \tag{8.6}$$

Observe that

$$\frac{\lambda_i}{\lambda}(1 - \eta_i) = \frac{\frac{\lambda_i}{\lambda}(\mu - \lambda)}{\mu - \lambda + \lambda_i} = \eta_i \frac{\mu - \lambda}{\lambda}$$

so that

$$f_i(s_i + 1) - f_i(s_i) = \frac{\mu - \lambda}{\lambda}\eta_i^{s_i+1}.$$

Thus to solve the problem with $a_i = 1$, we can use the following algorithm. Set $I = 0$, $s_i = 0$, for all $i$.

WHILE $I < C$

Find $i^*$ such that $\eta_{i^*}^{s_{i^*}+1} = \max_i \eta_i^{s_i+1}$

$s_{i^*} \leftarrow s_{i^*} + 1$

$I \leftarrow I + 1$

CONTINUE

Rather than maximizing the expected fill rate, suppose we want to minimize the expected number of outstanding backorders. Again assuming $a_i = 1$, for all $i$, we can state the problem as

$$\min \sum_{i=1}^{n} \sum_{j=s_i}^{\infty} (j - s_i)(1 - \eta_i)\eta_i^j$$

$$\sum_{i=1}^{n} s_i \leq C \qquad (8.7)$$

$$s_i = 0, 1, \ldots$$

Now let $g_i(s_i) = \sum_{j=s_i}^{\infty} (j - s_i)(1 - \eta_i)\eta_i^j$. The reduction in expected backorders by increasing the stock level from $s_i$ to $s_i + 1$ is given by

$$g_i(s_i + 1) - g_i(s_i) = -\eta_i^{s_i+1}.$$

Thus if you add stock to the item that decreases expected backorders most, you add stock to the product for which $\eta_i^{s_i+1}$ is largest. It is easy to see that the optimal solution to problems (8.5) and (8.7) must be the same, since the greedy algorithm used to solve each problem always selects the same product at each step as the one whose stock level should be incremented.

### 8.3.4   A Further Extension

To this point we have assumed that the order arrival process to our system is governed by a Poisson process. Furthermore, we assumed that the processing times are governed by an exponential distribution. Let's suppose that the arrival process of a single item to the single bottleneck repair center is a general process (finite expected time between arrivals) and independent interarrival times and service times come from a general distribution. Thus we assume that the repair system is analogous to a $GI/G/1$ queueing system. Let

$c_a$ = coefficient of variation of the interarrival time distribution

$c_s$ = coefficient of variation of the service time distribution.

Several approximations exist for calculating the expected number of units in the system (Buzacott and Shanthikumar [34]). When $c_a^2 \leq 2$ and $\rho$ is high ($> .9$) the following approximation is quite accurate.

$$E[N] \cong \left\{ \frac{\rho^2(1+c_s^2)}{1+\rho^2 c_s^2} \right\} \frac{c_a^2 + \rho^2 c_s^2}{2(1-\rho)} + \rho. \tag{8.8}$$

Next, let's approximate the steady-state probability distribution of the number of units in the production system, that is, $P\{N = j\}$. Recall that the probability that the production system is idle is $1 - \rho$. Hence $P\{N = 0\} = p(0) = 1 - \rho$. Next let us assume that the remaining probabilities have a geometric form, that is, $P\{N = j\} = p(j) = a\eta^j, j = 1, 2, \ldots$. This assumption is similar to the one we made earlier in this chapter. There we assumed a mass exponential distribution would accurately represent the shortfall distribution. Since $\sum_{j=0}^{\infty} p(j) = 1$,

$$1 - \rho + a\eta \sum_{j=1}^{\infty} \eta^{j-1} = 1 \quad \text{or}$$

$$1 - \rho + a\frac{\eta}{1-\eta} = 1$$

$$\text{and} \quad a = \rho \cdot \frac{(1-\eta)}{\eta}.$$

Then

$$E[N] = \sum_{j=1}^{\infty} j\rho\frac{1-\eta}{\eta}\eta^j = \frac{\rho}{\eta} \sum_{j=1}^{\infty} j(1-\eta)\eta^j = \frac{\rho}{\eta}\frac{\eta}{1-\eta} = \frac{\rho}{1-\eta}.$$

Now suppose we approximate $E[N]$ using equation (8.8). Call this approximation $\hat{N}$. Then

$$1 - \eta = \frac{\rho}{\hat{N}} \text{ or } \eta = \frac{1-\rho}{\hat{N}}.$$

This approximation has proven to be accurate, and hence quite useful in practice.

We can then approximate the distribution of the number of units awaiting repair for an item in the multi-item case. Assuming that $P\{N_i = k|N = j\}$ is binomially distributed, we can again construct the distribution for the number awaiting repair for item $i$. We can also find "optimal" stock levels using the same type of models described earlier.

We also observe that we can extend the preceding ideas to cases where more than one repair line is in the bottleneck station of the repair system. Again using queueing ideas we could construct the mathematical expression for $p_i(j)$. However, the analysis results in expressions that are not as simple as the ones we have found for the single workcenter cases we have examined in some detail.

## 8.4  Problem Set, Chapter 8

**8.1.** Show that the expression for $P[N = n]$ as given in Section 8.3.2 is correct.

**8.2.** In Section 8.3, we assumed repair times were exponentially distributed. Instead of this assumption, suppose repair times are distributed by an Erlang distribution. In this case the system is equivalent to an $M/E_k/1$ queueing system, where $k$ is the parameter of the Erlang distribution. In this case, find the probability distribution of the number of units in the system. Furthermore, find the steady state distribution of the number of units of item type $i$ in the repair system. How do inventory requirements change to minimize expected costs for the model presented in Section 8.3 as the parameter $k$ increases?

**8.3.** Suppose demands over an infinite horizon are independent and identically distributed from period to period. Furthermore, assume a production system operates in the manner described in Section 8.1. Assume the production capacity in a period is 3 units and the probability distribution of demand in each period is as follows: $P\{\text{demand} = k\} = 1/4$ for $k = 1, 2, 3, 4$, and 0 for other values of $k$. Compute the shortfall probability distribution for this situation.

**8.4.** Suppose that demand in each period is either exponentially or mass exponentially distributed, and that demands are independent from period to period. In these cases, prove that the shortfall random variable, $V$, has a mass exponential distribution. That is, show that the approximation given in equation (8.3) is exact in these cases.

**8.5.** Examine a variety of cases in which the following conditions hold. In each period of an infinite horizon, demand for a single product at a single location has a Negative Binomial distribution. Demands are independent from period to period. In each case, the expected demand per period remains unchanged; however, the variance of demand per period changes in different cases. Furthermore, the distribution of the per period demand changes from case to case. Capacity values are independent and identically distributed from period to period. The goal is to set the desired target stock level in each case so as to achieve a 98% fill rate.

The environment in which demands arise operates as follows. At the beginning of each period, demands occur at a location, which must be satisfied by the period's end. Next, production occurs. The amount produced in a period is available for shipment within that period. Production in a period is set to reach a desired target inventory level, if possible. If however, there is not enough available capacity to achieve the target level, then the amount produced equals the available capacity.

In all cases the expected demand per period is 50 units. However the variance to mean ratio of demand is either 1.01, 3 or 10. Additionally, the available capacity per period varies as follows. You will examine cases in which there are three different average per period capacity levels, and, for each capacity level, there are three possible probability distributions for the per period capacity. The three expected capacity levels are 58 units (normal capacity case), 55 (low capacity case), and 63 (high capacity case). The data in the following table are the probability distributions of capacity for each average capacity value.

| Normal Capacity (58 units per period) | | Low Capacity (55 units per period) | | High Capacity (63 units per period) | |
|---|---|---|---|---|---|
| Capacity | Probability | Capacity | Probability | Capacity | Probability |
| **Normal Mean Zero Variance** | | | | | |
| 58 | 1 | 55 | 1 | 63 | 1 |
| **Normal Mean Medium Variance** | | | | | |
| 50 | 0.2 | 47 | 0.2 | 55 | 0.2 |
| 54 | 0.2 | 51 | 0.2 | 59 | 0.2 |
| 58 | 0.2 | 55 | 0.2 | 63 | 0.2 |
| 62 | 0.2 | 59 | 0.2 | 67 | 0.2 |
| 66 | 0.2 | 63 | 0.2 | 71 | 0.2 |
| **Normal Mean High Variance** | | | | | |
| 46 | 0.2 | 43 | 0.2 | 51 | 0.2 |
| 50 | 0.2 | 47 | 0.2 | 55 | 0.2 |
| 58 | 0.2 | 55 | 0.2 | 63 | 0.2 |
| 66 | 0.2 | 63 | 0.2 | 71 | 0.2 |
| 70 | 0.2 | 67 | 0.2 | 75 | 0.2 |

Your goal is to compute the target inventory level to achieve a 98% fill rate for each combination of variance to mean ratio, average per period capacity level, and the variance of capacity corresponding to each average capacity level. In total 27 cases must be examined. Observe how difficult it is to establish exactly what the target inventory levels should be as the variances of demand and capacity increase. How does production vary from period to period in these different cases? What does your observation imply about the use of order-up-to policies in practice?

**8.6.** In Section 8.1.3 we discussed how a mass exponential approximation could be constructed for the distribution of the shortfall random variable, $V$. Suppose demand in each period is normally distributed and is independent from period to period. Prove that

$$\bar{p}_0 = \frac{\bar{F}_D(c)}{1 - \bar{F}_D(c)}, \text{ and } \gamma = \frac{2(c - E(D))}{\sigma^2}$$

in this case, where all nomenclature are defined in Section 8.1.3.

# 9

# Extension of Palm's Theorem to Nonstationary Demand Processes

Earlier, we constructed mathematical models based on two key assumptions. First, we assumed that demand was described by a stationary Poisson or compound process; second, we assumed resupply times are independent and identically distributed for all units that are resupplied. However, in many circumstances both the arrival and resupply processes are time dependent. Thus the applicability of the stationarity assumption is limited in certain dynamic environments. This is particularly true in military applications when flying activity and resupply of operating organizations are highly dynamic over relatively short periods of time. Time dependencies also occur in many commercial situations in which resupply times are so short that the two assumptions can substantially affect the accuracy of performance measure forecasts. For example, the demand process experienced at parts distribution centers differ substantially by day of the week for one automotive company we have studied.

Our goal in this chapter is to extend the results stated in Chapter 3. Specifically we will first extend Palm's Theorem when demands are nonstationary Poisson or compound processes and the resupply time distributions are time dependent. These extensions were first developed by Crawford [64] and Hillestad and Carrillo [129] and later summarized by Carrillo [42]. We will then demonstrate how to compute the probability distributions representing the number of units in resupply at each location in a two-echelon logistics system at any point in time.

## 9.1  The Nonstationary Poisson Demand Case at a Single Location

We will now extend Palm's theorem to the environment in which the demand process at a single location is a nonstationary Poisson process, and the resupply times for that location are independent from unit to unit but are time dependent.

Suppose $N(t)$ measures the number of arrivals (demands) that occur through time $t$. Assume that $N(0) = 0$; however, this is not a critical assumption. Also the rate at which the part is demanded at time $t$ is denoted by $\lambda(t) \geq 0$, where $\lambda(t)$

is an integrable function, and the expected number of demands through time $t$ is expressed as

$$m(t) = E(N(t)) = \int_0^t \lambda(s) \, ds.$$

Suppose a demand for a unit occurs at time $t$. We define $G(t, t+w) \equiv G_t(w)$ to be the probability that the resupply time for that unit is less than or equal to $w$, that is, the unit is resupplied by time $t + w$. The resupply times are assumed to be independent from unit to unit and are assumed to have finite, time dependent expectations.

To prove the extension of Palm's theorem, we will employ the following theorem.

**Theorem 12.** *Suppose demands occur according to a nonstationary Poisson process with the time dependent arrival rate given by $\lambda(t)$. Also, assume the resupply times are independent and time dependent and have distribution function $G_t(w)$, for all $t \geq 0$. Suppose $N(t) = n$. The arrival times of these demands have the same distribution as the order statistics for $n$ independent random variables each of which has a distribution function*

$$F(x) = \begin{cases} \frac{m(x)}{m(t)}, & 0 \leq x < t \\ 1, & x \geq t \end{cases}.$$

The proof of this theorem is similar to the one given in Chapter 3 for the case when all random variables are stationary. We are now ready to state and prove the following theorem.

**Theorem 13.** *Suppose demands and resupply times for a particular item are as stated in the hypothesis of the preceding theorem. Then the number of units in the resupply process at time t, which we denote by the random variable $X(t)$, is Poisson distributed with mean*

$$\alpha(t) = \int_0^t (1 - G_s(t - s))\lambda(s) \, ds.$$

The proof of this theorem is essentially the same as the one presented for the stationary case in Chapter 3; however, it is given for the sake of completeness.

*Proof.* Now suppose $N(t) = n$. We will first determine $P\{X(t) = k \mid N(t) = n\}$. The probability that an arbitrary unit of demand occurring during $[0, t)$ remains in resupply at time $t$ is given by

$$p = \int_0^t (1 - G_s(t - s)) \frac{\lambda(s)}{m(t)} \, ds$$

as a consequence of Theorem 12. Since this is the probability that each of the $n$ demanded units remains in the resupply system at time $t$,

$$P\{X(t) = k | N(t) = n\} = \binom{n}{k} p^k (1-p)^{n-k}.$$

The unconditional probability that $X(t) = k$ is given by

$$P\{X(t) = k\} = \sum_{n \geq k} P\{X(t) = k | N(t) = n\} P\{N(t) = n\}$$

$$= \sum_{n \geq k} \binom{n}{k} p^k (1-p)^{n-k} e^{-m(t)} \frac{m(t)^n}{n!}$$

$$= \sum_{n \geq k} \frac{n!}{(n-k)! k!} (p \cdot m(t))^k [(1-p) \cdot m(t)]^{n-k} \frac{e^{-m(t)}}{n!}$$

$$= \frac{(p \cdot m(t))^k e^{-m(t)}}{k!} \sum_{n \geq k} \frac{[(1-p) \cdot m(t)]^{n-k}}{(n-k)!}$$

$$= \frac{(p \cdot m(t))^k e^{-m(t)}}{k!} \sum_{j=0}^{\infty} \frac{[(1-p) \cdot m(t)]^j}{j!}$$

$$= \frac{(p \cdot m(t))^k e^{-m(t)}}{k!} \cdot e^{(1-p)m(t)}$$

$$= \frac{(p \cdot m(t))^k}{k!} e^{-p \cdot m(t)}.$$

Since

$$p \cdot m(t) = \int_0^t (1 - G_s(t-s)) \lambda(s) \, ds = \alpha(t),$$

$$P\{X(t) = k\} = e^{-\alpha(t)} \frac{(\alpha(t))^k}{k!}. \qquad \square$$

## 9.2 The Nonstationary Compound Poisson Process Case at a Single Location

The theorem presented in the last section stated that the probability distribution for the number of demands or orders that are in the resupply system at time $t$ is Poisson distributed with mean $\alpha(t)$. Suppose that each demand, or order, is for $j \geq 1$ units. Suppose that $u_j$ measures the probability that the order is for $j$ units. Furthermore, assume that this distribution does not change with time, that is, the order size distribution is stationary.

If $Y(t)$ is a random variable that measures the number of units demanded through time $t$, then it is clear, from our discussion in Chapter 3, that $Y(t)$ has a compound Poisson distribution. That is,

$$P[Y(t) = k] = \sum_{n=1}^{\infty} u_k^{(n)} e^{-m(t)} \frac{(m(t))^n}{n!},$$

where $u_k^{(n)}$ is the probability that $n$ orders result in a demand for $k$ units. Since the demands are independent, $u_k^{(n)}$ is the $n$-fold convolution of the compounding order size distribution $u_k$ with itself. It is easy to see that

$$E[Y(t)] = m(t) \cdot E[Q]$$

and

$$
\begin{aligned}
\mathrm{Var}[Y(t)] &= E\left[\mathrm{Var}(Y(t)|N(t))\right] + \mathrm{Var}\left[E[Y(t)|N(t)]\right] \\
&= E\left[N(t) \cdot (E(Q^2) - E(Q)^2)\right] + \mathrm{Var}\left[N(t) \cdot E(Q)\right] \\
&= m(t)[E(Q^2) - E(Q)^2] + E[Q]^2 \cdot m(t) \\
&= m(t)E[Q^2], \quad \text{since } \mathrm{Var}N(t) = m(t),
\end{aligned}
$$

where $Q$ is the random variable for the order size distribution.

Let us now extend the time dependent version of Palm's theorem to the case where the demand process is a compound Poisson process.

**Theorem 14.** *Assume the demand process is given by $Y(t)$ and that the resupply time for all units corresponding to an order arriving at time $t$ is given by $G_t(w)$. Then the probability distribution of the number of units in the resupply system at time $t$ is the compound Poisson distribution*

$$P[X(t) = k] = \sum_{n \geq 1} u_k^{(n)} e^{-\alpha(t)} \frac{(\alpha(t))^n}{n!}.$$

*Proof.* Since the number of orders in resupply at time $t$ has a Poisson distribution, the distribution of the number of units in resupply at time $t$ has a compound Poisson distribution.                                                                    □

## 9.3  A Two-Echelon Model when Demand is Described by a Nonstationary Poisson Process

The extension to Palm's theorem developed in the previous section provides the basis for developing a time dependent model for the probability distribution of the number of units in resupply in a two-echelon, depot and base system. This system is the same as the one studied earlier in Chapter 5. Figure 9.1 shows the flow of material in this system. Recall that demands (orders) for units occur at each operating base. In this case, we assume that the demand process is a nonstationary Poisson process for each part type at each base. Each demand corresponds to a failure that requires repair. Repairs occur at either the base or at the depot. The place at which repairs occur depends only on the nature of the failure. Hence there is a probability that a failed unit will be repaired at the base and 1 minus that probability that the unit will be repaired at the depot. Base stock levels are assumed to be known at each point in time. When a failure occurs at a base, a unit is withdrawn from base stock to satisfy the demand. This withdrawal will occur immediately if there is stock on-hand; otherwise, the demand will be satisfied whenever the base inventory is replenished.

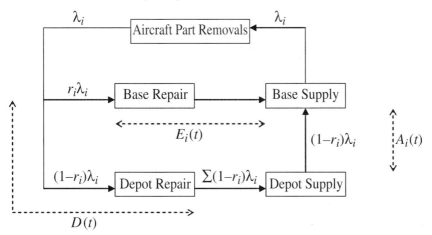

**Fig. 9.1.** Part Resupply Cycle

When a unit fails and is repaired at the base, the base's repair center is responsible for resupplying base stock. If the unit is sent to the depot, then the depot is responsible for satisfying the base's resupply request. If the depot has stock on-hand when the failure occurs, then a serviceable unit is sent to the base and the defective unit is sent to the depot for repair. If the depot does not have stock on-hand when the resupply request is made by the base, then that resupply request is satisfied on a first-come, first-serve basis by the depot. Again, refer to Figure 9.1 for a summary of resupply flows and flow times.

### 9.3.1 Notation

Let us now introduce notation that will be used to establish the probability distribution for the number of units in resupply for each base. Although we may have many item types for which we will construct these distributions, we will focus on only a single item type in our analysis to simplify the notation.

Let

$\lambda_i(t)$ represent the demand rate for the item at base $i$ at time $t$. Assume that this function is integrable.

$E_i(t)$ represents the base $i$ repair cycle time at time $t$, a constant. Furthermore, assume that $E_i(t) + t \geq E_i(s) + s, s < t$.

$A_i(t)$ represents the order and ship time from the depot for a failure occurring at base $i$ at time $t$, and is known and deterministic. Assume $A_i(t) + t \geq A_i(s) + s, s < t$.

$D(t)$ represents the depot repair cycle time for a failure occurring at a base at time $t$, a constant. Assume that $D(t) + t \geq D(s) + s$, $s < t$.

$r_i$ is the probability that a failure occurring at base $i$ is repaired there and $(1 - r_i)$ is the probability that it will be repaired at the depot.

$T_i(t)$ is the average resupply time for a part that fails at base $i$ at time $t$.

$\lambda_0(t) = \sum_{i=1}^{n}(1 - r_i)\lambda_i(t)$ is the depot demand rate at time $t$.

$s_i(t)$ is the stock level at base $i$ at time $t$.

$s_0(t)$ is the depot stock level at time $t$.

$\mathcal{B}_0(s_0(t))$ is the expected backorders at the depot at time $t$.

$\mathcal{B}_i(s_i(t))$ is the expected backorders at base $i$ at time $t$.

$X_0(t)$ is the random variable describing the number of units in resupply at the depot at time $t$.

$X_i(t)$ is the random variable describing the number of units in resupply at base $i$ that are backordered at the depot at time $t$.

Observe that we are assuming that the depot repair cycle time at time $t$, as well as the base repair cycle time and the depot to base order and ship times, are constant, but time dependent. While the assumption of deterministic repair cycle times and order and ship times is restrictive, we note that it is of significant value to have these times be time dependent. By allowing these times to be time varying, we can, for example, represent segments of times during when there is no repair capability, no ability to send failed units to the depot, or to receive serviceable units from the depot. Note also that the assumptions that $A_i(t) + t \geq A_i(s) + s$, $E_i(t) + t \geq E_i(s) + s$, and $D(t) + t \geq D(s) + s$, $s < t$, imply that there is no crossing of resupply times, that is, a unit entering either depot or base resupply at time $s$, $s < t$, completes its resupply cycle no later than a unit entering either of the corresponding resupply cycles at a later time $t$.

### 9.3.2  Depot Analysis

When the demand for an item at base $i$ is characterized by a nonstationary Poisson process and the probability that the failed unit is repaired at that base is $r_i$, then the demand process seen by the depot is a nonstationary Poisson process with rate $\lambda_0(t) = \sum_{i=1}^{n} \lambda_i(t)(1 - r_i)$. Our objective in this section is to compute the probability distribution of the random variable that measures the number of units in the resupply process at the depot at some time $t$ that correspond to resupply requests made by base $i$.

Let $t$ be an arbitrary point in time. Furthermore, let $\tilde{t} = \inf\{u : D(u) + u > t\}$. Then all demands for depot resupply made by bases prior to time $\tilde{t}$ will have been satisfied (i.e. shipped from the depot) by time $t$. Additionally, all entries into the depot resupply system occurring subsequent to time $\tilde{t}$ remain in the depot resupply (repair) system at time $t$, since, by assumption, if $\bar{u} > u \geq \tilde{t}$, then $D(\bar{u}) + \bar{u} \geq D(u) + u$. Thus the probability distribution for the number of units in the depot resupply system at time $t$, denoted by $X_0(t)$, is equal to $k$ is

$$P\{X_0(t) = k\} = e^{-m_0(\tilde{t}, t)} \frac{m_0(\tilde{t}, t)^k}{k!}, \quad \text{where}$$

$$m_0(\tilde{t}, t) = \int_{\tilde{t}}^{t} \sum_i \lambda_i(u)(1 - r_i) du.$$

When $k > s_0(t)$, then backorders exist at the depot at time $t$. The expected number of backorders at the depot at time $t$ is

$$\mathcal{B}_0(s_0(t)) = \sum_{x > s_0(t)} (x - s_0(t)) P[X_0(t) = x].$$

As stated, our goal is to determine the probability distribution for the number of these backorders that correspond to resupply requests from base $i$.

Recall that the random variable $X_i(t)$ measures the number of units of stock that are backordered at the depot that result from resupply requests made by base $i$. Suppose $X_i(t) = k, k > 0$. For this to occur, there exists a $u \in (\tilde{t}, t)$ such that the depot demand in the interval $(\tilde{t}, u)$, is equal to $s_0(t) - 1$, there is a demand placed on the depot at time $u$, and there are $k$ demands for depot resupply from base $i$ occurring during the interval $(u, t]$. All $s_0(t)$ demands for depot resupply that occurred in $(\tilde{t}, u]$ will have been shipped from the depot by time $t$, assuming a first-come, first-serve policy is followed. Furthermore, all resupply requests placed on the depot subsequent to time $u \in (\tilde{t}, t]$ will remain unsatisfied at time $t$. Hence, for $k > 0$,

$$P\{X_i(t) = k\} = \int_{\tilde{t}}^{t} e^{-m_0(\tilde{t}, u)} \frac{m_0(\tilde{t}, u)^{s_0(t-1)}}{(s_0(t) - 1)!}$$
$$\cdot \lambda_0(u) \cdot e^{-m_i(u,t)} \frac{m_i(u, t)^k}{k!} du \quad (9.1)$$

where $m_i(u, t) = \int_u^t (1 - r_i)\lambda_i(v) dv$.

There are two ways that $k$ can equal zero. First $X_0(t) \le s_0(t)$ and, second, depot demand in $(\tilde{t}, u]$ is $s_0(t) - 1$, a depot demand for resupply occurs at $u$, $u \in (\tilde{t}, t)$, and there are no base $i$ demands in $(u, t]$. Thus

$$P\{X_i(t) = 0\} = P\{X_0(t) \le s_0(t)\}$$
$$+ \int_{\tilde{t}}^{t} e^{-m_0(\tilde{t}, u)} \frac{m_0(\tilde{t}, u)^{s_0(t)-1}}{(s_0(t) - 1)!} \cdot \lambda_0(u) \cdot e^{-m_i(u,t)} du. \quad (9.2)$$

### 9.3.3  Base Analysis

Suppose $\bar{X}_i(t)$ is a random variable that represents the number of units in the resupply system at base $i$ at time $t$. Let $\bar{t} = \inf\{u : A_i(u) + u > t\}$ and $\bar{\bar{t}} = \inf\{u : E_i(u) + u > t\}$. Next, suppose $X_i^b(t)$ is a random variable that corresponds to the base $i$ demands occurring in $(\bar{\bar{t}}, t]$ that require base repair. Also, suppose $X_i^d(t)$

is a random variable that measures the number of units that require depot repair that fail during the interval $(\bar{t}, t]$ at base $i$. Then

$$\bar{X}_i(t) = X_i^b(t) + X_i^d(t) + X_i(\bar{t}).$$

Since these random variables are independent, $\bar{X}_i(t)$ is simply the convolution of these three random variables. The independent random variables $X_i^b(t)$ and $X_i^d(t)$ each has a nonstationary Poisson distribution, where

$$E[X_i^b(t)] = \int_{=\bar{t}}^{t} r_i \lambda_i(u) \, du$$

and

$$E[X_i^d(t)] = \int_{\bar{t}}^{t} (1 - r_i) \lambda_i(u) \, du.$$

The distribution of $X_i^b(t) + X_i^d(t)$ is a nonstationary Poisson distribution. Let

$$\tilde{m}_i(t) = \int_{\bar{t}}^{t} (1 - r_i)\lambda_i(v) \, dv + \int_{=\bar{t}}^{t} r_i \lambda_i(v) \, dv,$$

Then

$$P[X_i^b(t) + X_i^d(t) = k] = e^{-\tilde{m}_i(t)} \frac{\tilde{m}_i(t)^k}{k!},$$

and

$$P[\bar{X}_i(t) = k] = \sum_{j=0}^{k} P\left[X_i^b(t) + X_i^d(t) = j\right] \cdot P\left[X_i(\bar{t}) = k - j\right]$$

$$= \sum_{j=0}^{k} e^{-\tilde{m}_i(t)} \frac{\tilde{m}_i(t)^j}{j!} \cdot P\left[X_i(\bar{t}) = k - j\right],$$

where $P[X_i(\bar{t}) = k - j]$ is calculated using equations (9.1) and (9.2). Thus we may compute time dependent performance measures since we have shown how to compute $P[\bar{X}_i(t) = x]$. For example, the expected number of backorders at base $i$ at time $t$ is

$$\mathcal{B}_i(s_i(t)) = \sum_{x > s_i(t)} (x - s_i(t)) P[\bar{X}_i(t) = x].$$

The fill rate at time $t$ is $P[\bar{X}_i(t) < s_i(t)]$ when the demand process is a nonstationary Poisson process.

## 9.4  Problem Set, Chapter 9

**9.1.** Prove Theorem 12.

**9.2.** Let the random variable $Y$ measure the number of units demanded over a fixed length of time. Let $N$ be the random variable for the number of orders placed during the same period of time. Show that

$$\text{Var}[Y] = E[\text{Var}[Y|N]] + \text{Var}[E[Y|N]].$$

**9.3.** In Section 9.1 we assumed that $N(0) = 0$. Suppose that over the interval $(-\infty, 0]$ the demand process was a simple stationary Poisson process with rate $\lambda$. Modify Theorem 13 to account for this change and establish the new distribution for $X(t), t > 0$. Prove your result.

**9.4.** The analysis presented in Section 9.3 was based on the assumption that time was continuous. Suppose that time is divided into periods. Let $\lambda_{it}$ represent the expected demand at base $i$ in period $t$. Demand in each period is independent of demands in other periods and has a Poisson distribution with mean $\lambda_{it}$. For this discrete time case, develop the analysis required to compute the probability distributions for $X_{0t}$, the random variable for the number of units in the depot's resupply system at the end of period $t$, and $X_{it}$, the random variable for the number of units in resupply for base $i$ at the end of period $t$.

# 10

# Real-time Execution Systems

Real-time decision-making focuses on answering the question, "What do I do now?" The "what" activities differ by application, of course; however, every service parts resupply system faces this type of question. Let us begin this chapter by examining briefly the types of questions that arise in a few different settings. We will then explore one application in substantial depth.

The first example comes from the high technology sector. In one environment, machines such as photocopiers and printers are located at customer sites. Machines are installed and removed daily from these sites by field service engineers (FSE). Furthermore, a FSE also conducts repairs on these machines at customer locations. The following is a list of questions that must be addressed daily in such a system:

1. What should be done with a machine that is removed from a customer site? Should it be refurbished, torn down for critical parts, or scrapped?
2. As the population of installed machines changes over time, how should stocks of service parts be reallocated within the supply network? Is it cost effective to transship these parts from one location to another? Should the parts be scrapped?
3. When many types of parts are awaiting repair, which ones should be entered into the repair process now?
4. When many types of machines are available for refurbishing, which ones should be entered into the repair process today?
5. As parts and machines become available through the procurement and repair processes, where should they be stocked?
6. Given the level of stocks on hand, on order, in repair, and in the pipelines, what parts should be ordered today? Should new machines be purchased and torn down for parts?

A second example arises in the environment we described in Chapter 6. There we examined a situation in which service contracts were written for each customer, for specified groups of machines (or parts), and for each customer site. The tactical planning models we developed in Chapter 6 produced a set of target stock levels based on long run expected behavior of the system. But contracts pertain to short run behavior, and insuring that contractual obligations are met is of critical importance to long term customer satisfaction.

With that said, suppose a machine requires repair, and a particular part is known to be needed to complete the repair. Suppose further that the FSE has access to the required part. Should the FSE use the part to complete the repair? This question seems to have an obvious answer, but upon some reflection, it is not clear what action should be taken.

For instance, suppose there are only two active customer contracts that could require a particular part. The first contract stipulates that over the contract period of one year, 80% of the machine repair calls must be satisfied within 48 hours. The second contract stipulates that over the same contract period of one year, 98% of the machine repair calls must be satisfied within 8 hours. Suppose that over the first 9 months of these contracts, 92% of the service calls under the first type of contract have been satisfied within 48 hours, but only 93% of the service calls under the second type of contract have been satisfied within 8 hours.

A call now comes that falls under the first contract type, and there is only one unit in stock of the part type that is required to complete the service. Furthermore, it is likely that this part will be needed to repair a machine under the second contract type before another unit becomes available. What action should be taken? A complex model is likely required to assist decision makers that manage contracts in this environment. Most static decision rules need not perform well. Obviously there are significant information system and business process implications associated with implementing various types of decision support mechanisms in these situations. The question remains as to what the relationship is between the cost of creating the infrastructure required to capture and act on the data, and the value derived (in terms of customer satisfaction and retention) from the use of a complex modelling environment. This question must be asked in all situations.

A third example is found in automotive service parts environments. Decisions must be made daily as to: (1) what parts should be stocked at each location in a multi-echelon resupply network, (2) what part types should be ordered each day at each location, (3) what mode of transport should be employed to ship parts from one location to another, given the current state of the system, (4) what location should provide a part to a location that needs it, and (5) where should parts be stocked in warehouses as they arrive from a supplying location? Clearly there are many other types of decisions that must be made in this environment. Since such systems often contain many hundreds of thousands of part types stocked at many thousands of dealers/retailers and tens of stocking locations within an automotive company, the need for effective real-time decision support models is evident.

Considerable evidence exists showing that service and costs can be improved by constructing and implementing real-time operations models and information infrastructures in many settings. Several authors, including [214], [215], [126], [198], [128], [129], and Abell et al. [1], have shown how real-time approaches and models to managing repairs and parts in aviation settings can improve performance substantially. In the remainder of this chapter, we will develop a series of real-time models for managing expensive service parts in the aviation sector. We will also demonstrate the effectiveness of employing these models.

## 10.1   Real-time Capacity and Inventory Allocations for Reparable Items: Introduction

The real-time operational planning process we examine in this section focuses on day-to-day decision-making in an aviation service parts repair and distribution system. Each day, planners determine the best way to allocate available repair capacity among different items and the best way to allocate available inventories to a set of bases. These allocation decisions, which must be made in highly dynamic environments, are largely driven by the information available about the current state of the system, as well as the information available about the demands that are likely to occur in the time periods that the decisions will affect. Since the level and reliability of such information is limited, the operational planning horizon for such a system is finite. The usual goal of the operational planning activity is to satisfy operational requirements at minimum expected cost over the time periods impacted by the current decisions.

Our objective in the remainder of this chapter is to formulate this operational planning problem in several ways. Each of the models measures the economic consequences of the current operating decisions; however, the models differ in their complexity. We will examine three different systems in detail, and will demonstrate the value of employing integrated decision models over using decentralized allocation rules in a range of commonly encountered operating environments.

### 10.1.1   System Description

The system consists of a depot repair facility, a depot warehouse, and a set of bases. Figure 10.1 shows the cyclic flow of materials in this system. When a part on an aircraft fails, the defective unit is removed, and a replacement part of the same type is dispatched from the base's stock. The defective unit is shipped back to the depot, where it joins a queue of parts awaiting repair. After the part is repaired, it is sent to the depot's warehouse, where it is then available for redeployment to any base. If a required part is not on-hand at the base, a backorder occurs. The backordered part will be subsequently supplied to the base from the depot warehouse. Parts may be shipped from the depot warehouse to the bases via a regular or an expedited shipment mode.

At any given point in time, there are inventories of each part type in various places in the system. Specifically, a part may be:

1. In-transit from a base to the depot repair facility;
2. Awaiting repair at the depot repair facility;
3. In-repair at the depot repair facility;

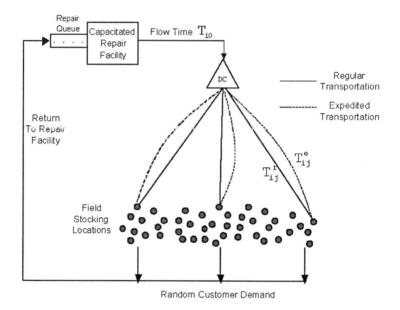

**Fig. 10.1.** The System Under Study

4. In-transit from the depot repair facility to the depot warehouse;
5. On-hand at the depot warehouse;
6. In-transit from the depot warehouse to a base; or
7. On-hand at a base.

Segments 1–4 constitute the *repair subsystem*, and segments 5–7 constitute the *distribution subsystem*. Given the quantities of each part type in each segment of the system, along with the knowledge of when parts in-transit will arrive at their destinations, decisions must be made in each period regarding what parts to repair and what parts to ship from the depot warehouse to the bases via each transportation mode.

We begin by constructing a dynamic programming model for making repair and allocation decisions for this system. Decisions are to be made each period regarding how many units of each item type to enter into the depot repair cycle, and how many units of each item type to ship from the depot warehouse to each base. The number of units entering the repair process each period is limited by the number of units available to be repaired, as well as by the maximum capacity of the repair process (which is described as a maximum number of units). The demand process at each base is described by a random variable whose distribution may vary from period to period. We will make some simplifying assumptions regarding lead times that will permit a relatively straightforward statement of the problem. Nonetheless, we will observe that the resulting dynamic program can not be solved for practical problems due to the size of the state space.

Since this multi-period dynamic programming based formulation of the problem is computationally intractable, we will concentrate on developing alternative formulations that are approximations, but are computationally tractable. Specifically, we will present discrete-time, periodic-review models for making allocation decisions in increasingly complex operating environments. The first model, although rather basic in its assumptions, provides both the framework and the insight for dealing with the second and third models, in which an additional shipment option and control over repair decisions, respectively, are captured in the formulation. It is the third model for which we are ultimately interested in developing practical solution algorithms.

Each approximating model is in essence a multi-period newsvendor problem, where the effects of the decisions made in one period are captured in subsequent periods over the effective horizon. Thus, when solving the problem, the model simultaneously makes decisions at the beginning of the current period for all periods in the planning horizon, implicitly assuming that these decisions will be implemented and that there is no opportunity to alter the plan once more information about the demand process is obtained. In practice, however, the models would be employed on a rolling horizon basis, so that new allocation decisions would be made every period based on new information about the system.

In describing each operational model, the difference between the *planning horizon* of the model and the *effective horizon* of the model must be noted. The *planning horizon* of a model encompasses the time periods in which allocation decisions will be made. The *effective horizon* of a model encompasses the time periods in which the effects of the allocation decisions will be measured. We now describe the dynamic programming model and each of the approximating models in greater detail.

In constructing the dynamic programming formulation of the decision problem, assume, for notational clarity, that there is only one mode of shipment available from the depot warehouse to the bases, that the shipment time is the same for all items and bases, and that the repair lead time plus the time to transport a repaired item to the depot warehouse is the same for all item types. (All three of these assumptions will be relaxed when we describe the approximate models.) We also assume that the decision-maker has visibility of all parts in the distribution subsystem as well as the repair subsystem, with the exception of parts that are in-transit to the repair facility (i.e., segment 1). It will be obvious how to extend the model when this assumption is relaxed; however, given this assumption, the length of the planning horizon is taken to be the repair lead time plus the transportation time from the repair facility to the depot warehouse. Thus, repair decisions in the dynamic programming model will be made in the first period only, and these decisions will affect the availability of parts at the depot warehouse in the last period of the planning horizon. Allocation decisions will be made over the entire planning horizon.

The first approximate model, which we call the *stock allocation model*, or **SAM**, addresses only the inventory allocation problem and, as in the dynamic programming model, allows only one mode of shipment from the depot warehouse to the bases (i.e, the regular shipment mode). The **SAM** determines how many parts to ship from the depot warehouse to each base in each period of the planning horizon. This model assumes that the decision-maker has complete visibility of parts in the distribution subsystem as well as those that are currently in the repair process and en route to the depot warehouse, but has *no* knowledge of or control over parts that are in the repair queue or en route to the repair facility (i.e., segments 1 and 2). Accordingly, the length of the planning horizon for the **SAM** (i.e., the number of periods for which allocation decisions will be made) is the repair lead time plus the transportation time from the repair facility to the depot warehouse. Since the current contents of this pipeline are visible to the decision-maker, this is the horizon over which the supply to the depot warehouse is known with certainty.

The *extended stock allocation model*, or **ESAM**, also addresses only the inventory allocation problem; however, in this model the decision-maker has the option of using an expedited shipment mode to transport parts from the depot warehouse to the bases. In making allocation decisions, the incremental benefit gained by having parts arrive at a base earlier than they would have using the regular shipment mode must be weighed against the incremental cost of expedited shipment. The visibility of the decision-maker and the length of the planning horizon for the **ESAM** are the same as for the **SAM**.

In the *extended stock allocation model with repair*, or **ESAMR**, capacitated repair decisions as well as inventory allocation decisions must be made. The repair decisions to be made are which parts (of those awaiting repair) should have their repair commenced in the first period of the planning horizon. Like the dynamic programming model, the **ESAMR** assumes that the decision-maker has visibility of all parts in the distribution subsystem as well as the repair subsystem, with the exception of parts that are in-transit to the repair facility (i.e., segment 1). The reason for this final restriction is that in practice, the decision-maker may have limited control over the timing and methods used by bases to return defective parts for repair. The length of the planning horizon for the **ESAMR** is the same as for the **SAM** and the **ESAM**; however, the repair decisions to be made in this model affect the availability of parts at the depot warehouse in the last period of the planning horizon, and hence, are integrated with the inventory allocation decisions.

The objective of all three approximate models is to minimize the *relevant* total expected system cost over the effective horizon, assuming that the current allocation decisions for future periods will be executed without change over the course of the planning horizon. For the **SAM**, this cost includes the *incremental* expected cost of holding parts at the bases (instead of holding them at the depot warehouse) and the expected backorder costs at the bases. For the **ESAM**, we also consider the *incremental* cost of shipping parts to bases via the expedited mode (instead of the regular shipment mode). For the **ESAMR**, in addition to the incremental shipping costs, we also allow for an *incremental* cost of holding parts at the depot warehouse (instead of holding them at the depot repair facility). Repair costs, depot repair facility holding costs, and regular shipment costs are not captured in these models since they are *not relevant*. That is, assuming that all demand placed on the system is to be satisfied eventually, these costs will be incurred *regardless* of the current capacity and inventory allocation decisions. The only relevant costs in the operating models are those associated with the *timing* with which the parts are made to flow through the system from the depot repair facility to the depot warehouse to the various bases.

As observed earlier, these approximate models would be implemented in a rolling horizon manner. That is, given the system's status at the beginning of a period, the model's recommendations would be followed for the current period only, recognizing that the solution considers the consequences of the current period decisions on future period decisions and costs.

The environment described in this section along with the models and algorithms presented in the following sections are based on [39].

## 10.2  Notation and Assumptions

In this section notation for the first two approximating models, **SAM** and **ESAM**, is defined as well as notation for the dynamic programming formulation . Additional notation will be defined as required.

The key to all of the model formulations lies in appropriately constraining the inventory allocation decisions by the part availability at the depot warehouse, as well as accurately capturing the relevant cost consequences of the inventory allocation decisions. Namely, for each part type, we must capture:

the current stock level at the depot warehouse and the quantities due to arrive at the depot warehouse over the planning horizon;

the current stock levels at the bases and the quantities due to arrive at the bases as a consequence of the inventory allocation decisions; and

the costs incurred each period as a consequence of the inventory allocation decisions; namely, the incremental expedited shipment costs to the bases, the incremental expected holding costs at the bases, and expected backorder costs at the bases.

Remember that repair costs, depot warehouse holding costs, and regular shipment costs are not relevant to the **SAM** or the **ESAM** since these costs will be incurred regardless of the current inventory allocation decisions. Since the flow of parts from the depot repair facility to the depot warehouse is predetermined in these models, the only relevant costs are those associated with the *timing* with which the parts are made to flow from the depot warehouse to the bases.

The following notation is used throughout this chapter:

**Network Parameters**
$I$    the set of items, indexed by $i$.
$J$    the set of bases, indexed by $j$.

**Time Parameters** (All are assumed to be an integer number of periods.)

$T_{i0}$     the repair lead time for item $i$, including transport to the depot warehouse from the repair facility. (The index 0 denotes the depot warehouse.)

$T_{ij}^r$     the regular transportation lead time for item $i$ to base $j$ from the depot warehouse.

$T_{ij}^e$     the expedited transportation lead time for item $i$ to base $j$ from the depot warehouse. We assume that $1 \leq T_{ij}^e < T_{ij}^r$.

**Decision Variables**

$y_{ijt}^r$     the number of units of item $i$ to ship via regular transport from the depot warehouse to base $j$ in time period $t$, $t = 0, \ldots, T_{i0}$.

$y_{ijt}^e$     the number of units of item $i$ to ship via expedited transport from the depot warehouse to base $j$ in time period $t$, $t = 0, \ldots, T_{i0}$.

$v_i$     the number of units of item $i$ to enter into repair in time period 0.

**Supply and Demand Parameters**

$\tilde{S}_{i0t}$     the *known* cumulative supply of item $i$ available at the depot warehouse through period $t$ (i.e., inventory on-hand at the beginning of the planning horizon plus stock arriving through period $t$). These parameters define the pipeline stock profile coming into the depot warehouse at the beginning of the planning horizon and are unaffected by current allocation decisions. They are defined for periods $t = 0, \ldots, T_{i0}$.

$\tilde{S}_{ijt}$     the *known* cumulative supply of item $i$ available at base $j$ through period $t$ (i.e., net inventory at the beginning of the planning horizon plus stock arriving through period $t$). These parameters define the pipeline stock profile coming into base $j$ from the depot warehouse at the beginning of the planning horizon and are unaffected by current allocation decisions. They are defined for periods $t = 0, \ldots, T_{ij}^r + T_{i0}$. Note, however, that $\tilde{S}_{ijt} = \tilde{S}_{ij(T_{ij}^r - 1)}$ for all $t = T_{ij}^r, \ldots, T_{ij}^r + T_{i0}$.

$S_{ijt}$     the cumulative supply of item $i$ available at base $j$ through period $t$. These parameters *are* affected by current depot warehouse allocation decisions, and are defined for periods $t = T_{ij}^e, \ldots, T_{ij}^r + T_{i0}$.

$X_{ijt}$     the cumulative demand of item $i$ at base $j$ through period $t$, a random variable, defined for periods $t = 0, \ldots, T_{ij}^r + T_{i0}$.

$R_i$     the number of units of item $i$ available for repair in time period 0.

$C_{max}$     the maximum number of units (over all item types) that can be entered into repair in time period 0 (i.e., the repair facility capacity).

$C_{min}$     the minimum number of units (over all item types) that must be entered into repair in time period 0, where $C_{min} \leq \sum_{i \in I} R_i$ by assumption. In many practical applications, this parameter would likely be set to $\min(\sum_{i \in I} R_i, C_{max})$ so that as many units as possible are repaired.

$C_i$     $\min(R_i, C_{max})$, the maximum number of units of item $i$ that can enter the repair process in time period 0.

**Cost Parameters and Functions**

$h_{ij}$    the incremental cost per period associated with holding a unit of item $i$ at base $j$ instead of at the depot warehouse.

$b_{ij}$    the unit shortage cost per period for item $i$ at base $j$.

$e_{ij}$    the incremental cost of shipping a unit of item $i$ via expedited mode from the depot warehouse to base $j$. This cost is assumed to include any incremental holding costs while in transit.

$G_{ijt}(\cdot)$  the function describing the expected incremental holding costs and backorder costs incurred in period $t$ for item $i$ at base $j$, defined for periods $t = T_{ij}^e, \ldots, T_{ij}^r + T_{i0}$. The function argument is $S_{ijt}$, the cumulative supply of item $i$ at base $j$ through period $t$. That is,
$$G_{ijt}(S_{ijt}) = h_{ij} E[S_{ijt} - X_{ijt}]^+ + b_{ij} E[X_{ijt} - S_{ijt}]^+.$$

$Q_{ij}(\cdot)$  the function describing the expected incremental holding costs incurred beyond the end of the effective horizon for item $i$ at base $j$. The function argument is $S_{ij(T_{ij}^r + T_{i0})}$, the cumulative supply of item $i$ at base $j$ at the end of the effective horizon. That is,
$$Q_{ij}(S_{ij(T_{ij}^r + T_{i0})}) = h_{ij} \sum_{t=T_{ij}^r + T_{i0}+1}^{\infty} E[S_{ij(T_{ij}^r + T_{i0})} - X_{ijt}]^+.$$

Some explanation should be given for the item-specific time periods over which the parameters and variables are defined. As mentioned earlier, the goal is to capture all of the incremental expected costs that are a *direct consequence* of the allocation decisions made during the planning horizon (i.e., in periods $t = 0, \ldots, T_{i0}$ for item $i$). Note that since different items may have different repair lead times, the number of periods for which supply information exists (i.e., $T_{i0}$, the pipeline length) may differ from item to item. Also, since the shipment lead times $T_{ij}^r$ and $T_{ij}^e$ to the various bases may differ across items and locations, the consequences of the allocation decisions made at the depot warehouse may be realized within time windows that vary by item and by base. Specifically, for a given item $i$ and a given base $j$:

  Incremental expected holding costs are realized in periods $t = T_{ij}^e, \ldots, T_{ij}^r + T_{i0}$.

  Expected backorder costs are realized in periods $t = T_{ij}^e, \ldots, T_{ij}^r + T_{i0}$.

  Incremental costs for expedited shipments from the depot warehouse are realized in periods $t = 0, \ldots, T_{i0}$.

In addition, the models capture the end-of-horizon implications of carrying inventory at each base for each item. Note that although the supply may make it possible to send extra inventory from the depot warehouse to bases, carrying inventory at a base at the end of the effective horizon is appropriate only if the expected future demand at that base warrants it. To address this, the model includes a cost term that reflects the expected incremental holding costs that would be incurred in periods *following* the end of the effective horizon. This cost function, first introduced in [43], measures the expected number of future periods worth of inventory that are on-hand at the end of the effective horizon and multiplies this figure by the item's incremental holding cost per period. That is, the expected future holding cost associated with having $[S_{ij(T_{ij}^r+T_{i0})} - X_{ij(T_{ij}^r+T_{i0})}]^+$ units of item $i$ on-hand at base $j$ at the end of the effective planning horizon is expressed as:

$$Q_{ij}(S_{ij(T_{ij}^r+T_{i0})}) = h_{ij} \sum_{t=T_{ij}^r+T_{i0}+1}^{\infty} E[S_{ij(T_{ij}^r+T_{i0})} - X_{ijt}]^+. \tag{10.1}$$

It is easily shown that $Q_{ij}$ is a convex function of its argument.

## 10.3  The Dynamic Programming Model

We now formulate the repair and inventory allocation problem as a dynamic program. Recall that there is only one mode of transportation from the depot warehouse to the bases in this model, and the repair and transportation times are the same for all items and bases. Thus, $T_{i0} = T_0$ for all $i \in I$, and $T_{ij}^r = T^r$ for all $i \in I, j \in J$. The planning horizon is $T_0$ for all items, with the effective horizon being $T_0 + T^r$, since the economic consequences of the decisions made in periods 0 through $T_0$ will be incurred through period $T_0 + T^r$. The incremental holding and backorder costs incurred at the bases through period $T^r - 1$ are beyond our control.

Given the notation of the previous section, the following constraints must be met as the optimization is carried out:

$$\tilde{S}_{i0t} \geq \sum_{j \in J} \sum_{t'=0}^{t} y_{ijt'}^r, \quad \forall i \in I, t = 0, \dots, T_0, \tag{10.2}$$

$$\tilde{S}_{i0(T_0)} = \tilde{S}_{i0(T_0-1)} + v_i, \quad \forall i \in I, \tag{10.3}$$

$$S_{ijt} = \tilde{S}_{ij(T^r-1)} + \sum_{t'=0}^{t-T^r} y_{ijt'}^r, \tag{10.4}$$

$$\forall i \in I, j \in J, t = T^r, \dots, T^r + T_0,$$

$$C_{\min} \leq \sum_{i \in I} v_i \leq C_{\max}, \tag{10.5}$$

$$0 \leq v_i \leq C_i \text{ and integer} \quad \forall i \in I, \tag{10.6}$$

$$y_{ijt}^r \geq 0 \text{ and integer} \quad \forall i \in I, j \in J, t = 0, \dots, T_0. \tag{10.7}$$

Let $\bar{S}_{t-T^r}$ be a vector whose components are the values of the $S_{ijt}$ quantities. Thus, $\bar{S}_{t-T^r}$ measures the cumulative amount allocated to each base location and available for allocation by the depot through period $t$ given the initial conditions and subsequent repair and allocation decisions. Also, let $\bar{X}_{t-T^r}$ be a vector of cumulative demands for each $(i, j)$ pair through period $t$. Define the state of the system at time $t$ to be $(\bar{S}_{t-T^r}, \bar{X}_{t-T^r-1})$. Furthermore, let $\bar{Y}_{t-T^r}$ be the vector of allocation decisions in period $t - T^r$, whose components are $y_{ij(t-T^r)}$ for each $(i, j)$ pair.

Given these definitions, one dynamic programming formulation of the simplified repair and allocation problem is as follows. The recursion is:

$$f_{T^r+T_0+1}(\bar{S}_{T_0}; \bar{X}_{T_0-1}) = \sum_{i \in I} \sum_{j \in J} Q_{ij}(S_{ij(T^r_{ij}+T_{i0})}; \bar{X}_{T_0-1}) \tag{10.8}$$

and

$$f_{T^r+t}(\bar{S}_t; \bar{X}_{t-1}) =$$
$$\min_{\bar{Y}_t} \sum_{i \in I} \sum_{j \in J} G_{ij(T^r+t)}(\bar{S}_t; \bar{X}_{t-1}) + \sum_{\bar{x}_t} f_{T^r+t+1}(\bar{S}_t + \bar{Y}_t; \bar{X}_t) P[\bar{X}_t | \bar{X}_{t-1}], \tag{10.9}$$

where $Q_{ij}(\cdot; \bar{X}_{T_0-1})$ and $G_{ij(T^r+t)}(\cdot; \bar{X}_{t-1})$ are conditioned on the vectors $\bar{X}_{T_0-1}$ and $\bar{X}_{t-1}$, respectively. $\bar{X}_0$ is the null vector. Decisions in period $t$ are subject to the constraints (10.2)-(10.7).

It should be obvious that the number of states that would exist for any reasonably-sized problem encountered in practice would be extraordinarily large. Hence, the dynamic programming model is not a practical one for generating repair and allocation decisions. We now turn our attention to alternative approximation models that allow computationally tractable methods for making repair and allocation decisions.

## 10.4   The Stock Allocation Model

In this section, the first approximation model is formulated, which we call the *stock allocation model*, or **SAM**. This model is a convex program, and we show that it is separable by item. We then propose two approaches for solving the item subproblems. In the first approach, we derive bounds on the optimal values of the decision variables and use these bounds to formulate and solve the subproblems as linear programs. Although the LPs can be quite large, this method is exact, and it provides a benchmark against which we can measure the performance of faster, greedy heuristics. Two such greedy algorithms are presented here.

### 10.4.1   Model Definition

Recall that the **SAM** addresses only the inventory allocation problem and allows only one mode of shipment (i.e., regular shipment) from the depot warehouse to the bases. Given the notation of the previous section, we can describe the **SAM** as follows:

$$\textbf{(SAM) minimize} \sum_{i \in I} \sum_{j \in J} \left\{ \sum_{t=T_{ij}^r}^{T_{ij}^r+T_{i0}} G_{ijt}(S_{ijt}) + Q_{ij}(S_{ij(T_{ij}^r+T_{i0})}) \right\} \quad (10.10)$$

subject to

$$\tilde{S}_{i0t} \geq \sum_{j \in J} \sum_{t'=0}^{t} y_{ijt'}^r, \forall i \in I, t = 0, \ldots, T_{i0}, \quad (10.11)$$

$$S_{ijt} = \tilde{S}_{ij(T_{ij}^r-1)} + \sum_{t'=0}^{t-T_{ij}^r} y_{ijt'}^r, \quad (10.12)$$

$$\forall i \in I, j \in J, t = T_{ij}^r, \ldots, T_{ij}^r + T_{i0},$$

$$y_{ijt}^r \geq 0 \text{ and integer } \forall i \in I, j \in J, t = 0, \ldots, T_{i0}. \quad (10.13)$$

Constraints (10.11) ensure that stock is available at the depot warehouse before it is allocated, and constraints (10.12) relate the quantities received at the bases with the corresponding quantities shipped from the depot warehouse.

Observe that no more than one item type occurs in any one constraint in this formulation. Hence, the problem is separable by item. Letting $Z^*$ denote the optimal objective function value to **SAM**, this means that we can write:

$$Z^* = \sum_{i \in I} Z_i^*,$$

where $Z_i^*$ denotes the optimal objective function of the subproblem **SAM**$_i$, given by:

$$\textbf{(SAM}_i) \textbf{ minimize} \sum_{j \in J} \left\{ \sum_{t=T_{ij}^r}^{T_{ij}^r+T_{i0}} G_{ijt}(S_{ijt}) + Q_{ij}(S_{ij(T_{ij}^r+T_{i0})}) \right\} \quad (10.14)$$

subject to

$$\tilde{S}_{i0t} \geq \sum_{j \in J} \sum_{t'=0}^{t} y_{ijt'}^r, \quad \forall t = 0, \ldots, T_{i0}, \quad (10.15)$$

$$S_{ijt} = \tilde{S}_{ij(T_{ij}^r-1)} + \sum_{t'=0}^{t-T_{ij}^r} y_{ijt'}^r, \quad (10.16)$$

$$\forall j \in J, t = T_{ij}^r, \ldots, T_{ij}^r + T_{i0},$$

$$y_{ijt}^r \geq 0 \text{ and integer } \quad \forall j \in J, t = 0, \ldots, T_{i0}. \quad (10.17)$$

The problem of solving **SAM** thus reduces to solving **SAM**$_i$ for each $i \in I$.

### 10.4.2  LP Formulation of SAM$_i$

The objective function of **SAM**$_i$ given in (10.14) has two types of terms, those involving single-period expected costs, and those involving end-of-horizon expected holding costs. For a given base $j$ and a given time period $t \in [T_{ij}^r, \ldots, (T_{ij}^r + T_{i0})]$, let us focus on the single-period cost function $G_{ijt}$. Since $G_{ijt}$ is convex in its argument, it is a simple matter to find the solution to the constrained newsvendor problem $CN_{ijt}$:

$$(CN_{ijt}) \text{ minimize } G_{ijt}(S_{ijt}) \tag{10.18}$$

$$\text{subject to } S_{ijt} \geq \tilde{S}_{ijt} \text{ and integer.}$$

Let $\hat{S}_{ijt}$ denote the *largest* optimal solution to this problem, so that

$$\hat{S}_{ijt} = \max(\tilde{S}_{ijt}, \; \arg\min_{S \in \mathcal{S}}(G_{ijt}(S))), \tag{10.19}$$

$$\text{where } \mathcal{S} = \{\lceil F_{X_{ijt}}^{-1}(\frac{b_{ij}}{b_{ij} + h_{ij}})\rceil, \lfloor F_{X_{ijt}}^{-1}(\frac{b_{ij}}{b_{ij} + h_{ij}})\rfloor\}.$$

Recall that $\tilde{S}_{ijt} = \tilde{S}_{ij(T_{ij}^r - 1)}$ for all $t \in [T_{ij}^r, \ldots, T_{ij}^r + T_{i0}]$. Thus, in order for the solutions $\hat{S}_{ijt}, t \in [T_{ij}^r, \ldots, (T_{ij}^r + T_{i0})]$ to be nondecreasing in $t$, it suffices for the corresponding distribution functions $F_{X_{ijt}}(x)$ to be nonincreasing in $t$ for all values $x$. Since $X_{ijt}$ denotes the *cumulative* demand of item $i$ at base $j$ through period $t$, this condition must hold. Hence,

$$\hat{S}_{ij(t-1)} \leq \hat{S}_{ijt} \quad \forall j \in J, t \in [T_{ij}^r, \ldots, (T_{ij}^r + T_{i0})], \tag{10.20}$$

and we have the following theorem to bound the optimal cumulative stock levels:

**Theorem 15.** *For all bases $j \in J$ and all time periods*
$$t \in [T_{ij}^r, \ldots, (T_{ij}^r + T_{i0})],$$
*let $\hat{S}_{ijt}$ denote the largest optimal solution to $CN_{ijt}$, and let $S_{ijt}^*$ denote the corresponding term within an optimal solution to $SAM_i$. Then:*

$$\tilde{S}_{ij(T_{ij}^r - 1)} \leq S_{ijt}^* \leq \hat{S}_{ijt} \quad \forall j \in J, t \in [T_{ij}^r, \ldots, (T_{ij}^r + T_{i0})]. \tag{10.21}$$

*Proof.* The first inequality must hold for any feasible solution. Only the second inequality requires proof. Suppose it does not hold for some base $j$, and let $k$ be the smallest index (i.e., the earliest time period) for which the base violates this condition. That is, $S_{ijk}^* > \hat{S}_{ijk}$, and $S_{ijt}^* \leq \hat{S}_{ijt}$ for all $t \in [T_{ij}^r, \ldots, (k-1)]$. (If $k = T_{ij}^r$, this means that $S_{ij(k-1)}^* = S_{ij(T_{ij}^r - 1)}^* = \tilde{S}_{ij(T_{ij}^r - 1)} \leq \hat{S}_{ij(T_{ij}^r - 1)}$.) By (10.20), we have that $\hat{S}_{ij(k-1)} \leq \hat{S}_{ijk}$. Thus, $S_{ijk}^* > \hat{S}_{ijk} \geq \hat{S}_{ij(k-1)} \geq S_{ij(k-1)}^*$, which implies that $S_{ijk}^* - S_{ij(k-1)}^* > 0$. Recall, however, that $\tilde{S}_{ijk} - \tilde{S}_{ij(k-1)} = \tilde{S}_{ij(T_{ij}^r - 1)} - \tilde{S}_{ij(T_{ij}^r - 1)} = 0$. Thus, the optimal solution must allocate at least one unit of item $i$ that arrives at base $j$ in time period $k$. That is, $y_{ij(k-T_{ij}^r)}^{r*} > 0$. Consider the following minor changes to the optimal solution to $SAM_i$:

$$y_{ij(k-T_{ij}^r)}^r \leftarrow y_{ij(k-T_{ij}^r)}^{r*} - 1 \quad \text{and}$$

$$y_{ij(k-T_{ij}^r+1)}^r \leftarrow y_{ij(k-T_{ij}^r+1)}^{r*} + 1.$$

The resulting solution is feasible. The shipment of one unit from the depot warehouse to base $j$ is delayed by one period, so that one *less* unit arrives in period $k$ and one *more* unit arrives in period $k + 1$. Moreover, this modified solution will have $S_{ijk} = S^*_{ijk} - 1$, but for all $t \neq k$, $S_{ijt} = S^*_{ijt}$. If $k \neq T^r_{ij} + T_{i0}$, then the only change to the objective function is that $G_{ijk}(S_{ijk})$ replaces $G_{ijk}(S^*_{ijk})$. Since $G_{ijk}$ is convex in its argument and $\hat{S}_{ijk}$ is the *largest* optimal solution to $\mathbf{CN}_{ijt}$, $G_{ijk}(\hat{S}_{ijk}) \leq G_{ijk}(S_{ijk}) = G_{ijk}(S^*_{ijk} - 1) < G_{ijk}(S^*_{ijk})$. If $k = T^r_{ij} + T_{i0}$, then the second term in the objective function will also change; but, since $Q_{ij}$ is a strictly increasing function of its argument, $Q_{ij}(S_{ij(T^r_{ij}+T_{i0})}) = Q_{ij}(S^*_{ij(T^r_{ij}+T_{i0})} - 1) < Q_{ij}(S^*_{ij(T^r_{ij}+T_{i0})})$. In either case, the modified solution will have an objective function value that is strictly less than the value achieved by the original solution. Hence, the original solution cannot be optimal, and in any optimal solution we must have $S^*_{ijt} \leq \hat{S}_{ijt}$ for all $t = T^r_{ij}, \ldots, (T^r_{ij} + T_{i0})$. □

Theorem 15 provides upper and lower bounds on the cumulative stock levels in any optimal solution to $\mathbf{SAM}_i$. This result is used to construct a linear programming formulation of $\mathbf{SAM}_i$. Letting

$$
\delta_{ijtk} = \begin{cases} 1 & \text{if } S_{ijt} = k, \\ 0 & \text{otherwise,} \end{cases} \tag{10.22}
$$

we can reformulate $\mathbf{SAM}_i$ as follows:

$$
\text{minimize} \sum_{j \in J} \Bigg\{ \sum_{t=T^r_{ij}}^{T^r_{ij}+T_{i0}} \sum_{k=\tilde{S}_{ij(T^r_{ij}-1)}}^{\hat{S}_{ijt}} \delta_{ijtk} G_{ijt}(k)
$$

$$
+ \sum_{k=\tilde{S}_{ij(T^r_{ij}-1)}}^{\hat{S}_{ij(T^r_{ij}+T_{i0})}} \delta_{ij(T^r_{ij}+T_{i0})k} Q_{ij}(k) \Bigg\} \tag{10.23}
$$

subject to

$$
\tilde{S}_{i0t} \geq \sum_{j \in J} \sum_{t'=0}^{t} y^r_{ijt'}, \quad \forall t = 0, \ldots, T_{i0}, \tag{10.24}
$$

$$
\sum_{k=\tilde{S}_{ij(T^r_{ij}-1)}}^{\hat{S}_{ijt}} \delta_{ijtk} \cdot k = \tilde{S}_{ij(T^r_{ij}-1)} + \sum_{t'=0}^{t-T^r_{ij}} y^r_{ijt'}, \tag{10.25}
$$

$$
\forall j \in J, t = T^r_{ij}, \ldots, T^r_{ij} + T_{i0},
$$

$$
\sum_{k=\tilde{S}_{ij(T^r_{ij}-1)}}^{\hat{S}_{ijt}} \delta_{ijtk} = 1 \quad \forall j \in J, t = T^r_{ij}, \ldots, T^r_{ij} + T_{i0}, \tag{10.26}
$$

$$\delta_{ijtk} \in \{0, 1\} \tag{10.27}$$

$$\forall j \in J, t = T_{ij}^r, \ldots, T_{ij}^r + T_{i0}, k = \tilde{S}_{ij(T_{ij}^r-1)}, \ldots, \hat{S}_{ijt},$$

$$y_{ijt}^r \geq 0 \text{ and integer} \quad \forall j \in J, t = 0, \ldots, T_{i0}. \tag{10.28}$$

Since the cost functions $G_{ijt}$ and $Q_{ij}$ are convex in their arguments, and since the cumulative stock levels $S_{ijt}$ and the parameters $\tilde{S}_{i0t}$ and $\tilde{S}_{ijt}$ assume only integer values, the integer restrictions on $\delta_{ijtk}$ and $y_{ijt}^r$ are unnecessary. That is, solving the LP relaxation of the preceding ILP will result in an integer optimal solution. Hence, the integer restrictions on $\delta_{ijtk}$ and $y_{ijt}^r$ in (10.27) and (10.28), respectively, can be dropped, and the solution to the resulting linear program will be an optimal solution to $\mathbf{SAM}_i$. Details can be found in [68].

The major drawback to solving $\mathbf{SAM}_i$ using the above LP formulation is that the number of variables can be very large. For practical purposes, it is possible to reduce the number of variables by limiting the number of values each $y_{ijt}^r$ can assume (e.g., multiples of 5, instead of 1). By making such restrictions, the LP will yield a solution that is only approximately optimal, but for large problems, this type of rescaling can be a very useful technique. Furthermore, since the second derivatives of the $G$ functions are not large in the neighborhood of their minimizers, the expected cost of the optimal solution to a scaled problem will, in most realistic instances, be very close to the expected cost of the optimal solution to the unscaled problem.

### 10.4.3   Greedy Algorithms for SAM$_i$

Instead of solving $\mathbf{SAM}_i$ as a linear program, let us consider two alternative greedy algorithms that myopically exploit the convexity of the objective function. Solutions resulting from these algorithms may be used in two ways. They may be used directly, or, since these algorithms provide basic feasible solutions for the preceding LP, they can be used to seed the LP with a near-optimal solution.

Before describing these algorithms, we define the following terms to capture incremental changes in the objective function. For all $t = T_{ij}^r, \ldots, T_{ij}^r + T_{i0}$, let:

$$\Delta G_{ijt}(S_{ijt}) = G_{ijt}(S_{ijt} + 1) - G_{ijt}(S_{ijt}) \tag{10.29}$$

be the incremental change in the expected holding and backorder costs realized in period $t$ when the allocation of item $i$ to base $j$ increases by one unit prior to period $t$ or in period $t$ (i.e., when one more unit of item $i$ *arrives* at base $j$ prior to period $t$ or in period $t$). Next, define

$$\Delta C_{ijt} = \sum_{k=t}^{T_{ij}^r + T_{i0}} \Delta G_{ijk}(S_{ijk}) \tag{10.30}$$

to be the incremental change in the expected holding and backorder costs over the entire effective horizon when the allocation of item $i$ to base $j$ increases by one unit in period $t$. Finally, let

$$\Delta Q_{ij} = Q_{ij}(S_{ij(T_{ij}^r+T_{i0})} + 1) - Q_{ij}(S_{ij(T_{ij}^r+T_{i0})})  \quad (10.31)$$

be the incremental change in the end-of-horizon expected holding costs when the allocation of item $i$ to base $j$ increases by one unit over the planning horizon.

Given these definitions, if in period $t$ an additional unit of item $i$ is sent from the depot warehouse to base $j$, the total incremental change in the objective function is as follows:

$$\Delta Z_i(j, t) = \Delta C_{ij(t+T_{ij}^r)} + \Delta Q_{ij} \quad \text{when } y_{ijt}^r \leftarrow y_{ijt}^r + 1.  \quad (10.32)$$

Each of the algorithms begins with $y_{ijt}^r$ equal to zero for all $t \in [0, \dots, T_{i0}]$ and iteratively assigns available stock at the depot warehouse to bases according to a greedy rule. For all $t \in [0, \dots, T_{i0}]$, let $A_{i0t}$ measure the total number of units of item $i$ that have been allocated in periods $[0, \dots, t]$ (i.e., sent to bases from the depot warehouse).

In the first algorithm, **GA**, the greedy rule is simple: Over all periods $t$ in which stock is available for allocation, find the $(j, t)$ combination that produces the largest objective function reduction $\Delta Z_i(j, t)$, and assign a unit of stock to be sent to base $j$ in time period $t$. We now formally state this heuristic:

### (GA) A Greedy Algorithm for SAM$_i$:

**Step 0:** Set $y_{ijt}^r \leftarrow 0$ for all $j \in J, t \in [0, \dots, T_{i0}]$. (Note that this implicitly sets $S_{ijt} \leftarrow \tilde{S}_{ijt}$ for all $t \in [T_{ij}^r, \dots, T_{ij}^r + T_{i0}]$.) Set $A_{i0t} \leftarrow 0$ for all $t \in [0, \dots, T_{i0}]$.
**Step 1:** If $A_{i0T_{i0}} = \tilde{S}_{i0T_{i0}}$, then **STOP** – no more allocations can be made. Otherwise, determine $t^* = \min\{t : \forall t' \geq t, A_{i0t'} < \tilde{S}_{i0t'}\}$, the earliest period in which a unit of stock is still available for allocation.
**Step 2:** For all $j \in J, k \in [t^*, \dots, T_{i0}]$, compute $\Delta Z_i(j, k)$, and determine $(j^*, k^*) = \arg\min_{(j,k)}(\Delta Z_i(j, k))$.
**Step 3:** If $\Delta Z_i(j^*, k^*) \geq 0$, then **STOP** – no further objective function reductions are possible. Otherwise, set $y_{ij^*k^*}^r \leftarrow y_{ij^*k^*}^r + 1$ (so that implicitly $S_{ij^*(T_{ij}^r+t)} \leftarrow S_{ij^*(T_{ij}^r+t)} + 1$ for all $t \in [k^*, \dots, T_{i0}]$). For all $t \in [k^*, \dots, T_{i0}]$, set $A_{i0t} \leftarrow A_{i0t} + 1$. Go to Step 1.

The primary benefit of **GA** is that it only requires $O(j)$ computations at each iteration. This is because in Step 2, it can be shown that for each $j \in J$, the largest reduction $\Delta Z_i(j, k)$ will be achieved in the first period $k \geq t^*$ such that $S_{ij(k+T_{ij}^r)} < \hat{S}_{ij(k+T_{ij}^r)}$. Hence, not all periods $k \in [t^*, \dots, T_{i0}]$ need to be checked. The total time required for the algorithm is $O(jT_{i0} + j\tilde{S}_{i0T_{i0}})$.

**GA** will not necessarily find the optimal solution to **SAM$_i$**. The reason is that the availability of the depot warehouse pipeline stock is spread out over time periods $0, \dots, T_{i0}$, but the algorithm allocates each unit of stock as if it were the very last unit available. That is, **GA** does not consider the fact that stock that will be available next period may be able to provide almost as much benefit to a base $j$

as stock that is available now, but for another base $j'$ waiting until next period may reduce the benefit significantly. Under certain conditions, however, **GA** *will* find the optimal solution to **SAM**$_i$. Namely, if for some integer $N$ we have $\tilde{S}_{i0t} = N$ for all $t \in [0, \ldots, T_{i0}]$, or if for some period $k \in [0, \ldots, T_{i0}]$ we have $\tilde{S}_{i0t} = 0$ for all $t \in [0, \ldots, (k-1)]$ and $\tilde{S}_{i0t} = N$ for all $t \in [k, \ldots, T_{i0}]$, then **GA** will find the optimal solution to **SAM**$_i$.

The second algorithm, **LGA**, is similar to the first, except that a look-ahead step is performed before an allocation decision is made. That is, the algorithm checks to see what the objective function reduction would be if the next *two* units of available stock were the last two units available.

### (LGA) A Look-Ahead Greedy Algorithm for SAM$_i$

Step 0: Set $y^r_{ijt} \leftarrow 0$ for all $j \in J, t \in [0, \ldots, T_{i0}]$. (Note that this implicitly sets $S_{ijt} \leftarrow \tilde{S}_{ijt}$ for all $t \in [T^r_{ij}, \ldots, T^r_{ij} + T_{i0}]$.) Set $A_{i0t} \leftarrow 0$ for all $t \in [0, \ldots, T_{i0}]$.

Step 1: If $A_{i0T_{i0}} = \tilde{S}_{i0T_{i0}}$, then **STOP** – no more allocations can be made. Otherwise, determine $t^*_1 = \min\{t : \forall t' \geq t, A_{i0t'} < \tilde{S}_{i0t'}\}$, the earliest period in which a unit of stock is still available for allocation, and $t^*_2 = \inf\{t \geq t^*_1 : \forall t' \geq t, \tilde{S}_{i0t'} - A_{i0t'} \geq 2\}$, the earliest period in which a *second* unit of stock is available for allocation. (Note that it is possible to have $t^*_2 = \infty$ if $t^*_1$ is the last period in which stock is available for allocation and only one unit remains.)

Step 2: For all $j \in J, k \in [t^*_1, \ldots, T_{i0}]$, compute $\Delta Z_i(j, k)$. If $\min_{(j,k)} \Delta Z_i(j, k) \geq 0$, then **STOP** – no further objective function reductions are possible. Otherwise, go to step 3.

Step 3: If $t^*_2 = \infty$, then for each $j \in J$, determine $k^j_1 = \arg \min_{k \in [t^*_1, \ldots, T_{i0}]} (\Delta Z_i(j, k))$, the locally best period in which to send another unit to base $j$, and set $Change(j) = \Delta Z_i(j, k^j_1)$. Otherwise, for each $j \in J$, determine $k^j_1 = \arg \min_{k \in [t^*_1, \ldots, T_{i0}]} (\Delta Z_i(j, k))$ and $(j_2, k_2) = \arg \min_{(j, k \in [t^*_2, \ldots, T_{i0}])} (\Delta Z_i((j, k^j_1), (j_2, k_2)))$, where $\Delta Z_i((j, k^j_1), (j_2, k_2))$ represents the change in the objective function $\Delta Z_i(j_2, k_2)$ *after* setting $y^r_{ij(k^j_1)} \leftarrow y^r_{ij(k^j_1)} + 1$ (this is only a look-ahead check – we are not actually augmenting $y^r_{ij(k^j_1)}$ in this step); then set $Change(j) = \Delta Z_i(j, k^j_1) + \min(0, \Delta Z_i((j, k^j_1), (j_2, k_2)))$.

Step 4: Determine $j^* = \arg \min_{j \in J} (Change(j))$. Set $y^r_{ij^*k^{j^*}_1} \leftarrow y^r_{ij^*k^{j^*}_1} + 1$ (so that implicitly $S_{ij^*(T^r_{ij}+t)} \leftarrow S_{ij^*(T^r_{ij}+t)} + 1$ for all $t \in [k^{j^*}_1, \ldots, T_{i0}]$). For all $k \in [k^{j^*}_1, \ldots, T_{i0}]$, set $A_{i0k} \leftarrow A_{i0k} + 1$. Go to Step 1.

**LGA** requires $O(j^2)$ computation time for each iteration, and the total time required for the algorithm is $O(jT_{i0} + j^2 \tilde{S}_{i0T_{i0}})$. Clearly, **LGA** can be generalized to be an $n$-step look-ahead algorithm for any $n$, but since the computation time for each iteration is $O(j^{(n+1)})$, such an algorithm is likely to be very computationally intensive for $n \geq 3$. Note, however, that such an $n$-step look-ahead algorithm will find the optimal solution to $\mathbf{SAM}_i$ if $\tilde{S}_{i0T_{i0}} \leq (n+1)$.

## 10.5 The Extended Stock Allocation Model

The *extended stock allocation model*, or **ESAM**, is a modification of the **SAM** in which the decision-maker has the option of using an expedited shipment mode to transport parts from the depot warehouse to the bases, in addition to the regular shipment mode. In this section, the formulation of the **SAM** is extended to allow for an expedited mode of shipment. The solution methods presented in the previous section to accommodate this extension are also modified. Unfortunately, Theorem 15 does not apply to the subproblems of the **ESAM**. Hence, new bounds are derived on the optimal values of the decision variables. Using these new bounds, the subproblems can still be formulated and solved as linear programs. Modifications of the two greedy algorithms presented in the previous section are presented for finding solutions to the **ESAM**.

### 10.5.1  Model Definition

Given the previously defined notation, we formulate the **ESAM** as follows:

(**ESAM**)

$$\text{minimize} \sum_{i \in I} \sum_{j \in J} \left\{ \sum_{t=T_{ij}^e}^{T_{ij}^r + T_{i0}} G_{ijt}(S_{ijt}) + Q_{ij}(S_{ij(T_{ij}^r + T_{i0})}) + \sum_{t=0}^{T_{i0}} e_{ij} y_{ijt}^e \right\} \quad (10.33)$$

subject to

$$\tilde{S}_{i0t} \geq \sum_{j \in J} \sum_{t'=0}^{t} (y_{ijt'}^r + y_{ijt'}^e), \ \forall i \in I, t = 0, \ldots, T_{i0}, \quad (10.34)$$

$$S_{ijt} = \tilde{S}_{ijt} + \sum_{t'=0}^{\min(t-T_{ij}^e, T_{i0})} y_{ijt'}^e, \ \forall i \in I, j \in J, t = T_{ij}^e, \ldots, T_{ij}^r - 1, (10.35)$$

$$S_{ijt} = \tilde{S}_{ij(T_{ij}^r-1)} + \sum_{t'=0}^{\min(t-T_{ij}^e, T_{i0})} y_{ijt'}^e + \sum_{t'=0}^{t-T_{ij}^r} y_{ijt'}^r, \quad (10.36)$$

$$\forall i \in I, j \in J, t = T_{ij}^r, \ldots, T_{ij}^r + T_{i0},$$

$$y_{ijt}^e, y_{ijt}^r \geq 0 \text{ and integer } \forall i \in I, j \in J, t = 0, \ldots, T_{i0}. \quad (10.37)$$

Note that the objective function (10.33) contains a new term to capture the cost of an expedited shipment. Also, note that two sets of constraints, (10.35) and (10.36), are used to relate the quantities received at the bases with the corresponding quantities shipped from the depot warehouse. When there was only a regular shipment mode, $t = T_{ij}^r$ was the earliest time period in which the allocation decisions could affect the cumulative stock level $S_{ijt}$ of item $i$ at base $j$ received through period $t$. However, with the option of using an expedited shipment, it is possible to affect $S_{ijt}$ for time periods $t = T_{ij}^e, \ldots, T_{ij}^r - 1$ as well.

As with **SAM**, **ESAM** is separable by item, and the problem reduces to solving **ESAM**$_i$, for each $i \in I$, given by:

(**ESAM**$_i$)

$$\text{minimize} \sum_{j \in J} \left\{ \sum_{t=T_{ij}^e}^{T_{ij}^r + T_{i0}} G_{ijt}(S_{ijt}) + Q_{ij}(S_{ij(T_{ij}^r + T_{i0})}) + \sum_{t=0}^{T_{i0}} e_{ij} y_{ijt}^e \right\} \quad (10.38)$$

subject to

$$\tilde{S}_{i0t} \geq \sum_{j \in J} \sum_{t'=0}^{t} (y_{ijt'}^r + y_{ijt'}^e), \quad \forall t = 0, \ldots, T_{i0}, \quad (10.39)$$

$$S_{ijt} = \tilde{S}_{ijt} + \sum_{t'=0}^{\min(t-T_{ij}^e, T_{i0})} y_{ijt'}^e, \quad (10.40)$$

$$\forall j \in J, t = T_{ij}^e, \ldots, T_{ij}^r - 1,$$

$$S_{ijt} = \tilde{S}_{ij(T_{ij}^r - 1)} + \sum_{t'=0}^{t-T_{ij}^r} y_{ijt'}^r + \sum_{t'=0}^{\min(t-T_{ij}^e, T_{i0})} y_{ijt'}^e, \quad (10.41)$$

$$\forall j \in J, t = T_{ij}^r, \ldots, T_{ij}^r + T_{i0},$$

$$y_{ijt}^e, y_{ijt}^r \geq 0 \text{ and integer} \quad \forall j \in J, t = 0, \ldots, T_{i0}. \quad (10.42)$$

### 10.5.2 LP Formulation of ESAM$_i$

As mentioned earlier, Theorem 15 does not hold for **ESAM**$_i$, and hence new bounds must be established in order to construct a meaningful linear program. The reason that Theorem 15 does not hold for **ESAM**$_i$ is that while the cumulative stock levels $S_{ijt}$ for $t \in [T_{ij}^e, \ldots, T_{ij}^r - 1]$ can be affected by sending expedited shipments in periods 0 through $(T_{ij}^r - 1) - T_{ij}^e$, we cannot *completely* control the cumulative stock pattern in these time periods. That is, the cumulative stock levels $S_{ijt}$ for $t \in [T_{ij}^e, \ldots, T_{ij}^r - 1]$ have lower bounds of $\tilde{S}_{ijt}$ because of what is in the pipeline at time 0, and while the ability to *augment* the pipeline with extra stock exists, stock that is already in the pipeline cannot be removed. This means that in addition to $S_{ijt} \geq \tilde{S}_{ijt}$ for every $t \in [T_{ij}^e, \ldots, T_{ij}^r - 1]$, every feasible solution also must have $S_{ij(t+1)} - S_{ijt} \geq \tilde{S}_{ij(t+1)} - \tilde{S}_{ijt}$. Because of this constraint, it is possible that $S_{ijt}^*$ is not less than or equal to $\hat{S}_{ijt}$.

In order to derive new bounds for the decision variables of **ESAM**$_i$, once again focus on the single-period expected cost function $G_{ijt}$ and the constrained newsvendor problem **CN**$_{ijt}$ given in (10.18). For each $j \in J$ and each $t \in [T_{ij}^e, \ldots, T_{ij}^r + T_{i0}]$, let $\hat{S}_{ijt}$ denote the *largest* optimal solution to **CN**$_{ijt}$, and define:

$$M_{jt} = \max_{k \in [T_{ij}^e, \dots, t]} \{\hat{S}_{ijk} - \tilde{S}_{ijk}\}. \tag{10.43}$$

We then have the following theorem:

**Theorem 16.** *For all bases $j \in J$ and all time periods $t \in [T_{ij}^e, \dots, (T_{ij}^r + T_{i0})]$, let $\hat{S}_{ijt}$ denote the largest optimal solution to $\textbf{CN}_{ijt}$, and let $S_{ijt}^*$ denote the corresponding term within an optimal solution to $\textbf{ESAM}_i$, and let $M_{jt} = \max_{k \in [T_{ij}^e, \dots, t]} \{\hat{S}_{ijk} - \tilde{S}_{ijk}\}$. Then:*

$$\tilde{S}_{ijt} \leq S_{ijt}^* \leq \tilde{S}_{ijt} + M_{jt} \quad \forall j \in J, t \in [T_{ij}^e, \dots, (T_{ij}^r + T_{i0})]. \tag{10.44}$$

*Proof.* The first inequality must hold for any feasible solution. The second inequality requires proof. Suppose it does not hold for some base $j$, and let $k$ be the smallest index (i.e., the earliest time period) for which the base violates this condition. That is, $S_{ijk}^* > \tilde{S}_{ijk} + M_{jk}$, and $S_{ijt}^* \leq \tilde{S}_{ijt} + M_{jt}$ for all $t \in [T_{ij}^e, \dots, (k-1)]$. (If $k = T_{ij}^e$, then note that $S_{ij(k-1)}^* = S_{ij(T_{ij}^e - 1)}^* = \tilde{S}_{ij(T_{ij}^e - 1)} \leq \tilde{S}_{ij(T_{ij}^e - 1)}$.) Then

$$
\begin{aligned}
S_{ijk}^* - S_{ij(k-1)}^* &> (\tilde{S}_{ijk} + M_{jk}) - S_{ij(k-1)}^* \\
&\geq (\tilde{S}_{ijk} + M_{jk}) - (\tilde{S}_{ij(k-1)} + M_{j(k-1)}) \\
&\geq (\tilde{S}_{ijk} + M_{jk}) - (\tilde{S}_{ij(k-1)} + M_{jk}) \\
&= \tilde{S}_{ijk} - \tilde{S}_{ij(k-1)}.
\end{aligned}
$$

Since the $\tilde{S}_{ijt}$ are nondecreasing, the optimal solution must allocate at least one unit of item $i$ that arrives at base $j$ in time period $k$. That is, we must have $y_{ij(k-T_{ij}^e)}^{e*} > 0$ or $y_{ij(k-T_{ij}^r)}^{r*} > 0$. Without loss of generality, assume the latter, and consider the following minor changes to the optimal solution to $\textbf{ESAM}_i$:

$$y_{ij(k-T_{ij}^r)}^r \leftarrow y_{ij(k-T_{ij}^r)}^{r*} - 1 \quad \text{and}$$

$$y_{ij(k-T_{ij}^r+1)}^r \leftarrow y_{ij(k-T_{ij}^r+1)}^{r*} + 1.$$

The resulting solution is feasible. The shipment of one unit from the depot warehouse to base $j$ is delayed by one period, so that one *less* unit arrives in period $k$ and one *more* unit arrives in period $k + 1$. Moreover, this modified solution will have $S_{ijk} = S_{ijk}^* - 1$, but for all $t \neq k$, $S_{ijt} = S_{ijt}^*$. If $k \neq T_{ij}^r + T_{i0}$, then the only change to the objective function is that $G_{ijk}(S_{ijk})$ replaces $G_{ijk}(S_{ijk}^*)$. But since $G_{ijk}$ is convex in its argument and $\hat{S}_{ijk}$ is the *largest* optimal solution to $\textbf{CN}_{ijt}$, we have that $G_{ijk}(\hat{S}_{ijk}) \leq G_{ijk}(\tilde{S}_{ijk} + M_{jk}) \leq G_{ijk}(S_{ijk}) = G_{ijk}(S_{ijk}^* - 1) < G_{ijk}(S_{ijk}^*)$. If $k = T_{ij}^r + T_{i0}$, then the second term in the objective function will also change; but, since $Q_{ij}$ is a strictly increasing function of its argument, $Q_{ij}(S_{ij(T_{ij}^r+T_{i0})}) = Q_{ij}(S_{ij(T_{ij}^r+T_{i0})}^* - 1) < Q_{ij}(S_{ij(T_{ij}^r+T_{i0})}^*)$. In either case, the modified solution will have an objective function value that is strictly less than the value achieved by the original solution. Hence, the original solution cannot be optimal, and in any optimal solution we must have $S_{ijt}^* \leq \tilde{S}_{ijt} + M_{jt}$ for all $t = T_{ij}^e, \ldots, (T_{ij}^r + T_{i0})$.                            □

Theorem 16 provides upper and lower bounds on the cumulative stock levels in any optimal solution to $\textbf{ESAM}_i$. As before, we can use these bounds to construct a linear programming formulation of $\textbf{ESAM}_i$. Letting

$$\delta_{ijtk} = \begin{cases} 1 & \text{if } S_{ijt} = k, \\ 0 & \text{otherwise,} \end{cases} \tag{10.45}$$

we can reformulate $\textbf{ESAM}_i$ as follows:

$$\text{minimize} \sum_{j \in J} \left\{ \sum_{t=T_{ij}^e}^{T_{ij}^r + T_{i0}} \sum_{k=\tilde{S}_{ijt}}^{\tilde{S}_{ijt} + M_{jt}} \delta_{ijtk} G_{ijt}(k) \right.$$

$$+ \sum_{k=\tilde{S}_{ij(T_{ij}^r-1)}}^{\tilde{S}_{ij(T_{ij}^r+T_{i0})} + M_{j(T_{ij}^r+T_{i0})}} \delta_{ij(T_{ij}^r+T_{i0})k} Q_{ij}(k) + \left. \sum_{t=0}^{T_{i0}} e_{ij} y_{ijt}^e \right\} \tag{10.46}$$

subject to

$$\tilde{S}_{i0t} \geq \sum_{j \in J} \sum_{t'=0}^{t} (y_{ijt'}^r + y_{ijt'}^e), \quad \forall t = 0, \ldots, T_{i0}, \tag{10.47}$$

$$\sum_{k=\tilde{S}_{ijt}}^{\tilde{S}_{ijt} + M_{jt}} \delta_{ijtk} \cdot k = \tilde{S}_{ijt} + \sum_{t'=0}^{\min(t-T_{ij}^e, T_{i0})} y_{ijt'}^e, \tag{10.48}$$

$$\forall j \in J, t = T_{ij}^e, \ldots, (T_{ij}^r - 1),$$

$$\sum_{k=\tilde{S}_{ij(T_{ij}^r-1)}}^{\tilde{S}_{ijt} + M_{jt}} \delta_{ijtk} \cdot k = \tilde{S}_{ij(T_{ij}^r-1)} + \sum_{t'=0}^{t-T_{ij}^r} y_{ijt'}^r + \sum_{t'=0}^{\min(t-T_{ij}^e, T_{i0})} y_{ijt'}^e, \tag{10.49}$$

$$\forall j \in J, t = T_{ij}^r, \ldots, T_{ij}^r + T_{i0},$$

$$\sum_{k=\tilde{S}_{ijt}}^{\tilde{S}_{ijt} + M_{jt}} \delta_{ijtk} = 1 \quad \forall j \in J, t = T_{ij}^e, \ldots, T_{ij}^r + T_{i0}, \tag{10.50}$$

$$\delta_{ijtk} \in \{0, 1\} \tag{10.51}$$

$$\forall j \in J, t = T_{ij}^e, \ldots, T_{ij}^r + T_{i0}, k = \tilde{S}_{ijt}, \ldots, \tilde{S}_{ijt} + M_{jt},$$

$$y_{ijt}^e, y_{ijt}^r \geq 0 \text{ and integer} \quad \forall j \in J, t = 0, \ldots, T_{i0}. \tag{10.52}$$

As with the $\textbf{SAM}_i$, solving the LP relaxation of the preceding ILP will result in an integer optimal solution. Hence, the integer restrictions in (10.51) and (10.52) can be dropped, and the solution to the resulting linear program will be an optimal solution to $\mathbf{ESAM}_i$.

### 10.5.3   Greedy Algorithms for $\mathbf{ESAM}_i$

Both of the greedy algorithms presented in Section 10.4.3 are easily modified to accommodate the $\mathbf{ESAM}_i$. The incremental changes in costs $\Delta G_{ijt}(S_{ijt})$, $\Delta C_{ijt}$, and $\Delta Q_{ij}$ are defined as before, but now, to capture the total change in the objective function, the mode of shipment used in the allocation must be taken into account. That is, we now define:

$$\Delta Z_i(j,t) = \begin{cases} \Delta Z_i^r(j,t) = \Delta C_{ij(t+T_{ij}^r)} \\ \qquad + \Delta Q_{ij} & \text{when } y_{ijt}^r \leftarrow y_{ijt}^r + 1, \\ \Delta Z_i^e(j,t) = \Delta C_{ij(t+T_{ij}^e)} \\ \qquad + \Delta Q_{ij} + e_{ij} & \text{when } y_{ijt}^e \leftarrow y_{ijt}^e + 1. \end{cases} \qquad (10.53)$$

The modified version of $\mathbf{GA}$, called $\mathbf{EGA}$, is given below. Similar changes are required to tailor the look-ahead algorithm $\mathbf{LGA}$ for the $\mathbf{ESAM}_i$, although the details of $\mathbf{ELGA}$ are omitted. As with the original algorithms, the solutions resulting from the modified greedy algorithms $\mathbf{EGA}$ and $\mathbf{ELGA}$ may be used directly, or they can be used to seed the $\mathbf{ESAM}_i$ LP with a near-optimal solution.

**(EGA) A Greedy Algorithm for $\mathbf{ESAM}_i$:**

**Step 0:** Set $y_{ijt}^m \leftarrow 0$ for all $j \in J, t \in [0, \ldots, T_{i0}], m \in [r, e]$. (Note that this implicitly sets $S_{ijt} \leftarrow \tilde{S}_{ijt}$ for all $t \in [T_{ij}^e, \ldots, T_{ij}^r + T_{i0}]$.) Set $A_{i0t} \leftarrow 0$ for all $t \in [0, \ldots, T_{i0}]$.

**Step 1:** If $A_{i0T_{i0}} = \tilde{S}_{i0T_{i0}}$, then **STOP** – no more allocations can be made. Otherwise, set $t^* = \min\{t : \forall t' \geq t, A_{i0t'} < \tilde{S}_{i0t'}\}$, the earliest period in which a unit of stock is still available for allocation.

**Step 2:** For all $j \in J, k \in [t^*, \ldots, T_{i0}]$, compute $\Delta Z_i^r(j,k)$ and $\Delta Z_i^e(j,k)$, and determine $(j^*, k^*, m^*) = \arg\min_{(j,k,m)}(\Delta Z_i^m(j,k))$.

**Step 3:** If $\Delta Z_i^{m^*}(j^*, k^*) \geq 0$, then **STOP** – no further objective function reductions are possible. Otherwise, set $y_{ij^*k^*}^{m^*} \leftarrow y_{ij^*k^*}^{m^*} + 1$ (so that implicitly $S_{ij^*(T_{ij}^{m^*}+t)} \leftarrow S_{ij^*(T_{ij}^{m^*}+t)} + 1$ for all $t \in [k^*, \ldots, T_{i0}]$). For all $k \in [k^*, \ldots, T_{i0}]$, set $A_{i0k} \leftarrow A_{i0k} + 1$. Go to Step 1.

We now turn our attention to an operating environment in which repair decisions as well as inventory allocation decisions must be made.

## 10.6   The Extended Stock Allocation Model with Repair

The premise for the *extended stock allocation model with repair*, or $\mathbf{ESAMR}$, is that a repair process, shared by all items, is the supply source for the depot warehouse. The repair allocation decisions made determine what items should be entered into repair in period $t = 0$.

Each item type $i$ is assumed to have a fixed repair time from the time the repair commences. Units of item type $i$ that are selected to enter repair in period 0 reach the depot warehouse in time period $T_{i0}$, the last period of the planning horizon for item $i$; hence, this supply will affect the costs realized over the effective horizon for item $i$.

When a decision is made to repair a part, there is a risk that the repaired part may subsequently sit in the depot warehouse for some time if the projected demand at the bases for the item is small. In such a case, it may have been better to wait to repair the part until the projected demand at the bases made its need more likely, since it is typically more costly to hold a serviceable part in storage than to hold a damaged part in the repair queue. Thus, since the **ESAMR** includes repair decisions, one of the additional economic considerations that is captured in this model is the incremental cost per period associated with holding serviceable units at the depot warehouse instead of at the repair facility. The following notation is used to model this consideration:

$h_{i0}$  the incremental cost per period associated with holding a unit of item $i$ at the depot warehouse instead of at the repair facility.

$Q_{i0}(\cdot)$  the function approximating the expected incremental holding costs incurred beyond the end of the planning horizon for item $i$ at the depot warehouse. The function argument is $\tilde{S}_{i0T_{i0}}$, the cumulative supply of item $i$ at the depot warehouse at the end of the planning horizon. (See below.)

The end-of-horizon expected incremental holding cost function in **ESAMR** is given by:

$$Q_{i0}(\tilde{S}_{i0T_{i0}}) = h_{i0} \sum_{t=T_{i0}+1}^{\infty} E\left[\left(\sum_{j \in J} \tilde{S}_{ij(T_{ij}^r-1)} + \tilde{S}_{i0T_{i0}}\right) - \sum_{j \in J} X_{ijt}\right]^+. \quad (10.54)$$

As with $Q_{ij}$, $Q_{i0}$ is a convex function of its argument. By incorporating $Q_{i0}$ into the objective function of **ESAMR**, a mechanism is provided for determining the most cost-effective use of repair capacity. That is, items will be prioritized for repair. The ones likely to be needed in the near future will be repaired instead of items that may not be needed for some time. Note that for each period beyond the end of the planning horizon, the expression (10.54) charges an incremental holding cost on the expected excess of the cumulative total supply (in the subsystem rooted at the depot warehouse) over the cumulative total demand. As such, it is only an *approximation* of the true expected incremental holding costs since it implicitly assumes that the cumulative total supply is fungible in satisfying the cumulative total demand.

Having defined these additional parameters, the *extended stock allocation model with repair*, or (**ESAMR**), is:

**(ESAMR)**

$$\text{minimize} \sum_{i \in I} \left[ \sum_{j \in J} \left\{ \sum_{t=T_{ij}^e}^{T_{ij}^r+T_{i0}} G_{ijt}(S_{ijt}) + Q_{ij}(S_{ij(T_{ij}^r+T_{i0})}) \right. \right.$$

$$\left. \left. + \sum_{t=0}^{T_{i0}} e_{ij} y_{ijt}^e \right\} + Q_{i0}(\tilde{S}_{i0T_{i0}}) \right] \quad (10.55)$$

subject to

$$\tilde{S}_{i0t} \geq \sum_{j \in J} \sum_{t'=0}^{t} (y_{ijt'}^r + y_{ijt'}^e), \quad \forall i \in I, t = 0, \dots, T_{i0}, \quad (10.56)$$

$$S_{ijt} = \tilde{S}_{ijt} + \sum_{t'=0}^{\min(t-T_{ij}^e, T_{i0})} y_{ijt'}^e, \quad (10.57)$$

$$\forall i \in I, j \in J, t = T_{ij}^e, \dots, T_{ij}^r - 1,$$

$$S_{ijt} = \tilde{S}_{ij(T_{ij}^r-1)} + \sum_{t'=0}^{t-T_{ij}^r} y_{ijt'}^r + \sum_{t'=0}^{\min(t-T_{ij}^e, T_{i0})} y_{ijt'}^e, \quad (10.58)$$

$$\forall i \in I, j \in J, t = T_{ij}^r, \dots, T_{ij}^r + T_{i0},$$

$$\tilde{S}_{i0(T_{i0})} = \tilde{S}_{i0(T_{i0}-1)} + v_i, \quad \forall i \in I, \quad (10.59)$$

$$C_{\min} \leq \sum_{i \in I} v_i \leq C_{\max}, \quad (10.60)$$

$$0 \leq v_i \leq C_i \text{ and integer} \quad \forall i \in I, \quad (10.61)$$

$$y_{ijt}^e, y_{ijt}^r \geq 0 \text{ and integer} \quad \forall i \in I, j \in J, t = 0, \dots, T_{i0}. \quad (10.62)$$

Since $Q_{i0}$ is a convex function of its argument, the objective function remains convex in each $S_{ijt}$. Constraints (10.59), (10.60), and (10.61) are new. Constraints (10.59) replace the previously deterministic values $\tilde{S}_{i0(T_{i0})}$ with $\tilde{S}_{i0(T_{i0}-1)}$ plus whatever gets entered into repair in period 0. Constraint (10.60) ensures that the minimum repair requirements are met without exceeding repair capacity, and constraints (10.61) ensure that only units available for repair may be repaired. Unlike the **SAM** and **ESAM**, **ESAMR** is *not* separable by item due to the capacity constraint (10.60). However, with a little more work **ESAMR** can be represented in a form that is easily solvable.

Given an instance of **ESAMR**, let $Z_i^*(s)$ denote the optimal objective function value of the subproblem **ESAM**$_i$, subject to the condition that $\tilde{S}_{i0T_{i0}} = s$. Suppose $Z_i^*(s)$ is computed for each $i \in I$ and for each $s \in \{\tilde{S}_{i0(T_{i0}-1)}, \dots, \tilde{S}_{i0(T_{i0}-1)} + C_i\}$. Letting

$$\delta_{ik} = \begin{cases} 1 & \text{if } v_i = k, \\ 0 & \text{otherwise,} \end{cases} \quad (10.63)$$

we can write **ESAMR** as follows:

$$\text{minimize} \quad \sum_{i \in I} \left[ \sum_{k=0}^{C_i} Z_i^*(\tilde{S}_{i0(T_{i0}-1)} + k) + Q_{i0}(\tilde{S}_{i0(T_{i0}-1)} + k) \right] \delta_{ik} \quad (10.64)$$

subject to

$$\sum_{k=0}^{C_i} \delta_{ik} = 1, \quad \forall i \in I, \quad\quad (10.65)$$

$$C_{\min} \le \sum_{i \in I} \sum_{k=0}^{C_i} k \delta_{ik} \le C_{\max}, \quad\quad (10.66)$$

$$\delta_{ik} \in \{0, 1\}, \quad \forall i \in I, k = 0, \dots, C_i. \quad\quad (10.67)$$

It is not hard to show that for each item $i$, the optimal **ESAM**$_i$ value $Z_i^*(\tilde{S}_{i0(T_{i0}-1)} + k)$ is nonincreasing and (discretely) convex in $k$. (This convexity depends critically on the fact that the $k$ additional units of available inventory all arrive at the depot warehouse in the *same* planning horizon period $T_{i0}$. If the inventory were to arrive in different periods, then convexity could not be guaranteed.) Moreover, $Q_{i0}(\tilde{S}_{i0(T_{i0}-1)} + k)$ is increasing and convex in $k$. Hence, the objective function (10.64) captures the tradeoff, among item types, of using repair capacity to make more inventory available for use at the depot warehouse and potentially incurring incremental holding costs on this additional inventory. Because of the convexity of (10.64), the following greedy marginal analysis algorithm, **EGAR**, can be used to solve **ESAMR** to optimality, provided that the optimal $Z_i^*$ values are used.

**(EGAR) A Greedy Algorithm for ESAMR**:

**Step 0:** For all $i \in I$, $k = 0, \dots, C_i$, determine

$$W_i(k) = Z_i^*(\tilde{S}_{i0(T_{i0}-1)} + k) + Q_{i0}(\tilde{S}_{i0(T_{i0}-1)} + k).$$

Set $v_i \leftarrow 0$ for all $i \in I$. Set $A \leftarrow 0$.

**Step 1:** If $A = C_{\max}$, then **STOP** – no more repair capacity is available. Otherwise, for all $i \in I$ with $v_i < C_i$, compute

$$\Delta W_i(v_i) = W_i(v_i + 1) - W_i(v_i)$$

and determine $i^* = \arg\min_{i \in I}(\Delta W_i(v_i))$.

**Step 2:** If $\Delta W_{i^*}(v_{i^*}) \ge 0$ and $A \ge C_{\min}$, then **STOP** – no further objective function reductions are possible. Otherwise, set $v_{i^*} \leftarrow v_{i^*} + 1$, and $A \leftarrow A + 1$. Go to Step 1.

Even if the greedy algorithm **EGA** is used to solve the **ESAM**$_i$ subproblems, so that the resulting values $Z_i^*(\tilde{S}_{i0(T_{i0}-1)} + k)$ for $k = 0, \dots, C_i$ are only approximate, it can be shown that these values will remain nonincreasing and convex

in $k$. Hence, **EGAR** can be used to make real-time repair and inventory allocation decisions using an exact or an approximate method for solving the **ESAM**$_i$
subproblems. Note also that since this approach is largely separable by item, the
computational effort required to perform a sensitivity analysis is minimal.

In the next section, we compare the solution quality and computational efficiency of the approximate method to the exact LP solution under a variety of
supply system configurations and operating system states.

## 10.7 Numerical Study

Two primary reasons exist for conducting numerical experiments. First, the quality of the solution resulting from the heuristic allocation techniques is compared
with the optimal solution to the **ESAMR** under a variety of different operating
conditions.

Second, and more important, the conjecture that integrated real-time decision
models are of significant operational value in dynamic environments where inventory imbalances exist should be tested, as is often the case in actual service
parts supply chains. As noted earlier, inventory levels found in service parts supply chains are often too high or too low for the current operating situation. This
can occur because an item's inventory position often must be decided far in advance of its availability. When demands eventually arise for the item, the available
quantity may not be appropriate for the current demand processes. See [199] for a
detailed description and example of this phenomenon for the C-5 Galaxy aircraft.

To address the first objective, the **EGAR** algorithm outlined in Section 10.6
is implemented, where the item subproblems are solved approximately using the
**EGA** heuristic from Section 10.5. The solution from this approach is compared
with the optimal solution to the **ESAMR**, which is obtained by solving a linear
programming version of the **ESAMR** formulation given in (10.55)-(10.62). A
simulator randomly generates problem instances for this comparison. Details are
given in Section 10.7.2.

To address the second objective, the continuous operation of a service parts
supply chain is simulated using two different methods for making allocation decisions. The long-run performance of these methods is then compared. The first
method uses the **EGAR** algorithm (with the **EGA** heuristic embedded) to jointly
make repair and inventory allocation decisions in a rolling-horizon manner. The
second method employs a decentralized approach in which the **EGAR** algorithm
is used for making inventory allocation decisions, but a first-come, first-served
rule is used to manage the repair queue. Details of the experiment are given in
Section 10.7.3.

### 10.7.1  Testing Environment

To facilitate the numerical study, a large-scale periodic-review service parts supply chain simulator was used to create operating system states for different supply

system configurations. A supply system configuration is defined by the following factors:

the number of items and bases;

the lead times for repair, regular transport, and expedited transport for each item and base;

the mean and variance of demand each period for each item and base;

the incremental holding, backorder, and transportation costs for each item and base;

the maximum repair capacity each period; and

the total system inventory level of each item.

In the numerical experiments for testing the value of the integrated approach, the following factors are held constant. There are 12 items demanded at 5 bases. Repair, expedited transport, and regular transport lead times are 3, 1, and 2 periods, respectively. The return transportation lead time for failed items to the depot repair facility is 5 periods. The incremental holding costs at the depot warehouse range between $0.14 and $0.27 per unit per period across the items. The incremental holding costs at the bases are twice that of the depot warehouse and range between $0.28 and $0.54 per unit per period across the items. The backorder cost is 9 times the holding cost rate for each item at each base. The incremental unit cost of expedited transport was $0.80. The factors that are varied include the repair capacity utilization (80%, 90% and 95%) and the demand process variance-to-mean ratios (1 and 10).

In the numerical experiments for testing the quality of the **EGAR** heuristic approach, the same factor values just described are used, except that two values of the regular transportation time are tested (2 and 5), and two backorder-to-holding-cost ratios are tested (9 and 20).

| Location | Item | | | Item Demand Rates | | | | | | | | |
|---|---|---|---|---|---|---|---|---|---|---|---|---|
| | 1 | 2 | 3 | 4 | 5 | 6 | 7 | 8 | 9 | 10 | 11 | 12 |
| 1 | 6.2 | 5.0 | 4.0 | 3.0 | 1.2 | 0.6 | 6.2 | 5.0 | 4.0 | 3.0 | 1.2 | 0.6 |
| 2 | 6.2 | 5.0 | 4.0 | 3.0 | 1.2 | 0.6 | 6.2 | 5.0 | 4.0 | 3.0 | 1.2 | 0.6 |
| 3 | 6.2 | 5.0 | 4.0 | 3.0 | 1.2 | 0.6 | 6.2 | 5.0 | 4.0 | 3.0 | 1.2 | 0.6 |
| 4 | 6.2 | 5.0 | 4.0 | 3.0 | 1.2 | 0.6 | 6.2 | 5.0 | 4.0 | 3.0 | 1.2 | 0.6 |
| 5 | 6.2 | 5.0 | 4.0 | 3.0 | 1.2 | 0.6 | 6.2 | 5.0 | 4.0 | 3.0 | 1.2 | 0.6 |
| Item System Inventory Levels | 515 | 410 | 325 | 240 | 115 | 60 | 255 | 220 | 155 | 85 | 40 | 15 |

**Table 10.1.** An example of the item demand rates and system inventory levels.

For both experiments, the demand processes for each item at each base are taken to be stationary and independent over time and location. While the methodology can handle more complicated nonstationary demand processes, the experiments are designed to demonstrate the value of making integrated decisions even with relatively stable demand processes. Two types of demand distributions were used, Poisson (to simulate a variance-to-mean ratio of 1) and negative binomial (to simulate a variance-to-mean ratio of 10).

The demand rates for the 12 items are set as follows. For the first 6 items, a mixture of high, medium, and low demand rates represent a composite set of items, as is typically found in practice. The demand rates for the second 6 items are identical to those of the first 6 items. The demand rates for each item are identical across the bases.

The system inventory levels for the first 6 items are set to values that are slightly above those required to minimize their steady-state expected holding and backorder costs. This case represents items that are in long supply. For the second 6 items, the system inventory levels are set to approximately one-half of the inventory levels of the first 6 items, representing items that are in short supply. An example is given in Table 10.1. (The inventory levels shown in Table 10.1 correspond to an instance in which the demand processes are Poisson processes.) Thus, the testing environment contains a mixture of high, medium, and low demand rate items, some of which have ample inventory and others of which do not have sufficient inventory in the supply chain.

### 10.7.2   Exact Approach vs EGAR Heuristic

For this comparison, we examine supply system configurations that exhibit conditions under which the performance of the methods is likely to differ. Twenty-four different configurations are tested in all. For each configuration, one thousand operating system states were randomly generated, using the simulator, to create problem instances. The **EGAR** algorithm and the exact approach are applied to each instance. The simulator created instances instead of creating them manually to avoid potentially biasing the experimental results with statistically unlikely system states.

| | | Regular Lead Time = 2 | | Lead Time 2 | Regular Lead Time = 5 | | Lead Time 5 | Average |
|---|---|---|---|---|---|---|---|---|
| Utilization | Demand VTMR | Backorder Cost Multiple = 9 | Backorder Cost Multiple =20 | Average | Backorder Cost Multiple = 9 | Backorder Cost Multiple =20 | Averag e | |
| 80% | 1 | 0.85% | 0.95 % | 0.90% | 0.68% | 0.91 % | 0.79% | 0.85% |
| | 10 | 0.93% | 0.79 % | 0.86% | 0.85% | 1.03 % | 0.94 % | 0.90% |
| 80% Average | | 0.89% | 0.87% | 0.88% | 0.77% | 0.97 % | 0.87 % | 0.87% |
| 90% | 1 | 0.83% | 0.95% | 0.89 % | 0.78% | 0.72% | 0.75% | 0.82% |
| | 10 | 0.84% | 1.02% | 0.93 % | 1.11% | 0.95% | 1.03% | 0.98% |
| 90% Average | | 0.84% | 0.98% | 0.91% | 0.95 % | 0.83% | 0.89% | 0.90% |
| 95% | 1 | 1.04% | 0.75 % | 0.90 % | 0.89% | 0.94% | 0.91 % | 0.90% |
| | 10 | 1.07% | 0.85 % | 0.96 % | 1.29% | 1.00% | 1.15 % | 1.05% |
| 95% Average | | 1.06% | 0.80 % | 0.93 % | 1.09 % | 0.97% | 1.03% | 0.98% |
| Average | | 0.93% | 0.88 % | 0.91 % | 0.93% | 0.92% | 0.93 % | 0.92% |

**Table 10.2.** Average deviations of the EGAR cost above the ESAMR optimal cost.

Table 10.2 contains the results for each of the twenty-four configurations examined. The numbers in the table represent the average percentage deviation of the expected effective horizon cost incurred by employing the **EGAR** heuristic from the expected effective horizon cost incurred by using the optimal **ESAMR** solution.

There are several questions of interest to discuss:

*1. How frequently did the decisions resulting from the two approaches match?*
Matches in every single allocation decision were observed for many of the problem instances (i.e., operating states) tested; however, complete matches were not a usual or frequent occurrence. Due to the combinatorial nature of the problem, small differences in repair decisions, shipment timing, and quantity were common, and hence makes it difficult to devise a meaningful measure of the "closeness" of two solutions directly. It is clear from Table 10.2, however, that these small differences did not greatly affect the resulting expected costs.

*2. How close are the expected costs over the effective horizon resulting from the two approaches?* As Table 10.2 shows, the average relative difference between the expected costs of the **EGAR** heuristic and the exact approach are very small over all supply system configurations tested. The largest overall cost differences occur when the demand process variance-to-mean ratios were high, and repair capacity utilization was high. This combination caused large and frequent shortages to occur for long periods of time, which, in turn, magnified the small deviations from the optimal inventory allocation that were made by the myopic **EGAR** heuristic. However, at the item level, the most significant factor in determining the quality of the **EGAR** solution relative to the exact solution was the choice of the system inventory level. The simple explanation for this is that the constrained newsvendor cost functions are relatively flat in the area of their minima. When there is sufficient inventory of an item to meet most demand requirements, small deviations (in terms of quantity shipped and shipment timing) from the optimal operating allocation will have relatively little impact on the overall system cost incurred for this item. In such cases, the performance of most reasonable allocation policies is likely to be good. Since the **EGAR** heuristic attempts to minimize expected costs, albeit myopically, it performs well when the system, as a whole, has enough stock. When the system inventory level for an item is far too low, small deviations from the optimal solution will produce much larger relative cost differences than when the system has sufficient inventory. It is in these cases that the logic and robustness of the allocation rule becomes extremely important. Our experiments show that the **EGAR** heuristic performs well even when there are significant inventory imbalances in the system.

*3. What is the magnitude of the computational advantage of using the EGAR heuristic (with EGA embedded) over the exact method?* On average, the exact method's solution time was approximately five times longer than that of the **EGAR** heuristic. The **EGAR** heuristic is much faster than the exact method because the necessary expected cost computations are done on an as-needed basis. The majority of the processing time for the exact method is devoted to computing all of the necessary cost coefficients found in the **ESAMR** model.

### 10.7.3   First-Come, First-Served vs Integrated Repair Allocations

As stated earlier, to test the value of an integrated approach over a decentralized one, two different methods for making allocation decisions are implemented. The first method uses the **EGAR** algorithm (with the **EGA** heuristic embedded) to jointly make repair and inventory allocation decisions. The second method employs a decentralized approach in which the **EGAR** algorithm makes inventory allocation decisions, but a first-come, first-served rule (FCFS) is used to manage the repair queue. That is, the repair decisions were made with no knowledge of downstream requirements and were based solely on the order in which items are returned for repair. It is not necessary to examine a FCFS rule for *both* inventory allocation and repair since such an approach will perform less well than FCFS for repair decisions and **EGAR** for inventory allocation decisions.

For each supply system configuration, a simulation was run for one thousand time periods, after a warmup phase. Five separate random seeds are employed, along with their antithetic random number streams, for a total of ten simulations runs for each configuration. The outcomes for different levels of repair capacity utilization and demand uncertainty are summarized in Table 10.3. The data in the table indicate the average per period cost observed across the ten runs and the standard deviation of these average per period cost values. These results show that by considering repair and inventory allocation decisions jointly, costs are reduced by 13.3% on average, with the reductions ranging between 9.2% and 16.0%.

| Repair Capacity Utilization | Demand Process VTMR | Allocation Rule | | | | Relative Cost Improvement of ESAMR |
|---|---|---|---|---|---|---|
| | | FCFS | | ESAMR | | |
| 80% | 1 | $1,712.71 | ±$11.70 | $1,443.95 | ±$13.05 | 15.7% |
| | 10 | $1,763.01 | ±$26.89 | $1,554.13 | ±$26.50 | 11.8% |
| 90% | 1 | $1,713.15 | ±$11.73 | $1,443.80 | ±$13.03 | 15.7% |
| | 10 | $1,773.05 | ±$29.25 | $1,569.92 | ±$29.61 | 11.5% |
| 95% | 1 | $1,716.65 | ±$11.71 | $1,442.60 | ±$12.92 | 16.0% |
| | 10 | $1,802.50 | ±$40.32 | $1,636.45 | ±$56.32 | 9.2% |
| Average | | $1,746.84 | ±$21.93 | $1,515.14 | ±$25.24 | 13.3% |

**Table 10.3.** Simulation results for FCFS vs ESAMR.

Note that these results are based on fairly benign stationary demand processes. Real demand data are typically nonstationary and exhibit attributes for which the integrated approach will likely outperform simple and/or decentralized allocation rules by a wider margin. Furthermore, these experiments reveal that by allowing lateral transshipments in the integrated model, costs are lowered by an additional 10.1%, on average. This demonstrates that there is potential for significant economic value in using an integrated repair and stock allocation decision process.

# 10.8 Problem Set, Chapter 10

**10.1.** Llenroc Industries provides service parts and repairs for a variety of types of computers. Demands for these parts arise in two distinct ways. First, there are requests for these parts from companies that provide on-site repair of computers. These companies order parts on a daily basis and expect that these parts will be shipped to them on the same day. Second, Llenroc also repairs and refurbishes computers as well as repairs certain component parts. Hence Llenroc also consumes parts to complete its repair and refurbishing activities.

There are four ways that Llenroc obtains serviceable parts. First, for many item types, Llenroc can purchase parts directly from the Original Equipment Manufacturer (OEM), such as IBM. Second, Llenroc can buy new machines from these OEMs and remove parts from these machines. This is called "stripping" the machines. Third, Llenroc can buy used machines on the open market. Once on hand, critical parts are removed from these machines and tested. If they pass the test, they are placed into serviceable inventory. Otherwise, they are either repaired or scrapped. Not all components are repairable, for example, transistors. Even items that are designed to be repairable are not always repairable. Certain failure modes result in parts that can not be repaired. Thus only a fraction of the units of repairable type items can be repaired. Fourth, certain defective parts removed from computers, which have entered into Llenroc's repair or refurbishment processes, are repaired by Llenroc. As mentioned, not all item types are repairable. Obviously, Llenroc is capable of repairing only a subset of the item types that are found to be defective in the computers that it is repairing or refurbishing.

At a given moment there is a certain amount of stock on hand in Llenroc's parts warehouse. Furthermore, there are new machines on order from OEMs with known due dates. There are also various amounts of individual item types on order with appropriate OEMs that are expected to be in Llenroc's warehouses at known times in the future. There are also used machines that are on-hand that have yet to be stripped down and to have their parts tested. There are also parts of various item types that are awaiting repair in Llenroc's repair facility.

There are costs associated with each of the four possible ways that Llenroc can obtain serviceable parts. There are current market prices for new and used machines. There are also current market prices for each new item type that could be purchased. There are also costs for stripping down both new and used machines, and these costs differ by machine type and whether the machine is new or used. Needed parts removed from used machines require testing, which has an associated cost, and these parts may fail the test. Finally, there is a cost incurred to repair individual parts.

Additionally, there are lead times associated with the purchase of new and used machines, and for the acquisition of new parts. There is also a time required to test parts removed from the used machines. There is also capacity required to repair defective parts. Certain types of parts require the same type of capacity. Thus items are grouped by the type of capacity required to complete their repair. Each item type is repaired in a single type of capacitated work center.

Every type of computer entered into the repair and refurbishment process has a probability of needing each part type to complete its service. Demand for parts occur for two reasons, as stated earlier. First, demands occur from external customers for each item type according to a time dependent Poisson process, with rate $\lambda_i(t)$ for item $i$ on day $t$ in the future.

Second, a computer repair and refurbishment schedule dictates the requirements for parts. This schedule indicates what type and quantity of that type of computer will be worked on during the next $T$ days. This plan is fixed for $T$ days on a rolling $T$ day horizon basis. That is, each day planners determine what computer types should be repaired and refurbished on the $T^{\text{th}}$ day in the future. For service parts planning purposes they look beyond this horizon so that they can plan external purchases of new and used machines and new parts from OEMs.

OEMs can always provide parts on a one day basis either from their stocks or through parts brokers. But, this cost is very high and is avoided if possible.

Your task is to develop a real-time parts acquisition model. That is, develop a model that minimizes the total cost of acquiring and repairing parts over the planning horizon. Consider acquisition costs, repair costs, lead times, and capacities in your model. Also indicate how you would compute your purchase and repair quantities.

**10.2.** At the beginning of this chapter, a situation was discussed in which service contracts are written for customers and for groups of machines at the customer's operating locations. These contracts are for service guarantees over a fixed period of time. Obviously, meeting these contractual obligations is of importance since future business with each customer is related to the service provided currently.

From the service provider's perspective, resource allocation decisions are made every day that affect service provided to these customers. A conjecture was made that the following situation can occur. Suppose there is stock on hand for a part type and a demand arises for that part type at a customer site. Given the level of service already provided to that customer at that site, and the service provided to other customers to that point in time, it may not always be best to satisfy that demand immediately. An illustration of such conditions was given in the beginning of this chapter.

Your assignment is to construct a model that could be used to determine whether or not a part should be dispatched to meet a demand arising at a given time. Among the factors you should consider in the model are the current level of satisfaction of contracts that could require the use of this particular part, the length of time until inventory replenishment will occur for that part at the stocking location, and the likelihood of requiring that part by other customers over the time until future replenishments will occur.

**10.3.** Consider the two echelon system depicted in the following figure, consisting of a depot and $n$ bases.

The depot orders units of a given part type from an external supplier. If the depot desires, it can also return stock to the supplier. Activity with external customers arises only at the bases. Demands for the item arise at a base and returns

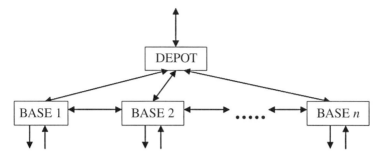

**Fig. 10.2.** System Structure

from customers also occur at bases. Hence net demand at a base in a period may be negative.

Assume that both regular and expedited shipments are permitted between any two bases or the depot in a period. Expedited shipments are assumed to be less expensive between bases than from the depot to a base. Although returns from customers may occur in a period at a base, only the depot may return stock to the external supplier.

Suppose we have a planning horizon of but one period in length for a single item. The following decisions must be made for that item.

(1) How much should the depot purchase from the external supplier or return to the external supplier.
(2) How much should the depot allocate to each base and how much should be shipped from a base back to the depot.
(3) How much stock of the item should be transshipped between bases (regular and expedited modes of transport).

Assume that events arise and costs are incurred in the period as follows. At the beginning of the period, we see how much stock is on hand at each base and the depot. At that point, we decide:

(1) the amount to purchase or return to the supplier by the depot,
(2) the quantity to ship back to the depot from each base,
(3) the amount allocated by the depot from its stock to each base (assume what is purchased from the supplier by the depot in the period is available for allocation), and
(4) the transshipment quantities between the bases.

Assume that the cost associated with each action is proportional to the quantity purchased, returned, or shipped. That is, there are no fixed costs. Also assume that all these transactions occur immediately.

Once the allocations of stock have been made, the net demand at each base is observed. Remember, the net demand could be negative. When demand at a base exceeds the stock on-hand at that base, an expedited or emergency resupply will occur, if possible, from another base. If no base has on-hand stock, an emergency shipment will be requested from the depot. (The emergency resupply cost from the depot is greater than that from another base.) A system shortage cost is charged on a per unit basis when there is no stock on hand anywhere in the system and backorders are incurred. Holding costs are charged at the end of the period. The holding costs are the same at all bases, but are lower at the depot.

Your goal is to construct a model and to present a method for finding purchase, return and transshipment quantities that minimize expected costs for the item in the period. Carefully state all assumptions.

**10.4.** Extend the model developed in Section 10.4.1 to consider transshipments among bases. Assume bases are located so that transshipments can be made only among certain groups of bases. Assume these transshipment times are shorter than $T_{ij}^{e}$.

# References

1. J.B. Abell, L.W. Miller, C.E. Neumann, and J.E. Payne. Drive (distribution and repair in variable environments). Report R-3888-AF, RAND Corporation, Santa Monica, CA, 1992.
2. P.K. Aggarwal and K. Moinzadeh. Order expedition in multi-echelon production/distribution systems. *IIE Transactions*, 26(2):86–96, March 1994.
3. V. Agrawal, M.A. Cohen, and Y.S. Zheng. Service parts logistics: A benchmark analysis. *IIE Transactions, Special Issue on Supply Chain Co-ordination and Integration*, 29(8):627–639, August 1997.
4. S.C. Albright. An approximation to the stationary distribution of a multi-echelon repairable-item inventory system. *Naval Research Logistics*, 36:179–195, 1989.
5. S.C. Albright and A. Soni. Markovian multi-echelon repairable inventory system. *Naval Research Logistics Quarterly*, 35:49–61, 1988.
6. P. Alfredsson and J. Verrijdt. Modeling emergency supply flexibility in a two-echelon inventory system. *Management Science*, 45(10):1416–1431, October 1999.
7. S.G. Allen. Redistribution of total stock over several user locations. *Naval Research Logistics Quarterly*, 5:51–59, 1958.
8. S.G. Allen. A redistribution model with set-up charge. *Management Science*, 8(1):99–108, October 1961.
9. S.G. Allen. Computation for the redistribution model with set-up charge. *Management Science*, 8(4):482–489, July 1962.
10. S.G. Allen and D.A. D'Esopo. An ordering policy for repairable stock items. *Operations Research*, 16(3):669–674, May-June 1968.
11. S.G. Allen and D.A. D'Esopo. An ordering policy for stock items when delivery can be expedited. *Operations Research*, 16(4):880–883, July-August 1968.
12. T.W. Archibald, Sassen S.A. E., and Thomas L.C. An optimal policy for a two depot inventory problem with stock transfer. *Management Science*, 43(2):173–183, February 1997.
13. K.P. Aronis, I. Magou, R. Dekker, and G. Tagaras. Inventory control of spare parts using a Bayesian approach: A case study. *European Journal of Operational Research*, 154:730–739, 2004.
14. K.J. Arrow, S. Karlin, and H.E. Scarf, editors. *Studies in the Mathematical Theory of Inventory and Production*. Stanford University Press, Stanford, California, 1958.
15. J. Ashayeri, R. Heuts, A. Jansen, and B. Szczerba. Inventory management of repairable service parts for personal computers: A case study. *International Journal of Operations and Production Management*, 16:74–97, 1996.

16. Yossi Aviv and Awi Federgruen. Stochastic inventory models with limited production capacity and periodically varying parameters. *Probability in the Engineering and Informational Sciences*, 11:107–135, 1997.

17. S. Axsäter. Modeling emergency lateral transshipments in inventory systems. *Management Science*, 36(11):1329–1338, November 1990.

18. S. Axsäter. Simple solution procedures for a class of two-echelon inventory problems. *Operations Research*, 38(1):64–69, January-February 1990.

19. S. Axsäter. Continuous review policies for multi-level inventory systems with stochastic demand. In S.C. Graves, A.H.G. Rinnooy Kan, and P.H. Zipkin, editors, *Handbook in OR and MS*, volume 4, pages 175–197. Elsevier Science Publishers B.V., North-Holland, Amsterdam, 1993.

20. S. Axsäter. Evaluation of unidirectional lateral transshipments and substitutions in inventory systems. *European Journal of Operational Research*, 149(2):438–447, 2003.

21. S. Axsäter. A new decision rule for lateral transshipments in inventory systems. *Management Science*, 49(9):1168–1179, 2003.

22. S. Axsäter. Note: Optimal policies for serial inventory systems under fill rate constraints. *Management Science*, 49(2):247–253, 2003.

23. M. Baganha. Feeney and Sherbrooke revisited. Working Paper, Department of Decisions Sciences, The Wharton School, University of Pennsylvania, 1985.

24. A.R. Balana, D. Gross, and R.M. Soland. Optimal provisioning for single-echelon repairable item inventory control in a time varying environment. *IIE Transactions*, 15:344–352, 1983.

25. A. Banerjee, J. Burton, and S. Banerjee. A simulation study of lateral shipments in single supplier, multiple buyers supply chain networks. *International Journal of Production Economics*, 81-82:103–114, 2003.

26. E.W. Barankin. A delivery-lag inventory model with an emergency provision. *Naval Research Logistics Quarterly*, 8:285–311, 1961.

27. M.N. Bartakke. A method of spare parts inventory planning. *Omega*, 9:51–58, 1981. Oxford.

28. S. Bessler and A.F. Veinott, Jr. Optimal policy for a dynamic multi-echelon inventory problem. *Naval Research Logistics Quarterly*, 13:355–389, 1966.

29. J.P. Bowman. A multi-echelon inventory stocking algorithm for repairable items with lateral resupply. Working Paper, 1986.

30. R. Brooks and A. Geoffrion. Finding Everett's Lagrange multipliers by linear programming. *Operations Research*, 14(6):1149–1153, November-December 1966.

31. R.B.S. Brooks, C.A. Gillen, and J.Y. Lu. Alternative measures of supply performance. Report RM-6094-PR, RAND Corporation, Santa Monica, CA, 1969.

32. Robert Goodell Brown. *Decision Rules for Inventory Management*. Holt, Rinehart, and Winston, New York, 1967.

33. M.D. Buyukkurt and M. Parlar. A comparison of allocation policies in a two-echelon repairable item inventory model. *International Journal of Production Economics*, 29:291, 1993.

34. J. Buzacott and J.G. Shanthikumar. *Stochastic Models of Manufacturing Systems*. Prentice Hall, Englewood Cliffs, NJ, 1993.

35. K.E. Caggiano, P.L. Jackson, J.A. Muckstadt, and J.A. Rappold. Optimal stocking in reparable parts networks with pooling. Under revision for Naval Research Logistics.

36. K.E. Caggiano and J.A. Muckstadt. A combinatorial, multi-indenture, multi-item inventory model for NASA's reusable launch vehicle program. Technical Report 1284, School of Operations Research and Industrial Engineering, Cornell University, 2001. Under revision for Operations Research.

37. K.E. Caggiano, J.A. Muckstadt, P.L. Jackson, and J.A. Rappold. A multi-echelon, multi-item inventory model for service parts management with generalized service level constraints. Under revision for Operations Research.

38. K.E. Caggiano, J.A. Muckstadt, and J.A. Rappold. Simple algorithms for pooling inventory to satisfy customer service requirements at minimum cost. Working paper.

39. K.E. Caggiano, J.A. Muckstadt, and J.A. Rappold. Real time capacity and inventory allocation decisions for reparables in a two-echelon system with expedited shipments. Technical report, School of Operations Research and Industrial Engineering, Cornell University, 2003.

40. C.L. Li Caglar, D. and D. Simchi-Levi. Two-echelon spare parts inventory system subject to a service constraint. *IIE Transactions*, 36(7):655–666, 2004.

41. H.S. Campbell and T.L. Jones, Jr. *A Systems Approach to Base Stockage–Its Development and Test*, page 3345. RAND Corporation, Santa Monica, CA, 1966.

42. Manuel J. Carrillo. Generalizations of Palm's theorem and dyna-METRIC's demand and pipeline variability. Report R-3698-AF, RAND Corporation, Santa Monica, California, 1989.

43. E.W. Chan, J.A. Rappold, and J.A. Muckstadt. Determining and allocating capacity-driven safety stock in multi-item, multi-echelon systems. Technical report, School of Business, University of Wisconsin, 1999.

44. F. Chen and J. Song. Optimal policies for multi-echelon inventory problems with Markov-modulated demand. *Operations Research*, 49(2):226–234, 2001.

45. F. Chen and Y. Zheng. Lower bounds for multi-echelon stochastic inventory systems. *Management Science*, 40(11):1426–1443, November 1994.

46. K.L. Cheung and W.H. Hausman. A multi-echelon inventory model with multiple failures. *Naval Research Logistics*, 40:593–602, 1993.

47. K.L. Cheung and W.H. Hausman. Multiple failures in a multi-item spares inventory model. *IIE Transactions*, 27(2):171–180, April 1995.

48. A. Clark. Experiences with a multi-indentured, multi-echelon inventory model. In L.B. Schwarz, editor, *Multi-Level Production, Inventory Control Systems: Theory and Practice*, volume 16 of *Studies in the Management Sciences*, pages 229–330. North-Holland Publishing, Amsterdam, 1981.

49. A. Clark and H. Scarf. Optimal policies for a multi-echelon inventory problem. *Management Science*, 6(4):475–490, July 1960.

50. A.J. Clark. An informal survey of multi-echelon inventory theory. *Naval Research Logistics Quarterly*, 19(4):621–650, 1972.

51. A.J. Clark. An allocation model. Report, CACI, Inc.-Federal, Arlington, Virginia, 1978.

52. A.J. Clark. Common item problems involved in the operation of the optimal ao inventory and allocation models. Report, CACI, Inc.-Federal, Arlington, Virginia, 1978.

53. A.J. Clark. Logistics support economic evaluation – An introduction. Report R7801, CACI, Inc.-Federal, Arlington, Virginia, 1978.

54. A.J. Clark. Optimal operational availability inventory model. Report R7806, CACI, Inc.-Federal, Arlington, Virginia, 1978.

55. A.J. Clark and H.E. Scarf. Approximate solutions to a simple multi-echelon inventory problem. In K. J. Arrow, S. Karlin, and H. Scarf, editors, *Studies in Applied Probability and Management Science*, chapter 5. Stanford University Press, Stanford, California, 1962.

56. M.A. Cohen and R. Ernst. Operations related groups (ORGs): A clustering procedure for production/inventory systems. *Journal of Operations Management*, 9(4):574–598, October 1990.

57. M.A. Cohen, P.R. Kleindorfer, and H.L. Lee. Optimal stocking policies for low usage items in multi-echelon inventory systems. *Naval Research Logistics Quarterly*, 33(1):17–38, 1986.

58. M.A. Cohen, P.R. Kleindorfer, and H.L. Lee. Service constrained (s, S) inventory systems with priority demand classes and lost sales. *Management Science*, 34(4):482–499, April 1988.

59. M.A. Cohen, P.R. Kleindorfer, and H.L. Lee. Near optimal service constrained stocking policies for spare parts. *Operations Research*, 37(1):104–117, January-February 1989.

60. M.A. Cohen and H.L. Lee. Strategic analysis of integrated production-distribution systems: Models and methods. *Operations Research*, 36(2):216–228, March-April 1988.

61. M.A. Cohen and H.L. Lee. Out of touch with customer needs? Spare parts and after sales service. *Sloan Management Review*, 31(2):55–66, 1990.

62. M.A. Cohen, Y-S. Zheng, and V. Agrawal. Service parts logistics: A benchmark analysis. *IIE Transactions*, 29(8):627–639, 1997.

63. M.A. Cohen, Y. S. Zheng, and Y. Wang. Identifying opportunities for improving Teradyne's service-parts logistics system. *Interfaces*, 29(4):1–18, July-August 1999.

64. G. B. Crawford. Palm's theorem for nonstationary processes. Report R-2750-RC, RAND Corporation, Santa Monica, CA, 1981.

65. G.B. Crawford. Variability in the demands for aircraft spare parts. Report R-3318-AF, RAND Corporation, Santa Monica, California, 1988.

66. M. Dada. A two-echelon inventory system with priority shipments. *Management Science*, 38(8):1140–1153, August 1992.

67. V. Daniel, R. Guide, and R. Srivastava. Repairable inventory theory: Models and applications. *European Journal of Operational Research*, 102:1–20, 1997.

68. G. Dantzig. *Linear Programming and Extensions*. Princeton University Press, Princeton, NJ, 1962.

69. C. Das. Supply and redistribution rules for two-location inventory systems: One-period analysis. *Management Science*, 21(7):765–776, March 1975.

70. J. Davidson. Initial provisioning of rotable spares for airlines: A case study. *INFOR*, 18:139–149, 1980.

71. W.S. Demmy. Allocation of spares and repair resources to a multi-component system. Report 70-17, Air Force Logistics Command, Wright Patterson Air Force Base, Ohio, 1970.

72. W.S. Demmy and V.J. Presutti. Multi-echelon inventory theory in the air force logistics command. In L. B. Schwarz, editor, *Multi-Level Production/Inventory Control Systems: Theory and Practice*, volume 16 of *Studies in the Management Sciences*, pages 279–297. North Holland Publishing, Amsterdam, 1981.

73. V. Deshpande, M.A. Cohen, and K. Donohue. An empirical study of service differentiation for weapon system service parts. *Operations Research*, 51(4):518–530, 2003.

74. T.S. Dhakar, C.P. Schmidt, and D.M. Miller. Base stock level determination for high cost low demand critical repairable spares. *Computers and Operations Research*, 21:411–420, 1994.

75. I. Duenyas. A simple release policy for networks of queues with controllable inputs. *Operations Research*, 42:1162–1171, 1994.

76. C. E. Ebeling. Optimal stock levels and service channel allocations in a multi-item repairable asset inventory system. *IIE Transactions*, 23(2):115–120, June 1991.

77. G. D. Eppen. Note: Effects of centralization on expected costs in a multi-location newsboy problem. *Management Science*, 25(5):498–501, May 1979.
78. G. D. Eppen and L. Schrage. Centralized ordering polices in a multi-warehouse system with lead times and random demand. In L.B. Schwarz, editor, *Multi-Level Production/Inventory Control Systems: Theory and Practice*, Studies in the Management Sciences, pages 51–67. North-Holland Publishing, Amsterdam, 1981.
79. N. Erkip, W. Hausman, and S. Nahmias. Optimal centralized ordering policies in multi-echelon inventory systems with correlated demands. *Management Science*, 36(3):381–392, March 1990.
80. H. Everett, III. Generalized Lagrange multiplier method for solving problems of optimal allocation of resources. *Operations Research*, 11(3):399–417, May-June 1965.
81. P. T. Evers. Hidden benefits of emergency transshipments. *Journal of Business Logistics*, 18(2):55–77, 1997.
82. P. T. Evers. Filling customer orders from multiple locations: A comparison of pooling methods. *Journal of Business Logistics*, 20(1):121–140, 1999.
83. A. Federgruen and P. Zipkin. Approximations of dynamic, multilocation production and inventory problems. *Management Science*, 30(1):69–84, January 1984.
84. A. Federgruen and P. Zipkin. Computational issues in an infinite horizon multi-echelon inventory model. *Operations Research*, 32(4):818–836, July-August 1984.
85. A. Federgruen and P. Zipkin. An inventory model with limited production capacity and uncertain demands i: The average-cost criterion. *Mathematics of Operations Research*, 11(2):193–207, 1986.
86. A. Federgruen and P. Zipkin. An inventory model with limited production capacity and uncertain demands ii: Discounted-cost criterion. *Mathematics of Operations Research*, 11(2):208–215, 1986.
87. G. J. Feeney and C. C. Sherbrooke. An objective Bayes approach for inventory decisions. Report RM-4362-PR, RAND Corporation, Santa Monica, CA, 1965.
88. G. J. Feeney and C. C. Sherbrooke. A system approach to base stockage of recoverable items. Report RM-4720PR, RAND Corporation, 1965.
89. G. J. Feeney and C. C. Sherbrooke. The (s-1,s) inventory policy under compound Poisson demand. *Management Science*, 12(5):391–411, January 1966.
90. W. W. Fisher and J. J. Brennan. The performance of cannibalization policies in a maintenance system with spares, repair, and resource constraints. *Naval Research Logistics Quarterly*, 33:1–15, 1986.
91. B. L. Fox and D. M. Landi. Optimization problems with one constraint. Report RM-5791, RAND Corporation, Santa Monica, CA, 1968.
92. B. L. Fox and D. M. Landi. Searching for the multiplier in one-constraint optimization problems. *Operations Research*, 18(2):253–262, March-April 1970.
93. Y. Fukuda. Optimal disposal policies. *Naval Research Logistics Quarterly*, 8:221–227, 1961.
94. G. Gallego and O. Ozer. Integrating replenishment decisions with advance demand information. *Management Science*, 47:1344–1360, 2001.
95. P. Glasserman. Bounds and asymptotics for planning critical safety stocks. *Operations Research*, 45(2):244–257, March-April 1997.
96. P. Glasserman and S. Tayur. The stability of a capacitated, multi-echelon production-inventory system under a base-stock policy. *Operations Research*, 42(5):913–925, September-October 1994.
97. P. Glasserman and S. Tayur. Sensitivity analysis for base-stock levels in multiechelon production-inventory system. *Management Science*, 41(2):263–281, February 1995.

98. P. Glasserman and S. Tayur. A simple approximation for a multistage capacitated production-inventory system. *Naval Research Logistics*, 43(1):41–58, February 1996.

99. S. C. Graves. A multi-echelon inventory model for a repairable item with one-for-one replenishment. *Management Science*, 31(10):1247–1256, October 1985.

100. S. C. Graves. Determining the spare and staffing levels for a repair depot. *Journal of Manufacturing and Operations Management*, 1:227–241, 1988.

101. D. Gross. Centralized inventory control in multilocation supply systems. In H.E. Scarf et al., editors, *Multistage Inventory Models and Techniques*, chapter 3. Stanford University Press, Stanford, California, 1963.

102. D. Gross. On the ample service assumption of Palm's theorem in inventory modeling. *Management Science*, 28(9):1065–1079, September 1982.

103. D. Gross, B. Gu, and R. Soland. Iterative solution methods for obtaining the steady-state probability distributions of Markovian multi-echelon repairable item inventory sytems. *Computers and Operations Research*, 20:817–828, 1993.

104. D. Gross and C. M. Harris. On one-for-one ordering inventory polices with state-dependent leadtimes. *Operations Research*, 19(3):735–760, May-June 1971.

105. D. Gross and C. M. Harris. *Fundamentals of Queuing Theory*, pages 415–419. John Wiley and Sons, New York, 1974.

106. D. Gross and J. F. Ince. Spares provisioning for a heterogeneous population. Technical Report T-376, Program in Logistics, The George Washington University, 1978.

107. D. Gross and J. F. Ince. Spares provisioning for repairable items: Cyclic queues in light traffic. *AIEE Transactions*, 10:307–314, 1978.

108. D. Gross, H. D. Kahn, and J.D. Marsh. Queuing models for spares provisioning. *Naval Research Logistics Quarterly*, 24(4):521–536, 1977.

109. D. Gross, L. C. Kioussis, and D. R. Miller. A network decomposition approach for approximating the steady-state behavior of Markovian multi-echelon reparable item inventory system. *Management Science*, 33(11):1453–1468, November 1987.

110. D. Gross, L. C. Kioussis, and D. R. Miller. Transient behavior of large Markovian multi-echelon repairable item inventory systems using a truncated state space approach. *Naval Research Logistics Quarterly*, 34:173–198, 1987.

111. D. Gross and D. R. Miller. Multi-echelon repairable-item provisioning in a time varying environment using the randomization technique. *Naval Research Logistics Quarterly*, 31:347–361, 1984.

112. D. Gross, D. R. Miller, and C. G. Plastiras. Simulation methodologies for transient Markov processes: A comparative study based on multi-echelon repairable item inventory systems. In W. D. Wade, editor, *Proceedings of the 1984 Summer Computer Simulation Conference*, pages 37–43, La Jolla, California, 1984. The Society for Computer Simulation.

113. D. Gross, D. R. Miller, and R. M. Soland. A closed queuing network model for multi-echelon repairable item provisioning. *IIE Transactions*, 15:344–352, 1983.

114. D. Gross and A. Soriano. On the economic application of airlift to product distribution and its impact on inventory levels. *Naval Research Logistics Quarterly*, 19:501–507, 1972.

115. O. Gross. A class of discrete-type minimization problems. Report RM-1655-PR, RAND Corporation, Santa Monica, CA, 1956.

116. A. Gunasekaran, C. Patel, and R.E. McGaughey. A framework for supply chain performance measurement. *International Journal of Production Economics*, 87(3):333–348, 2004.

117. A. Gupta. Approximate solution of a single-base multi-indentured repairable-item inventory system. *Journal of the Operational Research Society*, 44:701–710, 1993.

118. A. Gupta and S.C. Albright. Steady-state approximations for a multi-echelon multi-indentured repairable-item inventory system. *European Journal of Operational Research*, 62:340–353, 1992.

119. S.E. Haber and R. Sitgreaves. An optimal inventory model for the intermediate echelon when repair is possible. *Management Science*, 21(6):638–648, 1975.

120. S.E. Haber, R. Sitgreaves, and H. Solomon. A demand prediction technique for items in military inventory systems. *Naval Research Logistics Quarterly*, 16:297–308, 1975.

121. G. Hadley and T.M. Whitin. *Analysis of Inventory Systems*. Prentice-Hall, Englewood Cliffs, N.J., 1963.

122. G. Hadley and T.M. Whitin. An inventory transportation model with N locations. In Scarf, Gilford, and Shelly, editors, *Multistage Inventory Models and Techniques*, chapter 5. Stanford University Press, Stanford, California, 1963.

123. T. Hamann and J. Proth. Inventory control of repairable tools with incomplete information. *International Journal of Production Economics*, 31:543–550, 1993.

124. R. Hariharan and P. Zipkin. Customer-order information, leadtimes, and inventories. *Management Science*, 41(10):1599–1607, 1995.

125. W.H. Hausman. Communication: On optimal repair kits under a job completion criterion. *Management Science*, 28(11):1350–1351, November 1982.

126. W.H. Hausman and G. D. Scudder. Priority scheduling rules for repairable inventory systems. *Management Science*, 28(11):1215–1232, November 1982.

127. D.P. Heyman. Return policies for an inventory system with positive and negative demands. *Naval Research Logistics Quarterly*, 25(4):581–596, 1978.

128. R.J. Hillestad. Dyna-METRIC: Dynamic multi-echelon technique for recoverable item control. Report R-2785-AF, RAND Corporation, Santa Monica, CA, 1982.

129. R.J. Hillestad and M.J. Carrillo. Models and techniques for recoverable item stockage when demand and the repair processes are nonstationary - part i: Performance measurement. Report N-1482-AF, RAND Corporation, Santa Monica, CA, 1980.

130. B. Hoadley and D.P. Heyman. A two-echelon inventory model with purchases, dispositions, shipments, returns, and transshipments. *Naval Research Logistics Quarterly*, 24:1–19, 1977.

131. J.S. Hodges. Modeling the demand for spare parts: Estimating the variance-to-mean ratio and other issues. Report N-2086-AF, RAND Corporation, Santa Monica, CA, 1985.

132. D. Hoekstra, R.L. Deemer, and S. Gajadlo. Optimal procurement decisions for spare aircraft components. Report, Frankfort Arsenal, Philadelphia, Pennsylvania, 1965.

133. W. Hopp, M. Spearman, and R.Q. Zhang. Easily implementable inventory control policies. *Operations Research*, 45(3):327–340, May-June 1997.

134. J.V. Howard. Service exchange systems–the stock control of repairable items. *Journal of the Operational Research Society*, 35:235–245, 1984.

135. E.L. Huggins and T.L. Olsen. Supply chain management with guaranteed delivery. *Management Science*, 49(9):1154–1167, 2003.

136. K.E. Isacson, P. Boren, C.L. Tsai, and R. Pyles. Dyna-METRIC version 4: Modeling worldwide logistics support of aircraft components. Report R-3389-AF, RAND Corporation, Santa Monica, CA, 1988. 95-96.

137. P.L. Jackson. Stock allocation in a two-echelon distribution system or 'what to do until your ship comes in'. *Management Science*, 34(7):880–895, July 1988.

138. P.L. Jackson and J.A. Muckstadt. Risk pooling in a two-period, two-echelon inventory stocking and allocation problem. *Naval Research Logistics*, 36(1):1–26, 1989.

139. G. Janakiraman and J.A. Muckstadt. Inventory control in directed networks: A note on linear costs. *Operations Research*, 52(3):491–495, 2004.

140. G. Janakiraman and J.A. Muckstadt. Optimality of multi-tier base-stock policies for a class of capacitated serial systems. Technical Report 1361, School of Operations Research, Cornell University, Ithaca, NY, 2003. Submitted to Operations Research.

141. W. Jung. Recoverable inventory systems with time-varying demand. *Production and Inventory Management Journal*, 34:77–81, 1993.

142. M. Kalchschmidt, G. Zotteri, and R. Verganti. Inventory management in a multi-echelon spare parts supply chain. *International Journal of Production Economics*, 81-82:397–414, 2003.

143. A.J. Kaplan. Economic retention limits. Report, Inventory Research Office, U.S. Army Logistics Management Center, Fort Lee, Virginia, 1969.

144. A.J. Kaplan. Mathematics of SESAME model stocakge models. Report, AMSAA Army Inventory Research Office, Philadelphia, Pennsylvania, 1980.

145. A.J. Kaplan. Incorporating redundancy considerations into stockage models. *Naval Research Logistics*, 36:625–638, 1989.

146. R. Kaplan. A dynamic inventory model with stochastic lead times. *Management Science*, 16(7):491–507, March 1970.

147. S. Karlin and H. Scarf. Inventory models of the Arrow-Harris-Marschak type with time lag. In K. J. Arrow, S. Karlin, and H. Scarf, editors, *Studies in the Mathematical Theory of Inventory and Production*, chapter 10. Stanford University Press, Stanford, California, 1958.

148. U.S. Karmarkar and N.R. Patel. The one-period, N-location distribution problem. *Naval Research Logistics Quarterly*, 24:559–575, 1977.

149. J. Kim, R.C. Leachman, and B. Suh. Dynamic release control policy for the semiconductor wafer fabrication lines. *Journal of the Operational Research Society*, 47:1516–1525, 1996.

150. J. Kim, K. Shin, and H. Yu. Optimal algorithm to determine the spare inventory level for a repairable-item inventory system. *Computers and Operations Research*, 23:289–297, 1996.

151. K.S. Krishnan and V.R.K. Rao. Inventory control in N warehouses. *Journal of Industrial Engineering*, 16:212–215, 1965.

152. W.K. Kruse. Waiting time in an (s-1,s) inventory system with arbitrarily distributed leadtimes. *Operations Research*, 28(2):348–352, March-April 1980.

153. W.K. Kruse. Waiting time in a continuous review (s, S) inventory system with constant lead times. *Operations Research*, 29(1):202–207, January-February 1981.

154. W.K. Kruse and A.J. Kaplan. Comments on Simon's two echelon model. *Operations Research*, 21(6):1318–1322, 1973.

155. S.H. Lawrence and M.K. Schaefer. Optimal maintenance center inventories for fault-tolerant repairable systems. *Journal of Operations Management*, 4:175–181, 1985.

156. H.L. Lee. A multi-echelon inventory model for repairable items with emergency lateral transshipments. *Management Science*, 33(10):1302–1316, October 1987.

157. H.L. Lee and Billington C. Material management in decentralized supply chains (in OR Practice). *Operations Research*, 41(5):835–847, September-October 1997.

158. H.L. Lee and K. Moinzadeh. Operating characteristics of a two-echelon system for repairable and consumable items under batch operating policy. *Naval Research Logistics*, 34:365–380, 1987.

159. H.L. Lee and K. Moinzadeh. A repairable item inventory system with diagnostic and repair service. *European Journal of Operational Research*, 40:210–221, 1989.

160. LMI, A model to allocate repair dollars and facilities optimally, 1974. Task 74-9, Washington, D.C.

161. LMI availability system: Levels of indenture model, 1978. Washington, D.C.

162. LMI availability system: Procurement model, 1978. Washington, D.C.

163. M.C. Mabini and L.F. Gelders. Repairable item inventory systems: A literature review. *Belgian Journal of Operations Research, Statistics, and Computer Science*, 30:58–69, 1991.

164. J.W. Mamer and S.A. Smith. Optimizing field repair kits based on job completion rate. *Management Science*, 28(11):1328–1333, November 1982.

165. K.F. Matta. A simulation model for repairable items/spare parts inventory systems. *Computers and Operations Research*, 12:395–409, 1985.

166. W.L. Maxwell. Priority dispatching and assembly operations in a job shop. Technical Report RM-5370-PR, RAND Corporation, Santa Monica, California, 1969.

167. B. Miller. A real time METRIC for the distribution of serviceable assets. Report RM-5687-PR, RAND Corporation, Santa Monica, CA, 1968.

168. B.L. Miller. Dispatching from depot repair in a recoverable item inventory system: On the optimality of a heuristic rule. *Management Science*, 21(3):316–325, November 1974.

169. B.L. Miller and M. Modarres-Yazdi. The distribution of recoverable inventory items from a repair center when the number of consumption centers is large. *Naval Research Logistics Quarterly*, 25(4):597–604, 1978.

170. K. Moinzadeh and P.K. Aggarwal. An information based multiechelon inventory system with emergency orders. *Operations Research*, 45:694–701, 1997.

171. K. Moinzadeh and H.L. Lee. Batch size and stocking levels in multi-echelon repairable systems. *Management Science*, 32(12):1567–1581, December 1986.

172. K. Moinzadeh and S. Nahmias. A continuous review model for an inventory system with two supply modes. *Management Science*, 34(6):761–773, June 1988.

173. K. Moinzadeh and C.P. Schmidt. An (s-1,s) inventory system with emergency orders. *Operations Research*, 39(2):308–321, March-April 1991.

174. J.A. Muckstadt. A model for a multi-item, multi-echelon, multi-indenture inventory system. *Management Science*, 20(4):472–481, December 1973.

175. J.A. Muckstadt. The consolidated support model (CSM): A three-echelon, multi-item model for recoverable items. Report R-1923-PR, RAND Corporation, Santa Monica, CA, 1976.

176. J.A. Muckstadt. Navmet: a four-echelon model for determining the optimal quality and distribution of navy spare aircraft engines. Technical Report 263, School of Operations Research and Industrial Engineering, Cornell University, Ithaca, NY, 1976.

177. J.A. Muckstadt. Some approximations in multi-item, multi-echelon inventory systems for recoverable items. *Naval Research Logistics Quarterly*, 25(3):377–394, September 1978.

178. J.A. Muckstadt. A three-echelon, multi-item model for recoverable items. *Naval Research Logistics Quarterly*, 26(2):199–221, 1979.

179. J.A. Muckstadt. Comparative adequacy of steady-state versus dynamic models for calculating stockage requirements. Report R-2636-AF, RAND Corporation, Santa Monica, California, 1980.

180. J.A. Muckstadt. A multi-echelon model for indentured, consumable items. Technical Report 548, School of Operations Research, Cornell University, Ithaca, NY, 1982.

181. J.A. Muckstadt and M.H. Issac. An analysis of a single item inventory system with returns. *Naval Research Logistics Quarterly*, 28:237–254, 1981.

182. J.A. Muckstadt and R.O. Roundy. Heuristic computation of periodic-review base stock inventory policies. Technical Report 1176, School of Operations Research, Cornell University, Ithaca, NY, 1996.

183. J.A. Muckstadt and L.J. Thomas. Are multi-echelon inventory methods worth implementing in systems with low-demand-rate items? *Management Science*, 26(5):483–494, May 1980.

184. A. Muharremoglu and J.N. Tsitsiklis. Echelon base stock policies in uncapacitated serial inventory systems. http://web.mit.edu/jnt/www/publ.html, 2001.

185. S. Nahmias. Simple approximations for a variety of dynamic leadtime lost-sales inventory models. *Operations Research*, 27(5):904–924, September-October 1979.

186. S. Nahmias. Managing reparable item inventory systems: A review. In L.B. Schwarz, editor, *Multi-Level Production/Inventory Control Systems: Theory and Practice*, volume 16 of *Studies in the Management Sciences*, pages 253–278. North Holland Publishing, Amsterdam, 1981.

187. S. Nahmias and H. Rivera. A deterministic model for a repairable item inventory system with a finite repair rate. *International Journal of Production Research*, 17:215–221, 1976.

188. P.M. Needham. The influence of individual cost factors on the use of emergency transshipments. *Transportation Research, Part E, Logistics and Transportation Review*, 34:149–161, June 1998.

189. T.J. O'Malley. The aircraft availability model: Conceptual framework and mathmematics. Report, Logistics Management Institute, Washington, D.C., 1983.

190. O. Ozer. Replenishment strategies for distribution systems under advanced demand information. *Management Science*, 49(3):255–272, 2003.

191. C. Palm. Analysis of the Erlang traffic formulae for busy-signal arrangements. *Ericsson Technics*, 5:39–58, 1938.

192. A. Pena Perez and P. Zipkin. Dynamic scheduling rules for a multiproduct make-to-stock queue. *Operations Research*, 45:919–930, 1997.

193. E.S. Phelps. Optimal decision rules for the procurement, repair or disposal of spare parts. Report RS-2920PR, RAND Corporation, Santa Monica, California, 1962.

194. E. Porteus and Z. Landsdowne. Optimal design of a multi-item, multi-location, multi-repair type repair and supply system. *Naval Research Logistics Quarterly*, 21(2):213–238, 1974.

195. U. Prabhu. *Queues and Inventories*. Wiley, New York, 1965.

196. U. Prabhu. *Stochastic Storage Processes*. Springer-Verlag, New York, 1980.

197. V.J. Presutti and R.C. Trepp. More ado about economic order quantities (EOQ). *Naval Research Logistics Quarterly*, 17:243–251, 1970.

198. D.F. Pyke. Priority repair and dispatch policies for repairable-item logistics systems. *Naval Research Logistics*, 37:1–30, 1990.

199. T.L. Ramey. Lean logistics: High-velocity logistics infrastructure and the C-5 Galaxy. Report MR-581-AF, RAND Corporation, Santa Monica, California, 1999.

200. J.A. Rappold and J.A. Muckstadt. A computationally efficient approach for determining inventory levels in a capacitated multi-echelon production-distribution system. *Naval Research Logistics*, 47(5):377–398, June 2000.

201. F.R. Richards. A stochastic model of a repairable-item inventory system with attrition and random lead times. *Operations Research*, 24(1):118–130, January-February 1976.

202. L.W. Robinson. Optimal and approximate policies in multiperiod multilocation inventory models with transshipment. *Operations Research*, 38(2):278–295, March-April 1990.

203. M. Rose. The (s-1,s) inventory model with arbitrary backordered demand and constant delivery times. *Operations Research*, 20(5):1020–1032, September-October 1972.

204. B. Rosenman and D. Hoekstra. A management system for high-value army components. Report TR64-1, U.S. Army, Advanced Logistics Research Office, Frankfort Arsenal, Philadelphia, Pennsylvania, 1964.

205. S.M. Ross. *Stochastic Processes*. John Wiley and Sons, New York, 1983.

206. R.O. Roundy and J.A. Muckstadt. Heuristic computation of periodic-review base stock inventory policies. *Management Science*, 46(1):104–109, January 2000.

207. J.W. Rustenburg, G.J. van Houtum, and W.H.M. Zijm. Spare parts management for technical systems: resupply of spare parts under limited budgets. *IIE Transactions*, 32(10):1013–1026, 2000.

208. J. Sarkis. Quantitative models for performance measurement systems - alternate considerations. *International Journal of Production Economics*, 86(1):81–90, 2003.

209. M.K. Schaefer. A multi-item maintenance center inventory model for low-demand repairable items. *Management Science*, 29(9):1062–1068, September 1983.

210. M.K. Schaefer. Replenishment policies for inventories of recoverable items with attrition. *Omega*, 17:281–287, 1989. Oxford.

211. D.A. Schrady. A deterministic inventory model for repairable items. *Naval Research Logistics Quarterly*, 14(3):391–398, 1967.

212. C.R. Schultz. On the optimality of the (s-1,s) policy. *Naval Research Logistics*, 37:715–723, 1990.

213. L.B. Schwartz, editor. *Multi-Level Production/Inventory Systems: Theory and Practice*, volume 16 of *Studies in the Management Sciences*. North Holland Publishing, New York, 1981.

214. G.D. Scudder. Priority scheduling and spares stocking policies for a repair shop: The multiple failure case. *Management Science*, 30:739–749, June 1984.

215. G.D. Scudder and W. Hausman. Spares stocking policies for repairable items with dependent repair times. *Naval Research Logistics Quarterly*, 29:303–322, 1982.

216. S. Seshadri and J.M. Swaminathan. A componentwise index of service measurement in multi-component systems. *Naval Research Logistics*, 50(2):184–194, 2003.

217. K. Shanker. An analysis of a two-echelon inventory system for recoverable items. Technical Report 341, School of Operations Research and Industrial Engineering, Cornell University, Ithaca, NY, 1977.

218. K. Shanker. Exact analysis of a two-echelon inventory system for recoverable items under batch inspection policy. *Naval Research Logistics Quarterly*, 28:579–601, 1981.

219. C. Sherbrooke. Multi-echelon inventory systems with lateral supply. *Naval Research Logistics*, 39:29–40, 1992.

220. C.C. Sherbrooke. Discrete compound Poisson processes and tables of the geometric Poisson distribution. Report RM-4831-PR, RAND Corporation, 1966.

221. C.C. Sherbrooke. Generalization of a queuing theorem of Palm to finite populations. *Management Science*, 12(11):907–908, July 1966.

222. C.C. Sherbrooke. Discrete compound Poisson processes and tables of the compound Poisson distribution. *Naval Research Logistics Quarterly*, 15(2):189–204, 1968.

223. C.C. Sherbrooke. METRIC: A multi-echelon technique for recoverable item control. *Operations Research*, 16(1):122–141, 1968.

224. C.C. Sherbrooke. An evaluator for the number of operationally ready aircraft in a multilevel supply system. *Operations Research*, 19(3):618–635, May-June 1971.

225. C.C. Sherbrooke. Waiting time in an (s-1,s) inventory system-constant service time case. *Operations Research*, 23(4):819–820, July-August 1975.

226. C.C. Sherbrooke. VARI-METRIC: Improved approximations for multi-indenture, multi-echelon availability models. *Operations Research*, 34(2):311–319, March-April 1986.

227. C.C. Sherbrooke. *Optimal Inventory Modeling of Systems: Multi-Echelon Techniques*. John Wiley and Sons, New York, 1992.

228. E.A. Silver. Inventory allocation among an assembly and its repairable subassemblies. *Naval Research Logistics Quarterly*, 19:261–280, 1972.

229. R.M. Simon. The uniform distribution of inventory position for continuous review (s, Q) policies. Report 3938, RAND Corporation, Santa Monica, California, 1968.

230. R.M. Simon. Stationary properties of a two-echelon inventory model for low demand items. *Operations Research*, 19(3):761–773, May-June 1971.

231. R.M. Simon and D.A. D'Esopo. Comments on a paper by S.G. Allen and D.A. D'Esopo: 'an ordering policy for repairable stock items'. *Operations Research*, 19(4):986–988, July-August 1971.

232. K.E. Simpson, Jr. A theory of allocation of stocks to warehouses. *Operations Research*, 7(6):797–805, November-December 1959.

233. V.P. Simpson. An ordering model for recoverable stock items. *AIIE Transactions*, 2:315–320, 1970 or 1971, check.

234. V.P. Simpson. Optimum solution structure for a repairable inventory problem. *Operational Research*, 26(2):270–281, March-April 1978.

235. F.M. Slay. VARI-METRIC: An approach to modeling multi-echelon resupply when the demand process is Poisson with a gamma prior. Report AF301-3, Logistics Management Institute, Washington, D.C., 1984.

236. F.M. Slay. Lateral resupply in a multi-echelon inventory system. Report AF501-2, Logistics Management Institute, Washington, D.C., 1986.

237. C.H. Smith and M.K. Schaefer. Optimal inventories for repairable redundant systems with aging components. *Journal of Operations Management*, 5:339–349, 1985.

238. S.A. Smith. Optimal inventories for an (s-1,s) system with no backorders. *Management Science*, 23(5):522–528, January 1977.

239. S.A. Smith, J.C. Chambers, and E. Shlifer. Optimal inventories based on job completion rate for repairs requiring multiple items. *Management Science*, 26(8):849–852, August 1980.

240. M.J. Sobel. Fill rates of single-stage and multistage supply systems. *Manufacturing and Service Operations Management*, 6(1):41–52, 2004.

241. A. Svoronos and P. Zipkin. Estimating the performance of multi-level inventory systems. *Operations Research*, 36(1):57–72, January-February 1988.

242. A. Svoronos and P. Zipkin. Evaluation of one-for-one replacement policies for multi-echelon inventory systems. *Management Science*, 37(1):68–83, January 1991.

243. G. Tagaras. Effects of pooling on the optimization and service levels of two-location inventory systems. *IIE Transactions*, 21(3):250–258, September 1989.

244. G. Tagaras. Pooling in multi-location periodic inventory distribution systems. *Omega*, 27(1):39, February 1999. Oxford.

245. G. Tagaras and M.A. Cohen. Pooling in two-location inventory systems with non-negligible replenishment lead times. *Management Science*, 38(8):1067–1083, August 1992.

246. G. Tagaras and D. Vlachos. A periodic review inventory system with emergency replenishments. *Management Science*, 47(3):415–429, March 2001.

247. S. Tayur. Computing the optimal policy in capacitated inventory models. *Stochastic Models*, 9(4), 1993.

248. M.J. Tedone. Repairable part management. *Interfaces*, 19:61–68, 1989.

249. J. Verrijdt, I. Adan, and T. de Kok. A trade-off between emergency repair and inventory investment. *IIE Transactions*, 30:119–132, 1998.

250. Y. Wang, M.A. Cohen, and Y.S. Zheng. A two-echelon repairable inventory system with stocking-center-dependent depot replenishment lead times. *Management Science*, 46(11):1441–1453, November 2000.

251. L.M. Wein. Scheduling networks of queues: Heavy traffic analysis of a multistation network with controllable inputs. *Operations Research*, 40:S312–S334, 1992.

252. L.M. Wein and P.B. Chevalier. A broader view of the job-shop scheduling problem. *Management Science*, 38:1018–1033, 1992.

253. P.T. Evers, K. Xu, and M.C. Fu. Estimating customer service in a two-location continuous review inventory model with emergency transshipments. *European Journal of Operational Research*, 145(3):569–584, 2003.

254. S. Yanagi and M. Sasaki. Reliability analysis for a two-echelon repair system considering lateral resupply, return policy and transportation times. *Computers and Industrial Engineering*, 27(1-4):493–497, September 1994.

255. S. Zacks. A two-echelon multi-station inventory model for navy applications. *Naval Research Logistics Quarterly*, 17(1):79–85, 1970.

256. S. Zacks and J. Fennell. Bayes adaptive control of two-echelon inventory systems, I: Development for a special case of one-station lower echelon and Monte Carlo evaluation. *Naval Research Logistics Quarterly*, 19(1):15–28, 1972.

257. S. Zacks and J. Fennell. Distribution of adjusted stock levels under statistical adaptive control procedures of inventory systems. *Journal of American Statistical Association*, 68:88–91, 1973.

258. S. Zacks and J. Fennell. Bayes adaptive control of two-echelon inventory systems, II: The multi-station case. *Naval Research Logistics Quarterly*, 21(4):575–593, 1974.

259. W.H. Zijm and Z.M. Avsar. Capacitated two-indenture models for repairable item systems. *International Journal of Production Economics*, 81-82:573–588, 2003.

260. P. Zipkin. Critical number policies for inventory models with periodic data. *Management Science*, 35(1):71–80, 1989.

# Index